W9-AGA-416

# HOW TO DESIGN SOLID-STATE CIRCUITS

2ND EDITION

No. 2975
$24.95

# HOW TO DESIGN SOLID-STATE CIRCUITS

## 2ND EDITION

MANNIE HOROWITZ AND DELTON T. HORN

TAB BOOKS Inc.
Blue Ridge Summit, PA

SECOND EDITION
FIRST PRINTING


Printed in the United States of America



Library of Congress Cataloging-in-Publication Data

Horowitz, Mannie.
How to design solid-state circuits.

Rev. ed. of: How to design circuits using semiconductors / by Mannie Horowitz. 1st ed. c1983.
Includes index.
1. Semiconductors 2. Electronic circuit design.
I. Horn, Delton T. II. Horowitz, Mannie. How to design circuits using semiconductors. III. Title.
TK7871.85.H66 1988 621.3815'2 88-8572
ISBN 0-8306-9075-1
ISBN 0-8306-2975-0 (pbk.)

Questions regarding the content of this book should be addressed to:

Reader Inquiry Branch
TAB BOOKS Inc.
Blue Ridge Summit, PA 17294-0214

# Contents

# Introduction

This book is directed to the circuit designer, the engineer, the technician, and the student. It provides information on semiconductors, their performance in various applications and their operation in a manner that will ensure reliability.

The transistor and other semiconductor devices are treated here as circuit components exhibiting specific characteristics and limitations. As a general practice, the equivalent circuit will be shown or derived and equations for the circuit will be presented. The manipulations of these equations may be omitted in the reading without loss of continuity. Because of this, much of the text serves as a useful handbook for the imaginative and inventive designer.

The discussion in the text starts with devices made of individual semiconductors, leading into a description of the semiconductor diode and its many functions in the modern circuit. Power supplies, filter circuits, and diode characteristics are among the topics. Zener diode regulators, tunnel diodes, and other diodes are also detailed.

In the transistor circuit sections, the dc bias and stabilization conditions for bipolar and field effect transistor (FET) devices are considered. The use of semiconductors in audio and radio-frequency (rf) amplifiers with a varying input signal is discussed. A chapter is devoted to power amplifiers because of the light this discussion throws on all transistor applications.

Pulse and switching circuits are described. This type of circuit is useful in many transistor applications. Special emphasis is placed on digital-circuit design along with its many variations. Designs using both ICs and discrete components are described.

A chapter is devoted to a discussion of silicon-controlled rectifiers and other lesser-known thyristors. Descriptions of the devices, their limitations, and their circuit applications are presented.

This revised edition features information on a number of devices that were previously unavailable to the experimenter. New devices are constantly appearing on the market. Initially, they tend to be quite expensive and often unavailable to individuals, but as they are increasingly used in commercial or industrial equipment, costs tend to come down and the availability goes up. New semiconductor devices open up many exciting new design applications for the circuit designer or experimenter. New devices also often simplify older design techniques.

In this book, a basic knowledge of electronics is assumed. Where it is deemed necessary, however, a review of this information is presented in a concise manner. A special section is devoted to Boolean algebra, along with methods of applying this type of mathematics to digital designs.

# 1

# Semiconductors and Diodes

Solid-state (semiconductor) devices range in complexity from the simple temperature-sensitive resistor (thermistor) and the conventional bipolar transistor, up to high density integrated circuits (ICs) composed of many semiconductor functions on one structure or substrate. Current, flowing through the different types of semiconductor devices, ranges from just a few microamperes up to several tens of amperes.

To facilitate design calculations covering such a wide range, you should become familiar with conversions from one measurement unit to another—for example, from amperes to milliamperes—as well as with exponential notation.

As for exponential notation, numbers are conveniently expressed as powers of 10. $10^n$ means 10 multiplied by itself n times.

$$10^2 = 10 \times 10 = 100$$
$$5.7 \times 10^5 = 5.7 \times 10 \times 10 \times 10 \times 10 \times 10 = 570{,}000$$

$10^{-n}$ denotes a number divided by 10, n times.

$$10^{-2} = 1/(10 \times 10) = 1/100$$
$$2.7 \times 10^{-5} = 2.7/(10 \times 10 \times 10 \times 10 \times 10) = 2.7/100{,}000$$

When numbers with exponents are multiplied by each other, the exponents are added.

Table 1-1. Commonly Used Prefixes Are Widely Encountered in Electronics Work.

| Prefix | Multiplier | Abbreviation for Prefix |
|---|---|---|
| pico- | $10^{-12}$ (0.000000000001) | p |
| nano- | $10^{-9}$ (0.000000001) | n |
| micro- | $10^{-6}$ (0.000001) | $\mu$ |
| milli- | $10^{-3}$ (0.001) | m |
| (no prefix) | $10^{0}$ (1) | – |
| kilo- | $10^{3}$ (1000) | k |
| mega- | $10^{6}$ (1,000,000) | M |
| giga- | $10^{9}$ (1,000,000,000) | G |

If numbers with exponents are divided one into the other, subtract the exponent in the denominator from the exponent in the numerator.

$$10^2 \times 10^3 = 10^5$$
$$(8 \times 10^3)/(2 \times 10) = 4 \times 10^2$$
$$(9 \times 10^{-3})/(3 \times 10^{-1}) = 3 \times 10^{-2}$$

Several of the exponents have been given special notations. Prefixes added to the basic unit indicate the multiplier of that unit. This multiplier can be expressed as an exponent of 10. Thus, 1 milliamp = $10^{-3}$ amps; 7 millivolts = $7 \times 10^{-3}$ volts. Table 1-1 lists prefixes used in this book.

Since there are 1000 milliamperes (mA) for each ampere, amperes must be multiplied by $10^3$ to be converted to milliamperes. Similarly, a millivolt (mV) must be multiplied by $10^3$ if it is to be converted to microvolts ($\mu$V).

## SEMICONDUCTOR MATERIAL

All materials are artificially grouped in categories of conductors, semiconductors, and insulators. Frequently used materials that conduct electricity well are copper and silver. Plastic, cotton and enamel are often used as insulating materials, because these substances tend to resist the flow of electrons.

Semiconductors are in a category somewhere between true insulators and true conductors. Germanium and silicon are the two basic elements used in most modern semiconductor devices. Materials such as arsenic, antimony, or phosphorus (called *impurities*) are normally added to the pure (or intrinsic) semiconductor. The process of adding impurities to form the extrinsic crystal is referred to as doping. Material formed when the impurity is added to the silicon or germanium is characterized by a lower resistance than the pure germanium or silicon. It has an excess of negative charge and is referred to as an n-type semiconductor.

Similarly, if indium, aluminum, or gallium is added to the basic element, a shortage of electrons (positive charge) exists in the combined material. This shortage of electrons is referred to as an excess of holes and the material is a p-type semiconductor.

In most applications, two or more slabs of semiconductor material are combined to form diodes, transistors, silicon-controlled rectifiers, and many other devices. Characteristics exhibited by the individual semiconductors are also frequently taken advantage of.

## THERMISTORS

Semiconductors are generally temperature sensitive. This means their characteristics (especially resistance) fluctuate in response to changes in temperature.

Semiconductors can have either a negative or positive temperature coefficient of resistance. The factor determining whether the resistance of a particular semiconductor will decrease or increase as its temperature rises is the material used in the device. Semiconductors designed specifically to exhibit particular resistance variations with temperature are referred to as *thermistors*. Although thermistors ordinarily decrease in resistance as the temperature rises, *sensistors* produced by Texas Instruments and *posistors* produced by Murata are heavily doped semiconductors exhibiting a positive temperature coefficient. Generally, both negative and positive temperature coefficient devices are referred to as thermistors.

In some designs, the bead is mounted in a vacuum or gas-filled glass bulb. In this arrangement, even slight temperature variations caused by the presence of minute amounts of current can readily be sensed. It can also be used to detect extremely small ambient temperature variations.

Like all other devices, the thermistor has maximum voltage, current, power, and temperature operating limits. Recommended body temperatures on specification sheets should never be exceeded. Power dissipation limits at the various temperatures must be observed if the life of the device is not to be terminated prematurely.

### Useful Characteristics

Four characteristics of the thermistor are used in modern circuits.

1. Resistance changes with temperature. For many negative resistance coefficient thermistors, the change can be calculated by multiplying the resistance of the device at 25°C by 0.96 for each degree Celsius that the temperature increases. As for the sensistor or posistor, multiply the resistance of the device at 25°C by 1.007 for each degree Celsius that the temperature increases. These rules assume that the thermistor

in question is being heated by external sources only, and that no heat is generated by power being dissipated in the device.

2. There is a time delay for the temperature of the body of the thermistor to change from one value to another. The time it takes for the body temperature to increase to 63.2 percent of its final value is referred to as the *thermal time constant* of the device.

3. Assume that a voltage is placed across a series combination of a thermistor and a fixed resistor. Current flowing through the circuit heats the components. As the temperature of the thermistor rises, its resistance decreases (assuming a negative temperature coefficient device), allowing more current to flow. The temperature of the device increases further because of the added current, until the circuit constants, the voltage and the fixed resistor, limit the amount of current and hence power available to the thermistor. The time required to reach the minimum thermistor resistance and maximum circuit current levels can be used in practical circuits where a time-delay characteristic is essential.

4. Volt-ampere characteristics follow Ohm's law at low voltages. As it increases, the thermistor is heated, its resistance drops, and the voltage across the thermistor decreases as the current increases. The transition point between the Ohm's law portion of the characteristic and the negative resistance portion of the characteristic is referred to as the *self-heating voltage.*

Heat produced from the power dissipated in the thermistor is proportional to the dissipation constant of the device. If, for example, the dissipation constant of a specific thermistor is 0.7 mW/°C indicating that 0.7 milliwatts must be dissipated to increase the body temperature of the semiconductor by 1°C, 7 mW must be dissipated to increase body temperature by 10°C.

## Modifying the Thermal Characteristics

There are only a select number of cold resistances and temperature characteristics of commercially available thermistors. If the required characteristic in your design differs from that of the standard device, a circuit as in Fig. 1-1A or 1-1B can be used.

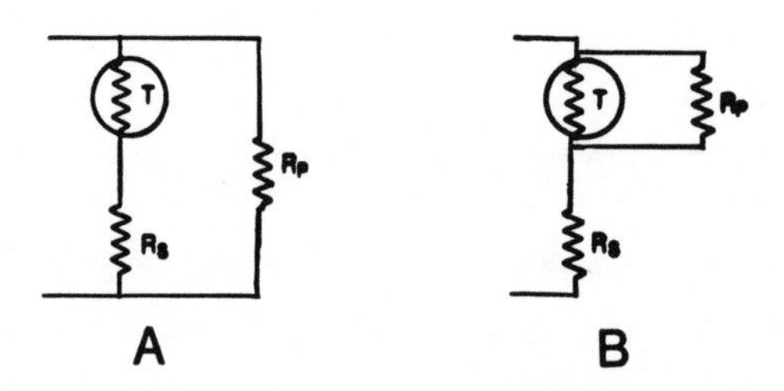


Fig. 1-1. Thermistor in circuit with resistors to establish the required resistance - temperature relationship.

As an example of how to select the proper thermistor and calculate $R_s$ and $R_p$ resistor values for one of the compensating circuits, first select a thermistor whose cold-to-hot resistance ratio, $R_{TC}:R_{TH}$, is greater at the temperature extremes than is required from the overall circuit resistance changes. Although $R_{TC}$, the cold resistance of the thermistor, must be above that required for the overall circuit, the hot resistance, $R_{TH}$, of the device should be less than that of the series-parallel circuit. Note that this is only a rule of thumb, and need not be adhered to strictly. In some designs the hot resistance of the thermistor may be more than that required of the overall circuit. If impossible combinations are chosen, it will show up as a negative resistance, $-R_s$ or $-R_p$, when calculating these resistors for the compensating circuit.

As you recall, the resistance of the combination of any two resistors, R1 and R2, when connected in parallel, is R = (R1)(R2)/(R1 + R2). Substituting $R_p$ from Fig. 1-1A for R1 in this relationship and $(R_T + R_S)$ for R2, the resistance of the parallel network in the figure becomes:

$$R_C = \frac{R_p\,(R_{TC} + R_s)}{R_p + (R_{TC} + R_s)} \qquad (1\text{-}1)$$

$$R_H = \frac{R_p\,(R_{TH} + R_s)}{R_p + (R_{TH} + R_s)} \qquad (1\text{-}2)$$

where $R_C$ is the cold resistance of the overall network, $R_{TC}$ is the cold resistance of the thermistor, $R_H$ is the hot resistance of the overall network, and $R_{TH}$ is the hot resistance of the thermistor. Solve the relationships simultaneously for $R_s$ and $R_p$.

If the circuit in Fig. 1-1B is used, the equations become:

$$R_C = R_s + \frac{R_{TC}R_p}{R_{TC} + R_p} \qquad (1\text{-}3)$$

$$R_H = R_s + \frac{R_{TH}R_P}{R_{TH} + R_p} \qquad (1\text{-}4)$$

Select the circuit that will supply you with the temperature characteristic that best suits your needs. To determine this, you should calculate resistance at several points between the temperature extremes for both circuits, and see which circuit provides you with resistances that most closely follow your compensation requirements.

### Thermistor Applications

Among the more common applications, temperature measurement, control, and compensation, are the most obvious. Two circuits showing these applications are in Figs.

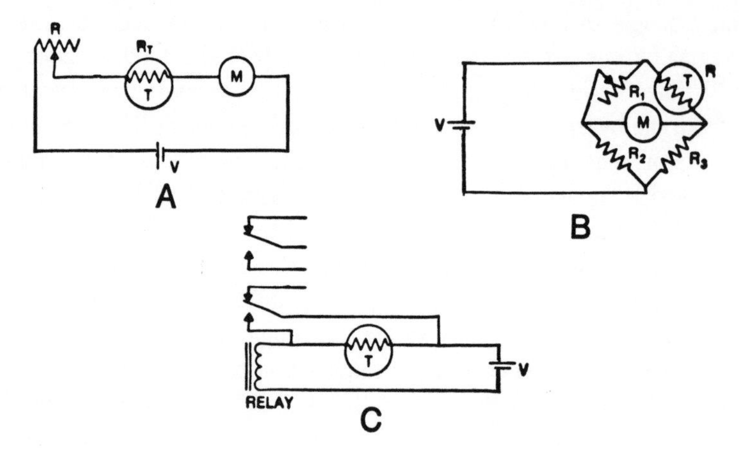


Fig. 1-2. Three practical circuits using thermistors.

1-2A and 1-2B. In Fig. 1-2A, the amount of current flowing through meter M depends upon the resistance of the thermistor, $R_T$, which in turn is contingent on the temperature of the semiconductor.

A more sensitive arrangement used in medical and industrial thermometers, involves the bridge circuit in Fig. 1-2B. Here, the meter can be used either to indicate temperature directly or as a device to show when the bridge is nulled. In the latter instance, R2 is usually made equal to R3. R1 is a variable resistor used to balance the bridge. When R1 is adjusted so that no current flows through the meter, the resistance of R1 is equal to that of the thermistor.

If two temperatures are to be compared, R1 can be replaced by a thermistor. Now the meter can be used to reveal directly the temperature difference of the two thermistors R1 and $R_T$. If this circuit is to be used to indicate the ratio of the temperatures of thermistors R1 and $R_T$, the meter can be used as a nulling indicator. R2 and R3 are made variable and adjusted for a null reading on the meter. The ratio of R2 to R3 is proportional to the resistance ratio of the thermistors, R1 to $R_T$. The temperature ratio is then determined from temperature-resistance curves of the thermistors.

By substituting a relay for either meter in Fig. 1-2A or 1-2B, a negative temperature coefficient thermistor can be used to control temperature. If temperature rises above a predetermined value, enough current flows through the relay to trip it. When connected in an oven circuit, it is turned off by the relay contacts.

Current surges through bulbs, power rectifiers, and tube filaments can damage these devices. They can be protected by including negative temperature coefficient thermistors in their circuits. When a switch is thrown, the thermistor does not allow the surge current to flow at the instant the switch is set to the ON position. Current buildup through the device is slow. As the current flows, the resistance of the thermistors drops gradually. The reduced resistance permits the desired amounts of current to flow through the entire series circuit. This time delay in the buildup of current and the absence of the initial surge extends the life of the switch and the other devices in the circuit.

When connected in series with a relay, as in Fig. 1-2C, the resistance of a cold negative temperature coefficient thermistor may be large enough so as not to allow sufficient current to flow to activate the relay. At elevated temperatures, the resistance of the thermistor drops. Sufficient current will then flow through the coil to trip the relay. Functionally, this circuit behaves as a time-delay relay. Note in the figure that the thermistor is shorted by the relay contacts after it has been tripped. This allows the thermistor to cool and be ready for the next time-delay cycle.

Metals such as copper increase in resistance as their temperatures rise. Copper is used as the coil in meter movements. Because the temperature characteristic of the negative temperature coefficient thermistor is opposite to that of copper, it can be used to compensate for the temperature characteristic of the coil. Use a circuit similar to one of those in Fig. 1-1.

Any positive temperature coefficient thermistor is useful in circuits such as those shown in Fig. 1-3. In Fig. 1-3A, the resistance of the semiconductor increases as current continues to flow through the circuit, heating the device. Should the voltage V rise excessively or should a short develop, the sensistor will limit circuit current to safe values. In this type of circuit, it acts as a thermostat. If the sensistor were placed in parallel with the resistor rather than in series, the device would serve to stabilize the current flowing through the circuit despite slow variations in supply voltage.

In Fig. 1-3B, the sensistor serves to suppress arcing across the switch when it opens the circuit. The function of the time-delay relay circuit in Fig. 1-3C is similar to that of the relay circuit in Fig. 1-2C. Upon closing the switch, large amounts of current flow through the cold sensistor. A voltage, $V - V_R$, is available for the relay. This voltage is insufficient to trip the relay. As the sensistor heats up because of the current flowing

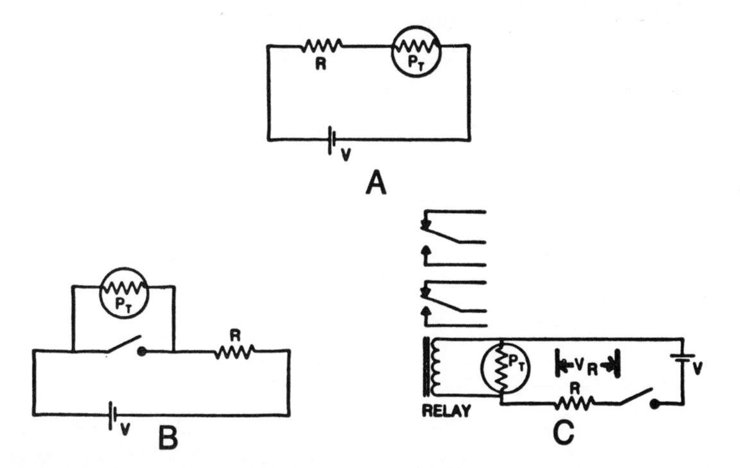


Fig. 1-3. Three practical circuits using a sensor or posistor.

through it, its resistance increases. Because less current flows through the hot sensistor than through the sensistor when it was cold, the voltage available for the relay is now sufficient to trip it.

A similar time-delay effect can be achieved by connecting the proper sensistor in series with the relay circuit. Now the delay is in the opening of the relay switch rather than in its closing. This period of time is known as the *time delay* of the circuit.

## VARISTORS

The prime importance of the varistor is the variation of resistance with the applied voltage. Two types of drawings for the current-to-voltage relationships, are shown for the same or similar varistors in Fig. 1-4. Resistance at any voltage is determined by extending a line from the voltage axis to a point on the curve. Draw a line from this point on the curve to the current axis. Applying Ohm's law, R = V/I, determine the resistance of the varistor at the voltage involved.

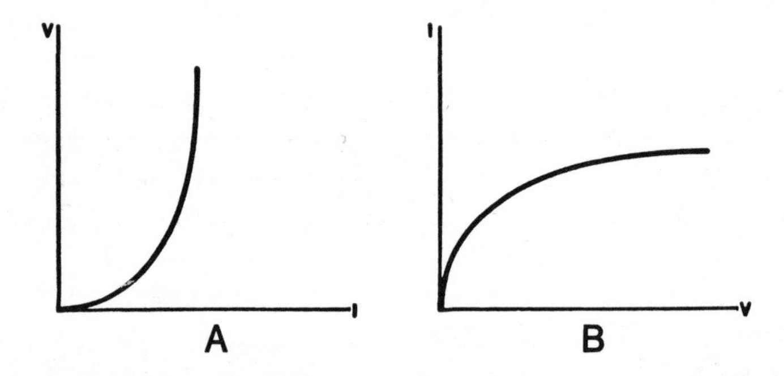


Fig. 1-4. Manufacturers use two different curves to show how current varies with the voltage applied to a varistor.

One of the prime applications of the varistor is as a spark suppressor to protect switch contacts. Protection is especially necessary when a series circuit inadvertently includes an inductor as one of its elements, such as in Fig. 1-5. An inductor is singled out here because when there is a change in the amount of dc current flowing through the coil, voltage developed across the inductor may be many times the voltage supplied by the battery. This peak voltage is commonly referred to as the *inductive kick*.

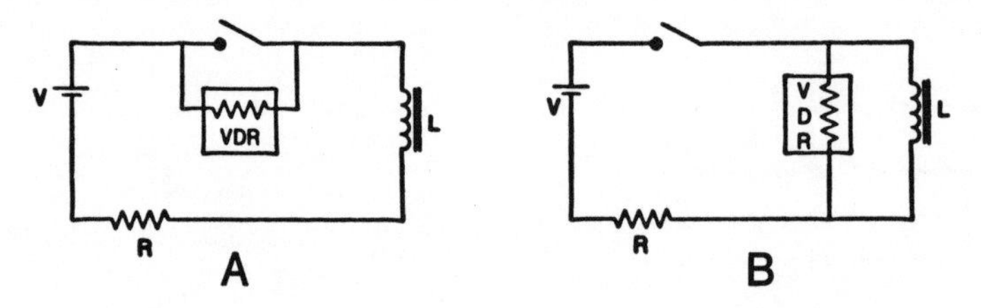


Fig. 1-5. A VDR or varistor can be used to protect contacts in a switching circuit.

In Fig. 1-5, a resistor is placed in series with the circuit to limit steady-state current to desirable levels. When the switch is opened, arcing may occur because of the inductive kick despite the presence of resistor R or the low voltage source V. Because of its small size, the resistor has little effect on the transient current. Arcing can, however, be suppressed by the presence of a varistor in the circuit.

In Fig. 1-5A, the varistor is across the switch, whereas in Fig. 1-5B it is across the inductor. In neither case does it affect circuit operation while dc current flows under steady-state conditions. Although the varistor is not in the circuit because it is shorted by the switch in Fig. 1-5A, its effect on the circuit in Fig. 1-5B is negligible because it is across an inductance. An inductance has very low dc resistance when compared to the resistance of the varistor.

Resistance of the varistor across the switch contacts in Fig. 1-5A is reduced by the voltage across the contacts while the switch is being opened. Current passes through the varistor rather than through the switch contacts. When the switch is opened in Fig. 1-5B, voltage across the inductor and its paralleling varistor produces a drop in the resistance of the semiconductor. Current from the inductor passes through the low resistance of the varistor rather than through the balance of the circuit and switch.

As was the case with the thermistor, here too, various characteristics can be generated through the use of circuits like those in Fig. 1-1. Varistors can be connected in series or parallel with resistor circuits to modify the voltage-resistance characteristics of the device being used.

Should you use several varistors in parallel, the current division between the semiconductors cannot be controlled precisely. Because of this, parallel varistor circuits cannot be used where more power dissipation capability is necessary than can be provided by only one of the devices. The dissipation ability of only one of the devices in the circuit should be considered as the reliable power dissipation limits of the overall parallel combination.

## PHOTORESISTORS

Semiconductors are sensitive to light. To avoid any degradation of the circuit in which a semiconductor is used, the device is normally placed in a light-proof package. Photoresistors utilize the effect of light striking the semiconductor. Its resistance drops radically as the intensity of light hitting it increases.

### Optics and the Photoresistor

The wavelength of ordinary visible light varies from about 0.33 μm (micrometers) to 0.8 μm. In optical language, these wavelengths are expressed in angstrom units, abbreviated Å. The number in Å is 10,000 times the magnitude of the number in μm. Visible wavelengths are from about 3,300 Å to 8,000 Å. There is a specific group of wavelengths for each color. For example, the range of wavelengths for violet is 3,300

Å to 4,300 Å. Only one wavelength in this range is generally considered as the wavelength of pure violet—namely 4,100 Å. As for the wavelengths of the other relatively pure colors, blue is 4,800 Å, green is 5,100 Å, yellow is 5,700 Å, orange is 5,900 Å, and red is 6,400 Å. Ultraviolet is below 3,800 Å and infrared is above 8,000 Å, but these are not visible to the naked eye.

Photoresistors are sensitive to selected groups of colors. The range of wavelengths to which a particular photoresistor is sensitive depends upon the material used to construct the device. Curves are available to show the response of photoresistors made from the different materials. The response of the device not only depends on the material used and the structure, but also on the concentration of the impurity in the mix and the size of the semiconductor particles.

### Applications

Photoresistors have found many uses in electronic and other equipment. Among these are light meters to measure light intensity for photographic applications, in conjunction with a relay to be turned on and off by the alternate presence and absence of light, and in movie projectors where the photoresistor responds to the rapid light intensity variations on the sound track.

A basic circuit involves a battery, V, applied to the series circuit of a photoresistor and an ordinary resistor, R. It is a voltage divider. In the dark, the resistance of the photoresistor is high and most of the voltage from V is developed across it. Very little voltage remains for R. More voltage is across R when light strikes the photoresistor reducing its resistance considerably. Specification sheets usually have curves showing the variations of resistance with light intensity. As a rule of thumb, resistance = 1/(light intensity).

Another useful characteristic of photoresistors is the time it takes for the device to respond to changes in light intensity. This is termed the *recovery rate* and is specified as the minimum amount the resistance will change each second as the light intensity drops. Time-delay circuits can make use of this characteristic.

## THE HALL EFFECT

Should a magnetic field be set up perpendicular to the direction of electric current flow in a semiconductor, a voltage is established across the semiconductor. This voltage, due to the *Hall effect*, is perpendicular to both the direction of the electric current and magnetic field. The Hall effect is demonstrated using Fig. 1-6 where I is the electron current and B is the magnetic field, due to current flowing through the inductor, L, that is supplied by battery, V. Voltage $V_H$ occurs across the remaining two surfaces of the slab.

The Hall generator can be used to determine if an n-type or p-type semiconductor is being used in a device. Assume the magnetic field is in the direction shown by the

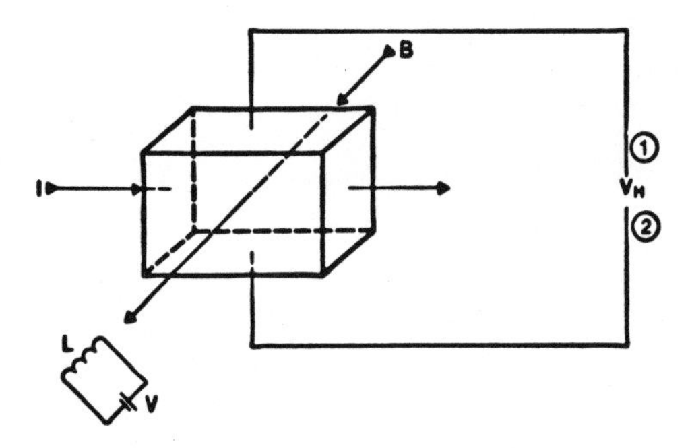


Fig. 1-6. The Hall effect.

B arrow in Fig. 1-6 and the electric current is in the direction depicted by the I arrow. If the slab is made of n-type semiconductor material, terminal 2 of $V_H$ is positive with respect to terminal 1. Should a p-type slab be used, the polarity is reversed.

The Hall effect can also be used to give the strength of a magnetic field. Because $V_H$ is proportional to the product of I and B, a meter measuring the Hall voltage can be calibrated to indicate strength of the magnetic field B rather than the voltage.

The relationship between $V_H$ and the product of B and I can also be applied to what is known as the *Hall effect multiplier*. Because this device indicates the product of B and I, a meter connected to indicate $V_H$ can be calibrated to show this product.

As $V_H$ can exist only if both I and B are present, the Hall effect can be applied to logic gates. The two-input AND gate works on the principle that there is an output from a circuit only if two inputs to that circuit exist simultaneously. Because I and B must be present if there is to be an output voltage, $V_H$, the Hall effect generator can be used as an AND gate.

## THE JUNCTION DIODE

The diode is the simplest commonly used semiconductor device involving more than one slab of semiconductor material. Its applications seem to be without bound, but each diode has specific characteristics and limitations. To extend the versatility, special diodes have been designed for power supply, af, rf, switching, and regulating circuits, as well as for other applications. The junction diode is the most popular diode in use today. It consists of pieces of n-type and p-type semiconductor materials in contact with each other.

Some of the free electrons from the n-material diffuse or spread across the junction to the p-material. Similarly, free holes enter the n-region and concentrate around the junction as shown in Fig. 1-7. The diffusion continues until there is so great a negative charge built up at the junction in the p-region that it repels the efforts of more electrons from the n-region to wander across the junction. A similar barrier is set up in the n-region to oppose the migration of an excess number of holes. The region at the junction

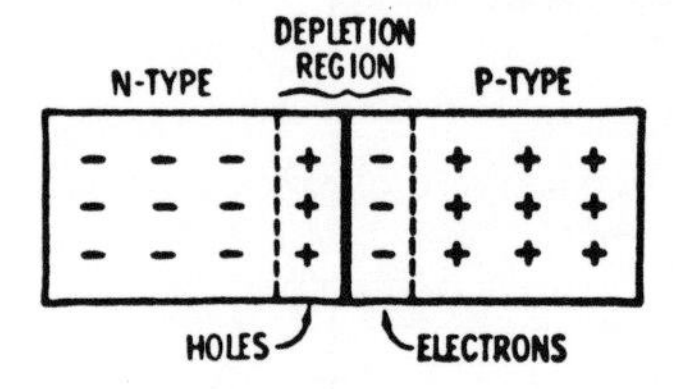

Fig. 1-7. This diagram of a PN junction diode shows the depletion region with holes and electrons that have migrated across the junction.

exists in a state of nonconducting equilibrium. This is known as the depletion region.

If a dc voltage source, such as a battery, were connected across the diode with the n-region connected to the positive terminal and the p-region connected to the negative terminal, there would be no conduction. By reversing the connections to the battery so that the n-region is connected to the negative terminal and the p-region to the positive terminal, current will flow through the device.

A diode is shown symbolically in Fig. 1-8A. The arrow represents the p-type material and the bar represents the n-type slab. In Fig. 1-8B, the battery is connected so that the diode conducts, whereas in Fig. 1-8C, the diode resists the flow of current.

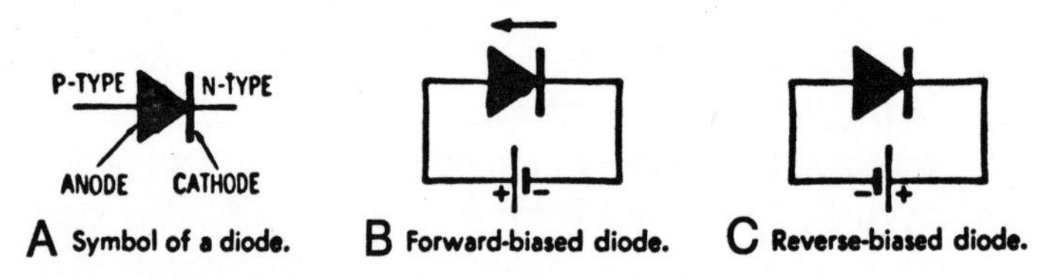

Fig. 1-8. Diode symbol and bias conditions.

## Junction Diode's Reverse-Bias Characteristics

The curve in Fig. 1-9A is a plot of the reverse characteristics of a conventional silicon junction diode. Since the diode is reverse biased (Fig. 1-8C), the curve is in the third

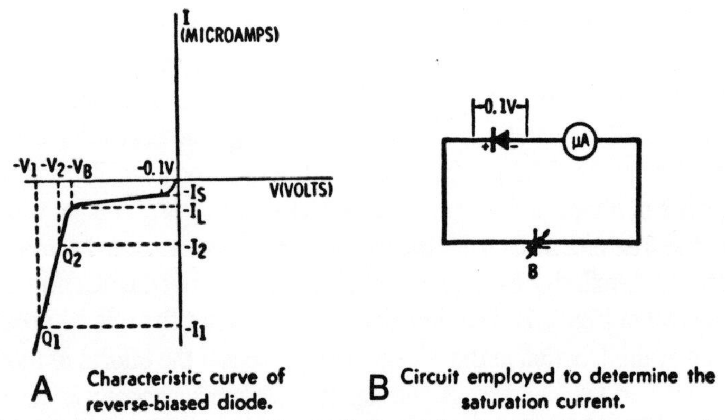

Fig. 1-9. The characteristics of the reverse-biased diode can be described and measured with a graph like this.

quadrant. Here, the voltage applied to the diode, as well as the current flowing through the diode, are negative.

To determine the amount of dc current flowing through the diode for a specific applied voltage such as V1 across the diode, first draw a vertical line from −V1 volts on the horizontal axis to the diode curve. The line intersects the diode curve at Q1. Next draw a horizontal line from Q1 to the vertical axis. It intersects this axis at −I1 microamps. Thus, with −V1 volts across the diode, there will be −I1 microamps of current. This construction may be repeated for any applied voltage. It is repeated in the drawing for a reverse voltage of −V2 volts.

The diode has definite dc and ac resistances, which vary with the portion of the curve under consideration. The dc resistance of the diode is calculated using Ohm's law. At point Q1, $R_{dc} = -V1/-I1$. At Q2, $R_{dc} = -V2/-I2$.

Ac resistance is different from dc resistance. Ac current and voltage vary with time through one or more cycles. Dynamic or ac resistance must be determined from the two points on the curve reached at extreme positive and negative gyrations of the cycle. Ac resistance is the resistance presented to a signal with amplitude variations throughout the cycle.

Assume an ac voltage is placed across the diode in such a manner as to cause the voltage across the diode to vary from −V1 to −V2. This can be done if the diode is biased by a dc supply to a point midway between −V1 and −V2. The ac resistance is actually the reciprocal of the slope of the diode curve. It can be determined from the formula:

$$r_{ac} = \frac{V1 - V2}{I1 - I2} \qquad (1\text{-}5)$$

The ac resistance will vary considerably at different parts of the curve. Several points on the curve characterize the diode when it is reverse biased. $I_S$ is the saturation current in the reverse direction and is dependent upon the temperature of the material. This current remains unchanged no matter what reverse voltage is applied across the diode. Saturation current approximately doubles for every 10°C rise in temperature although that figure can fall anywhere between 6°C and 16°C for silicon diodes. Because the saturation current of silicon diodes is low even at high temperatures, silicon diodes are frequently preferred over germanium devices.

Saturation current can be read from the curve as the current flowing when there is an applied reverse voltage of −0.1 or −0.2 volt. The test setup in Fig. 1-9B can be used in the laboratory to measure $I_S$. Here, the supply voltage B is adjusted until −0.1 or −0.2 volt is across the diode. The microammeter will indicate the saturation current.

Reverse current does increase in magnitude above $I_S$ as the reverse voltage is raised due to a leakage current caused by surface conditions and impurities. This leakage current, $-I_L$, rather than saturation current, is usually stated on diode specification sheets.

As the reverse voltage is raised, a value $-V_B$ is reached at which the current begins to increase rapidly in the negative direction ($-V_B$ is known as the breakdown voltage). Effectively, the current increases rapidly while the applied voltage remains relatively constant. This is the basis of operation of the zener or avalanche voltage-regulator diode, to be discussed later in this chapter. Breakdown voltage is most abrupt or sharp above $-6$ volts.

Many diodes used as power rectifiers do not exhibit sharp breakdown characteristics. Diodes that *do* have these attributes are referred to as *sharp avalanche breakdown* devices. They boast low ac resistance. Because a diode can dissipate as much power when current flows in the reverse direction as when it is forward biased, more reverse current can flow through a sharp avalanche breakdown device before its power dissipation capabilities are exceeded than can flow through a device without this sharp reverse breakdown characteristic. Consequently, sharp avalanche diodes are most desirable in any circuit where high transient voltage peaks are present. They should be used in power supplies.

### Forward-Biased Diode

The drawing in Fig. 1-10 shows the diode's curve when it is forward biased and conducting more readily than in the reverse-biased mode. Up to V1, the increase in current is very slow, and may be considered at 0 mA. Above V1, the current rises rapidly. The forward current, $I_F$, is related to the reverse saturation current, $I_S$. Voltage across the diode with a specific forward current flowing through the diode is:

$$V = \frac{(T + 273)}{5{,}000} \log_{10} \frac{(I_F + I_S)}{I_S} \qquad (1\text{-}6)$$

It is important to remember that the voltage across a forward-biased germanium diode is about 0.1 or 0.2 volt. For the silicon device, it ranges from 0.6 to 0.8 volt in most instances.

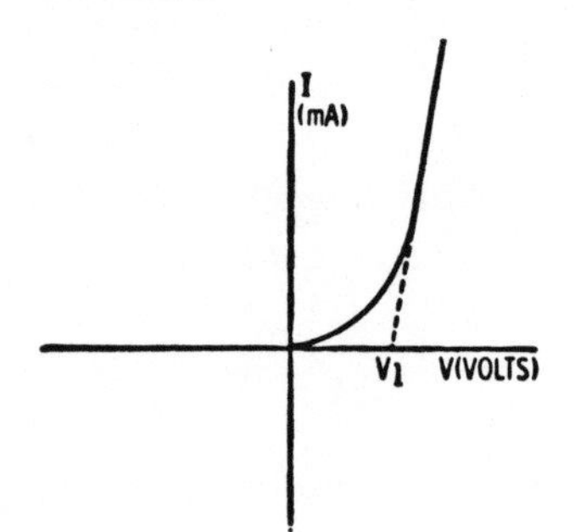


Fig. 1-10. This graph shows the characteristic curve of a diode in the forward-bias mode.

The ac forward impedance of the junction diode can be derived using Equation 1-5 and the curve in Fig. 1-10. A close approximation to the ac impedance at room temperature (about 25°C) is:

$$r_{ac} = \frac{26}{I_F} \quad (1\text{-}7)$$

where $I_F$ is expressed in milliamperes. Forward ac impedance increases linearly with temperature.

The discussion and equations stated above apply only to junction diodes. The shunt capacitance of this type of device is low enough so that it can be applied to many rf circuits. Detection or rectification applications at frequencies above 10 MHz should use the lower capacity point-contact diode that is composed of a metallic point as the anode. It is in contact with a semiconductor behaving as the cathode. Point-contact diodes are characterized by relatively high forward resistance and high reverse leakage currents.

### Diodes Connected in Series

In some applications, several diodes may be connected in series so that a greater peak inverse voltage may be applied to the combination than may be placed across any one of the individual devices. The resistors and capacitors shown in Fig. 1-11 should be connected across the diodes to swamp differences in individual devices.

Resistors are connected across each diode so that the applied voltage is divided equally between them. C1, C2, and C3 assure that transients are shared equally. Resistors R1, R2, and R3 should each be approximately half the reverse resistance of each diode. The value of the resistors is determined from Equation 1-8.

$$R \leq 500 \frac{V_B}{I_R(max)} \quad (1\text{-}8)$$

where,

$V_B$ is the peak voltage to appear across each diode,
$I_R(max)$ is the maximum specified reverse diode current in mA,
R is the value of the resistor across any one diode: R1, R2, or R3.

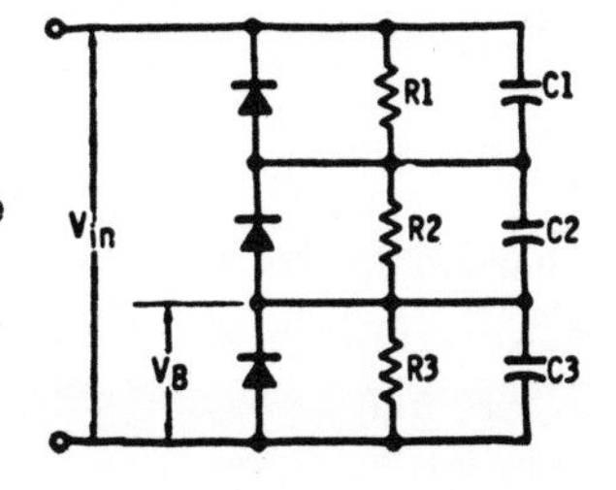


Fig. 1-11. In some circuits multiple diodes are connected in series.

The power dissipated by R is:

$$P_R = \frac{(V_B/\sqrt{2})^2}{R} \qquad (1\text{-}9)$$

where,

$P_R$ is the power dissipated by R,
$V_B$ is peak voltage across R.

The capacitor across a particular diode is usually made somewhat more than three times the rated capacitance of the zero-biased diode.

A diode may be switched rapidly in and out of conduction, and the switching may generate some instability in the form of a decaying oscillation. A 2- or 3-ohm resistor is frequently placed in series with each diode to assure rapid dissipation of the oscillating energy and let it settle quickly to a steady state.

## Diodes Connected in Parallel

Current required in a circuit may exceed the forward current capacity of an individual diode. Diodes may be connected in parallel to increase the overall current-carrying ability of the semiconductor portion of a circuit. The forward curves of the diodes (forward ac resistances) should be matched to assure that current is divided equally between the parallel-connected devices. To assure that any characteristic differences will not allow more current to flow through any one diode than it can pass safely, derate each device by at least 20 percent. Unless the ac resistances of all diodes are identical, one diode will pass more current than the other and it will be more subject to breakdown. This situation can be alleviated if a resistor is added in series with each diode. It can be done only if the voltage drop across the resistor is insignificant when compared to the voltage across the balance of the circuit. The resistance should be as small as possible to minimize any waste of power.

It is obvious that the reverse breakdown voltage of a circuit with diodes connected in parallel, is that of the device with the lowest rating. The leakage current of the circuit is the sum total of the leakage currents of all parallel-connected diodes. The general description of junction diodes can be applied to many specialized devices. These range from diodes used for power-rectifying purposes to switching and high-frequency rf devices.

## Rectifiers

Silicon junction diodes used as power rectifiers must exhibit several distinct characteristics. Reverse breakdown voltage must be sufficiently high so that the diode will not breakdown with the highest peak voltage that may be applied to it. Reverse leakage current must be low so as not to affect the average current levels in the load. Average and surge current ratings must be sufficient for the diode to perform its function properly for a long period of time. These factors are all detailed in Chapter 13.

## Bridge Rectifiers

Rectifiers are often used in a bridge configuration, as illustrated in Fig. 1-12. Special dedicated bridge rectifier units are available. They look like a small black box with four leads, as shown in Fig. 1-13. Such a bridge rectifier unit is nothing more than four ordinary rectifier diodes encapsulated in a single convenient housing. The unit is used in exactly the same way as a bridge made up of four separate diodes.

The chief advantages of a bridge rectifier unit are the convenience of using a single component in place of four, and the assurance that the four rectifiers making up the bridge will be well matched.

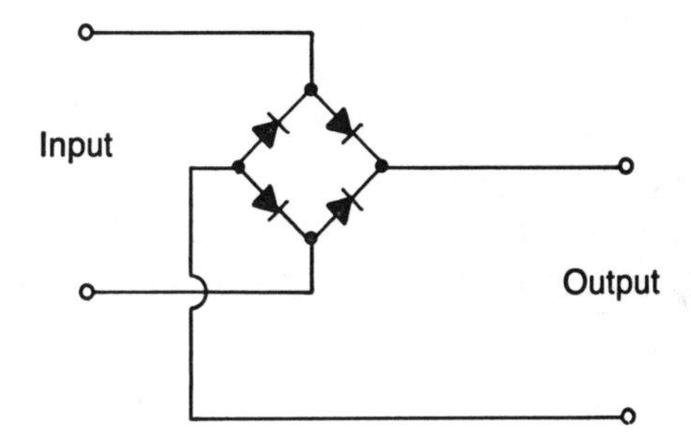


Fig. 1-12. A full-wave bridge rectifier network is made up of four diodes.

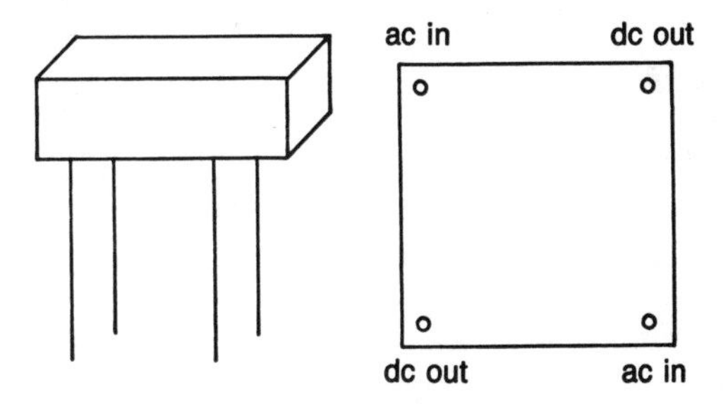


Fig. 1-13. Four separate diodes are often encapsulated in a single bridge rectifier package.

## Diodes in White-Noise Generators

Some p-n junction diodes, when biased in the reverse direction, generate noise as the avalanche breakdown voltage is approached. This is true of both standard diodes and zener diodes (which will be discussed later in this chapter).

Specially selected diodes can be used to produce true white noise when they are operated well into the breakdown region. White noise, used in audio tests as a "hiss," consists of frequencies of identical energies over the entire audible spectrum.

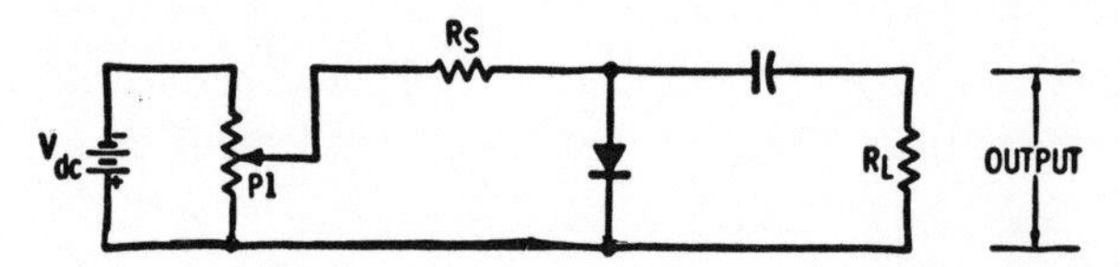


Fig. 1-14. The PN junction of a diode can be forced to act as a white-noise generator.

A typical white-noise generator circuit is shown in Fig. 1-14. The voltage is adjusted by the potentiometer, P1, until the diode is operating in its white-noise region. The noise across the diode is also developed across the high-impedance output load, $R_L$. $R_S$ limits diode current to safe values.

## Stabistor Voltage Reference Device

The voltage across the forward-biased germanium diode was accepted above as being 0.2 volt while the forward-biased silicon device was stated as having a forward voltage of 0.6 or 0.7 volt. These are more or less the voltages at which the respective devices start conducting a significant amount of current. Realistically, the voltage increases considerably with the amount of current flowing through the device. Single-junction germanium diodes can, at high current, have as much as 0.6 volt developed across the forward-biased junction, while more than 1 volt may be measured across a forward-biased silicon diode when there is a large amount of current passing through this device.

Over a limited current range, the voltage across the forward-biased diode varies little with changes in current. Should a load be placed across the diode, the voltage across the load and the stabistor is held relatively constant with variations of current. The action is not unlike that of the reverse-biased zener diode.

## Diode Switches

Diodes can serve as switches because they are ideally open circuits when reverse biased and closed circuits when forward biased. An arrangement using a diode as a switch is shown in Fig. 1-15. Here, the diode is in series with a radio-frequency or ac supply.

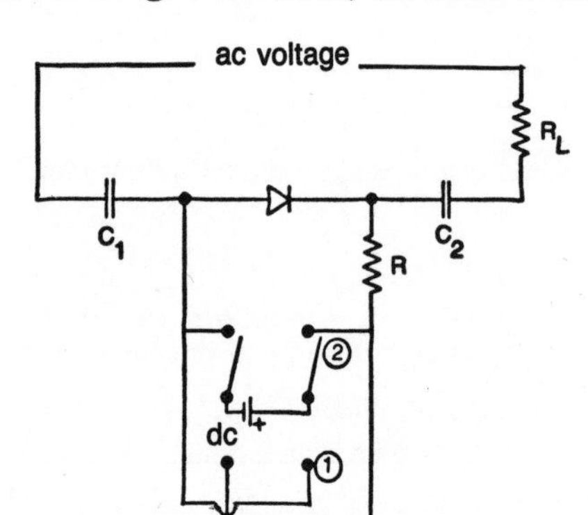


Fig. 1-15. Many applications use a diode as an electronic switch.

The device is isolated from the ac source by capacitors C1 and C2 so that dc is applied only to the diode. Ac is conducted when the switch is set to position 1, biasing the anode of the diode positive with respect to the cathode. Now the diode is turned on. There is no conduction when the switch is in position 2. The diode is reverse biased because the anode is made negative with respect to the cathode. Resistor R limits forward diode current to safe values.

The diode can also act to limit only a portion of the ac cycle. In this category, it behaves either as a clipping or clamping device, depending upon the design of the circuit. A typical clipping circuit is in Fig. 1-16. It can also be thought of as a clamping circuit, as the negative peak is limited to $-V_{BB}$.

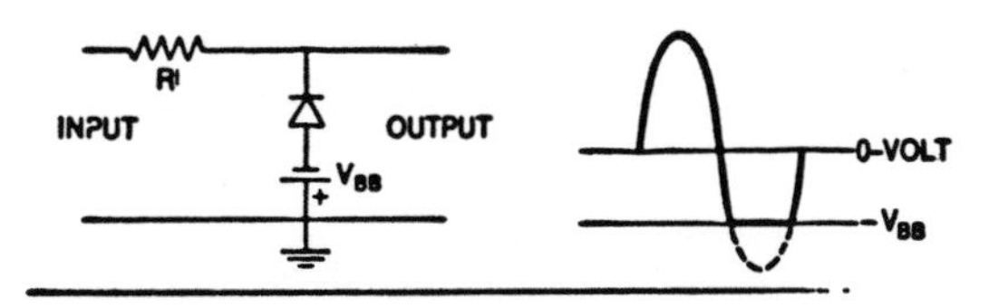


Fig. 1-16. Diodes can be used in clipping circuits.

A sine wave is applied to the input. The shape of the signal at the output is drawn at the right of the circuit. If the polarity of the battery and terminals of the diode were reversed, clipping would be in the alternate half of the cycle. The input signal supply is considered as having 0 ohms resistance for dc. Because of this, it is used to complete the circuit for the dc supplied by the battery. The battery itself presents 0 ohms to the ac signal voltage.

As can be noted in the drawing, the input and output signal is referenced to a 0-volt level. The battery voltage at the output is shown to be negative with respect to the 0-volt axis due to the polarity of the battery with respect to the common grounds. Should the polarity of the battery be reversed, its voltage would be represented by a horizontal line on the opposite side of the 0-volt axis.

The diode is reverse biased during the portion of the cycle that the upper input terminal is positive with respect to ground. During this half of the cycle, no portion of the signal is bypassed by the open circuit consisting of the diode and battery. All of the signal passes to the output in this positive half of the cycle. The diode is also reverse biased during part of the negative half of the input cycle. It is forward biased only after the applied ac input is of sufficient amplitude to make the cathode of the diode about 0.7-volt more negative (if a silicon device is used) than the anode. This is due to the presence and orientation of the battery, $V_{BB}$. When the diode is forward biased, it presents a short circuit to the input signal. During this portion of the cycle the signal is not permitted to reach the output. As for the balance of the cycle, the sinusoidal voltage is once again too positive to forward bias the diode. It is reverse biased and an open circuit exists where formerly there was a shunting path. Now the signal passes freely to the output. The shape of the signal at the output, shown as bold lines, is the input modified by the clipping circuit.

This description of a switching circuit presupposed that switching is instantaneous. This is not generally true. There are usually transients and time delays or time lags before the diode reaches the quiescent condition necessary for the particular mode of operation.

Let us examine the reaction of a series circuit consisting of a diode and resistor, when a square wave (see Fig. 1-17A) is applied to its input. Before $t_0$, the time when the diode is switched on, the input is at $v_r$ (min). The diode does not conduct. Any current flowing through the diode is shown in Fig. 1-17D. In the state when the diode supposedly does not conduct, there is a minute leakage current. A small negative voltage is developed across the series resistor (load) because of this current, as shown in Fig. 1-17C. The voltage across the load at all portions of the switching cycle is shown in this curve, while the curve in Fig. 1-17B describes the voltage across the diode itself.

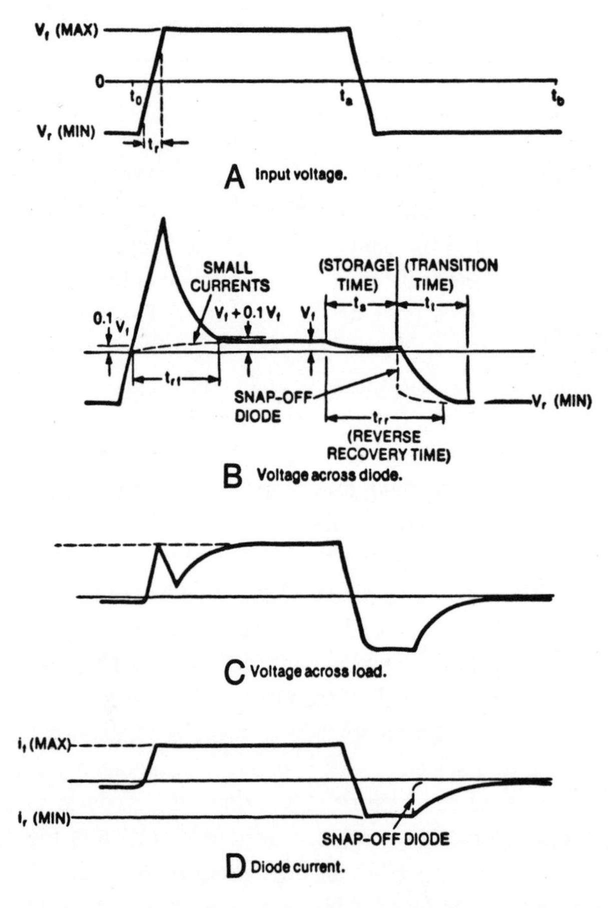


Fig. 1-17. The performance of a diode can be described via several current and voltage curves.

When the diode does not conduct, it is effectively an open circuit and the entire supply voltage (less the voltage drop across the load caused by the leakage current) is across the diode.

After time $t_0$, the diode is put into the forward-bias condition by the square wave. The maximum forward current flowing through the diode and circuit is determined by the size of the load resistor and is equal to $i_f(max) = V_f(max)/R_L$, by Ohm's law. As shown in Fig. 1-17D, the diode is conducting during this portion of the cycle, and the voltage across the diode rises to a peak value. This voltage, shown in Fig. 1-17B, is due to the diode behaving as a resistor rather than as a semiconductor while it is being switched. The size of this hypothetical resistor determines the size of the peak voltage across the diode, since $v = i_f(max)\ (R_b)$, where $R_b$ is the resistance of the diode at the instant it is switched on. Overshoot increases as the forward diode current ($i_f$) is increased and the rise time ($t_r$) of the input signal is decreased. (Rise time is the time it takes for the input of the square wave to rise from 10 percent to 90 percent of its final value, it is illustrated in Fig. 1-17A.) If the forward current is low, there is no peak. The signal takes the shape of the broken line shown in the forward-bias portion of the curve in Fig. 1-17B.

The voltage across the diode then drops to $v_f$. This is the approximate 0.6 volt for silicon diodes and the 0.2 volt for germanium devices, discussed above.

The voltage across the load rises to its peak value at the instant the diode is turned on. This is equal to $R_L$ multiplied by the steady-state forward current in the circuit, $i_f(max)$. The voltage then drops and once again rises to a constant level. The voltage across $R_L$ is the difference between the supply and diode voltages.

The forward recovery time, $t_{rf}$, covers the period of time from the instant the diode starts to conduct until it reaches some point within 10 percent of $v_f$. The latter point is 10 percent greater than $v_f$ if there is overshoot and 10 percent less than $v_f$ if the diode voltage rises slowly as shown by the dotted line curve marked "small currents."

Several factors affect the transients when the diode is first turned on. To reduce $t_{rf}$ as well as the instantaneous peak voltage across the diode, keep the reverse voltage $v_r$ (min) of the input signal to a minimum. Because only small currents will flow, $v_r(min)$ can be made equal to 0 volt or even made slightly positive. If $v_r(min)$ is made slightly positive, it must still be kept below the cut-in voltage of the diode. Other procedures that help minimize the peak voltage and $t_{rf}$ are: keep the rise time, $t_r$, of the input voltage as small as possible; keep the forward diode current, $i_f(max)$, to a minimum by limiting the applied circuit voltage; use the largest load resistor, $R_L$, compatible with circuit requirements; and keep the ambient temperature near the diode as high as possible, consistent with proper operation of the device.

The diode voltage remains at $v_f$ until time $t_a$, when the input voltage switches to a negative value. The diode current drops rapidly to $i_r(min) = v_r(min)/R_L$, as seen in Fig. 1-17D. The diode voltage remains positive for a short period, known as storage time, $t_s$. It then gradually drops to a constant value, $v_r(min)$. This interval of time, $t_t$, is known as the transition time. The reverse recovery time ($t_{rr}$) is the period of time from $t_a$ to the time the diode is at a specific high impedance, or conducts only a low leakage current.

Practically the entire supply voltage is then across the open circuit diode, as before time zero. Here, too, the voltage across the load is the difference between the input voltage and the voltage across the diode.

The reverse recovery time and the forward recovery time behave in similar fashions. If you use the suggestion noted above to achieve a low forward recovery time, the reverse recovery time will also be relatively fast. It follows that when choosing a switching diode, select one specified as having the lowest $t_{rr}$ in the group. Other factors indicative of a diode with good switching characteristics are low junction capacity and low forward voltage drop.

Special diodes have been developed to surpass the switching characteristics of ordinary junction diodes. As can be seen from the broken line in Figs. 1-17B and 1-17D, the "snap-off" or charge-storage diode provides a tremendous reduction in the transition time when compared to most other devices.

## Snap-Off and Hot-Carrier Diodes

Snap-off diodes behave as they do because of the stored charge. When forward biased, charge is stored in the diode. Should the diode be reverse biased after the charge has been stored, it conducts until all charge has been removed. Ideally, the diode turns off completely at the instant the charge is depleted. To approach the ideal in turn-off curve characteristics, the capacities and inductances in the circuit and device should be kept to a minimum. Turn-off time is also enhanced by minimizing the amount of charge stored in the device while maximizing the reverse current.

Besides serving in switching circuits, charge-storage diodes can be used in frequency-multiplier circuits in the GHz region. A signal at the fundamental frequency is fed to the circuit with the diode. During the positive portion of the cycle, charge is stored in the diode because it is forward biased. When the input signal is in the negative portion of its cycle, the diode instantly conducts large amounts of reverse current. It immediately reverts to the reverse-current characteristic of conventional diodes. Many high-order harmonics are in the transient signal at the output of the diode.

Another excellent switching device is the hot-carrier diode. Unlike the snap-off device, it stores no charge. Its primary advantage as a switch rests in the fact that there is practically no overshoot when the diode is switched to a reverse state. Consequently, the reverse recovery time is extremely low. This characteristic makes it very useful in applications in the microwave devices.

Even though the hot-carrier diode is not a junction device, it is mentioned here because of its excellent switching characteristics. The junction of the hot-carrier diode is a metal and a semiconductor. It generates little noise. When it is forward biased, there is about 0.25 volt across the device.

## PIN DIODES

Unlike other devices, PIN diodes will switch at extremely high rf frequencies above 200 or 300 MHz. They are composed of three sections. There are the usual p- and n-type semiconductor materials. Sandwiched between the p- and n-slabs is an intrinsic or pure layer of high resistivity material. The breakdown voltage is a function of the width of the intrinsic region.

The operation of the PIN diode can be divided into two bands separated, more or less, at a frequency $f_0$. Below $f_0$, the diode performs as if it were the conventional junction device with all its inherent nonlinearity and rectification properties. Above $f_0$, the diode is a pure resistor. It does not rectify at these frequencies. The resistance is dependent upon the dc flowing through the device and, at room temperature, is approximately equal to 48/I, where I is the dc flowing through the diode and expressed in milliamperes. Resistance may vary from 1 ohm to over 10,000 ohms. The frequency at the dividing point, $f_0$, is equal to $1/6.28\tau$, where $\tau$ is referred to as the *recombination lifetime* of the device.

Changing the dc across the device changes the rf resistance. It will thus behave as a switch acting as a short circuit when the resistance is low, and act as a relatively open circuit when the resistance is high. If an audio signal is applied across the diode, the variation in resistance will conform to the applied intelligence. Hence, the PIN diode can serve in a modulating function, as well as an rf switch.

## ZENER DIODES

Approximately 0.6 volt is developed across silicon diodes biased in the forward direction. If a variable load (with a high impedance relative to that of the diode) is placed across the forward-biased semiconductor, the voltage across the load will be maintained at about 0.6 volt, even if the current required by the load varies considerably.

Should the voltage applied to the semiconductor be reversed, the diode will break down at a specified voltage, $-V_B$. As the voltage is made more negative than $-V_B$, the current increases at an extremely rapid rate. If the current is limited by external circuitry so that it does not exceed the maximum specified value for the diode, it will not dissipate more than its rated power and will not disintegrate. Despite the increased current flowing through the diode, the voltage across the semiconductor will remain fairly constant. A relatively constant voltage will also be maintained across any load connected in parallel with the diode.

A drawing of a zener diode and a typical curve describing its characteristics are shown in Fig. 1-18. Only the reverse characteristic to be discussed here has been drawn. The forward characteristic is the same as for any other forward-biased junction diode. Ideally, $-V_B$ is a constant while I can vary from zero to minus infinity. I must be limited so that the power dissipation rating of the diode is not exceeded.

Several zener characteristics should be noted here. First, the zener breakdown voltage is temperature dependent. When the breakdown voltage is in the zener region, or

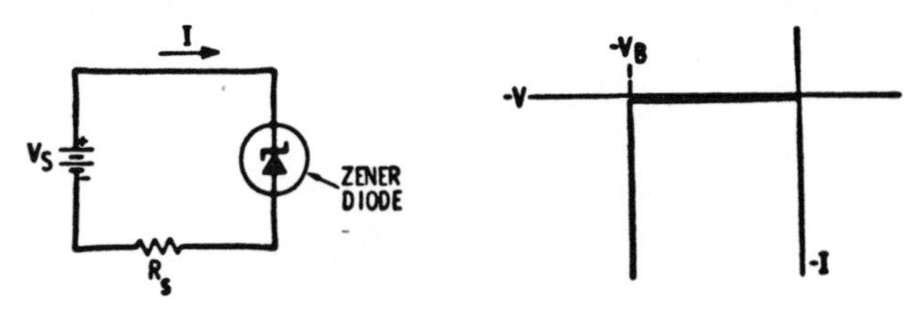

Fig. 1-18. Here we see a basic zener diode circuit with its ideal curve.

less than 5 to 6 volts, the zener voltage decreases 0.1 percent for every °C the temperature increases. In the 5- to 6-volt region, the zener voltage remains relatively unchanged despite temperature variations. In the avalanche region above about 6 volts, the zener voltage increases 0.1 percent for every °C the temperature rises.

The forward-biased diode voltage decreases as the temperature rises (−2.5 mV/°C). The voltage increase across the reverse-biased diode in the avalanche region can be compensated for by connecting a forward-biased diode of sufficient current-carrying capability in series with the reverse-biased zener. Do not forget to add the drop in the forward-biased diode (about 0.6 volt) to the zener voltage to determine the total voltage across the combination. Examples of this are shown in Fig. 1-19. In Fig. 1-19A, the diodes are individual units, and in Fig. 1-19B, they are combined into a single package.

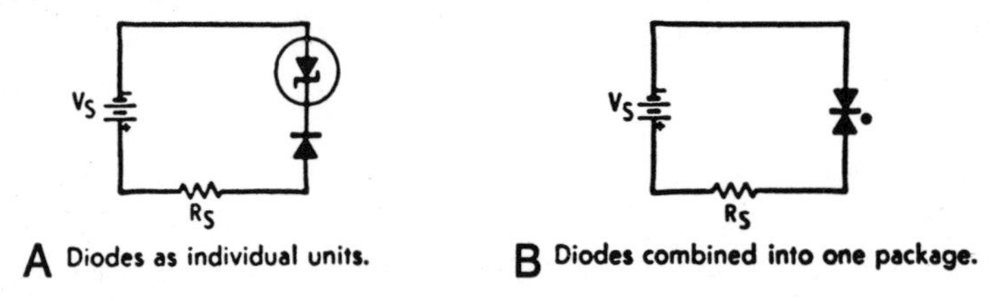

Fig. 1-19. In some applications, temperature compensation should be used with zener diodes.

A second important characteristic can best be described if you refer to the curve in Fig. 1-18. A sharp breakdown point is shown at $V_B$. This is reasonably true of diodes working in the avalanche region. For diodes operated in the zener region, this breakdown point is rounded and quite indecisive. The designer can bypass the region by choosing a minimum zener idling current equal to about 10 percent of the maximum current rating of the diode. Another factor guiding the idling current choice is that zener diodes are noisy in the low current region.

The third zener characteristic of importance is the ac resistance derived using Fig. 1-9A. The zener-diode resistance can be derived in a similar fashion.

The circuit in Fig. 1-18 will be used to derive the information required to make use of the zener diode as a regulator, and, for demonstration purposes, the curve in Fig. 1-20 will be used to represent the diode. It is assumed that the zener diode used here has a relatively high impedance.

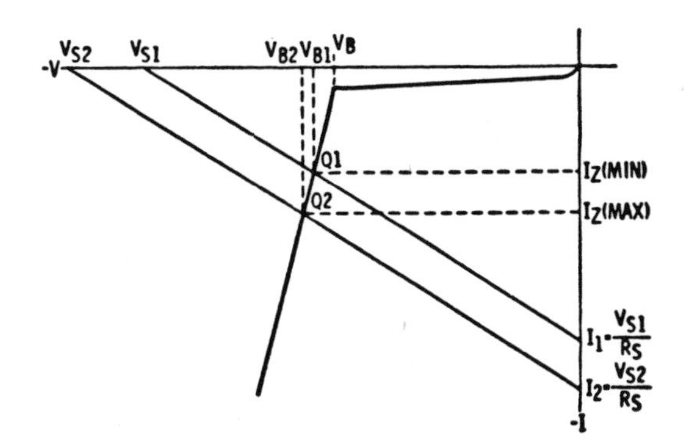


Fig. 1-20. Zener diode curve including load lines.

The supply voltage is nominally $V_S$. It varies from $V_{S1}$ to $V_{S2}$ with power-line voltage changes. The diode impedance is assumed to be negligible compared to that of the series resistor in the circuit, $R_S$. The required information to be derived from this data is how much the output voltage across the diode will change when the input supply jumps from $V_{S1}$ to $V_{S2}$ volts.

Using supply voltages $V_{S1}$ and $V_{S2}$, draw two load lines for $R_S$ over the diode curve. First, assume the supply voltage is $V_{S1}$. With no current, the entire supply voltage is across the diode. The intersection of $-V_{S1}$ volts and 0 mA is one point on the load line.

Next, assume there is 0 voltage across the diode. The current flowing through the circuit is $I1 = V_{S1}/R_S$. Mark I1 on the vertical axis. Connect the two points just located with a straight line. The line crosses the diode curve at Q1. Draw a vertical line from Q1 to the axis at $V_{B1}$. This is the voltage across the diode when the input voltage is $V_{S1}$. Draw a horizontal line from Q1 to the axis at $I_Z$(min). This is the current flowing through the diode under the same conditions.

Repeat this procedure, assuming that the supply voltage shifted to $V_{S2}$. Note that $V_{B2} - V_{B1}$ is much less than $V_{S2} - V_{S1}$. For a large supply-voltage variation, $V_{S2} - V_{S1}$, there is only a small voltage change, $V_{B2} - V_{B1}$, across the diode. However, the current flowing through the diode will vary a considerable amount from $I_Z$(min) to $I_Z$(max).

The voltage across a load, $R_L$, connected in parallel with the diode, will be subjected to the same good voltage regulation as is the diode. A circuit of this type is shown in Fig. 1-21. The equation for the voltage drops around the circuit in the drawing is:

$$V_S = R_S(I_{R_L} + I_Z) + V_B$$

where,

$I_{R_L}$ is the current flowing through the load,

$I_Z$ is the current flowing through the zener through the zener diode.

$V_B$ is the nominal reverse breakdown voltage of the diode.

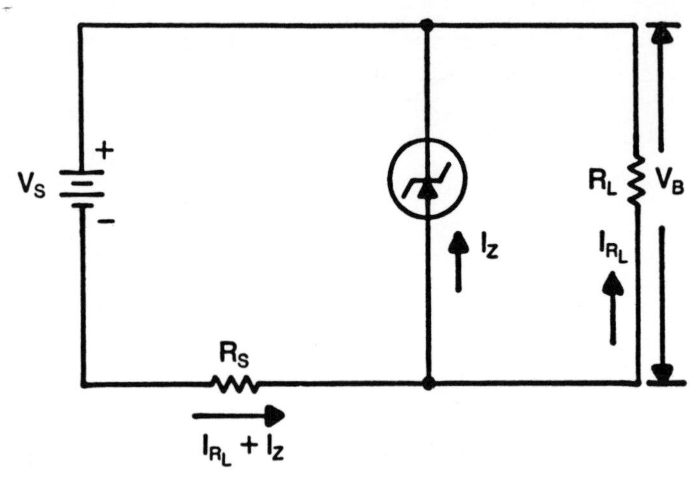

Fig. 1-21. Simplified circuit showing a zener diode with an output load.

Two factors may vary in a circuit of this type. These are the supply voltage, the load resistor (or load current, $I_{R_L}$), or both. First assume that only the supply voltage varies. This case has been depicted graphically in Fig. 1-20. In order to determine the components in the circuit, it is necessary to start with the circuit equation stated above. The zener idling current is assumed to be about 0.1 $I_{R_L}$ so that the total current in the circuit is $0.1\ I_{R_L} + I_{R_L}$, or $1.1 \times I_{R_L}$. The maximum size of the series resistor is:

$$R_S = \frac{V_{S1} - V_B}{1.1\ I_{R_L}} \text{ ohms} \qquad (1\text{-}10)$$

where,

$V_{S1}$ is the minimum value of the supply voltage.

The maximum current flowing through $R_S$ is:

$$I_{R_L} + I_Z(\text{max}) = \frac{V_{S2} - V_B}{R_S} \text{ amps} \qquad (1\text{-}11)$$

where,

$V_{S2}$ is the highest supply voltage,

$I_Z(\text{max})$ is the maximum zener current.

The power dissipated by the series resistor is derived from the general power equation, $P = I^2R = V^2/R$:

$$P_S = \frac{(V_{S2} - V_B)^2}{R_S} \text{ watts.} \qquad (1\text{-}12)$$

The maximum power dissipated by the zener is $I_Z$ (max) $V_B$, or, from Equation 1-11:

$$P_Z \text{ (max)} = \left( \frac{V_{S2} - V_B}{R_S} - I_{R_L} \right) V_B \text{ watts} \qquad (1\text{-}13)$$

If the input voltage is constant at $V_S$, and the load current varies from a maximum $I_{R_L}$ (max) to a minimum $I_{R_L}$ (min) due to a variation of load resistance, $R_L$, the maximum allowable series resistor that can be used is:

$$R_S = \frac{(V_S - V_B)}{1.1\ I_{R_L} \text{ (max)}} \text{ ohms} \qquad (1\text{-}14)$$

The power dissipated by the resistor is:

$$P_S = \frac{(V_S - V_B)^2}{R_S} \text{ watts} \qquad (1\text{-}15)$$

and the maximum power dissipated by the diode is:

$$P_Z \text{ (max)} = \left[ \frac{(V_S - V_B)}{R_S} - I_{R_L} \text{ (min)} \right] V_B \text{ watts} \qquad (1\text{-}16)$$

Should $I_L$ vary from $I_{R_L}$ (max) to $I_{R_L}$ (min) while the supply voltage also varies, use the following equation to calculate the maximum size of the series resistor:

$$R_S = \frac{V_{S1} - V_B}{1.1\ I_{R_L} \text{ (max)}} \text{ ohms} \qquad (1\text{-}17)$$

The power dissipated by $R_S$ is:

$$P_S = \frac{(V_{S2} - V_B)^2}{R_S} \text{ watts} \qquad (1\text{-}18)$$

Then zener diode dissipates a maximum of:

$$P_Z \text{ (max)} = \left[ \frac{(V_{S2} - V_B)}{R_S} - I_{R_L} \text{ (min)} \right] V_B \text{ watts} \qquad (1\text{-}19)$$

When zener diodes are connected in series, the breakdown voltage of the combination is equal to the sum of the breakdown voltages of the individual zener diodes. When connected in parallel, all current flows through the diode with the lowest breakdown voltage, after it has been exceeded, and all the power is dissipated by that device.

## VARACTORS

Referring to Fig. 1-7, the depletion region located at the junction of the n-type and p-type slabs acts as a dielectric. This is similar to the insulator between the plates of a capacitor. Hence, there is a capacity between the connections to the anode and cathode of all diodes. As the anode is made more negative with respect to the cathode, the width of the depletion region increases, effectively increasing the space between the anode and cathode and reducing the capacity of the junction diode. We consequently have a device whose capacity can be changed merely by altering the applied voltage.

Capacitance of a junction diode may be as high as 2,000 pF. The relationship of the capacity to the applied voltage is somewhere between $C \approx k(1/V_r^{1/3})$ and $C \approx k(1/V_4^{1/2})$, where $V_r$ is the reverse voltage applied across the diode and $k$ is a constant. The Q (the ratio of the capacitive reactance to the resistance) can be as low as 50, but is more likely to be near 300.

When using ordinary diodes in rf circuits, capacitance may become an important factor. High frequencies are conducted through that capacity. The net effect is a diode that acts like a conductor of signal.

Varactors are specifically designed to provide the maximum capacitance range with changes in applied voltage. The equivalent circuit is a capacitor, C, in series with a voltage-dependent resistor, R. A limiting factor, the cutoff frequency, $f_c$, can be derived from these two values:

$$f_c = \frac{1}{2\pi RC} \qquad (1\text{-}20)$$

$f_c$ is the frequency at which the Q of the diode is equal to 1, because:

$$Q = \frac{1}{2\pi fRC} \qquad (1\text{-}21)$$

where f is any frequency in Hz at which the diode is used. The diode cannot be used as a capacitor at frequencies above $f_c$.

The capacitor diode (varactor) has been used to replace mechanical tuning elements in rf circuits. The frequency of a resonant circuit is adjusted by varying the voltage across the diode. A typical tank circuit is shown in Fig. 1-22. The varactor can also be used in switching, limiting, harmonic generation, and parametric amplification applications.

## TUNNEL DIODES

Diodes discussed above, can be thought of as resistors. When forward biased, resistance is low; when reverse biased, resistance is high. In either case, resistance is

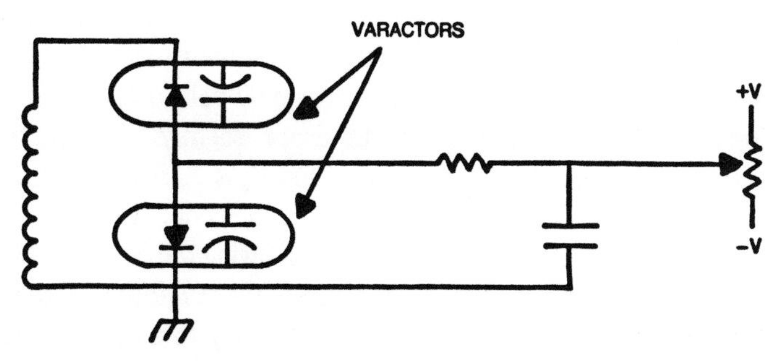


Fig. 1-22. A typical tank circuit using varactors.

positive. This is essentially true of the resistance of the tunnel diode. It is the ratio of two positive values—positive going voltage and positive going current. What makes the tunnel diode unique is that over part of its forward-biased characteristics, the current flowing through the device is reduced or negative, while the voltage across the device continues to go positive. If you divide a positive going voltage by a negative going current, you have a negative resistance. This negative resistance allows you to use the tunnel diode as a switch, as an amplifier, or as an oscillator.

## Diode Characteristics

In the tunnel diode, both semiconductor slabs are heavily doped. Because of this, electrons and holes will flow through or tunnel through the depletion region, despite the potential barrier. With a forward voltage across the diode, tunneling continues until about 0.07 volt is across the device. This voltage is symbolized $V_p$. (See Fig. 1-23.) The depletion region remains at its initial width however narrow it is, despite the forward

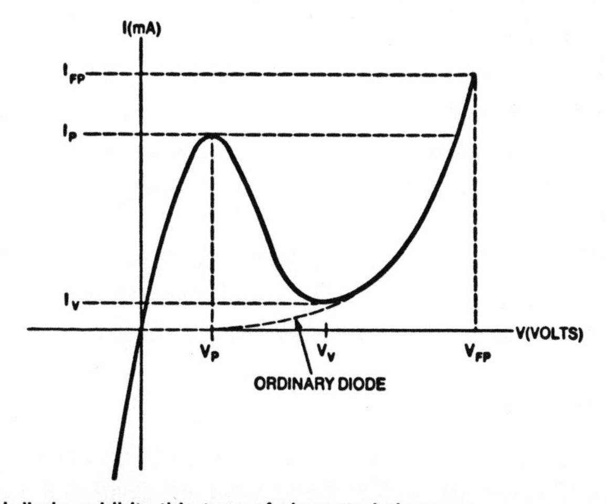


Fig. 1-23. A tunnel-diode exhibits this type of characteristic curve.

voltage. It can get narrower only after the cut-in voltage (0.3 to 0.6 volt) is across the diode. Refer to this cut-in voltage as $V_v$. Because of this, more electrons cannot flow through the diode despite the increase of voltage from 0.07 volt to 0.3 or 0.6 volt. Voltage increases between $V_p$ and $V_v$, but current drops slightly. The section of the characteristic curve of the tunnel diode between $V_p$ and $V_v$ is its negative resistance region. Above $V_v$ the device acts as an ordinary junction diode. This voltage-current relation is illustrated in Fig. 1-23. Note the broken line showing the curve for the ordinary diode. Several specific points on the tunnel diode curve are of interest to the designer.

At zero voltage, the current is zero. It rises immediately to a peak current, $I_p$ mA, when the voltage rises to $V_p$ volts. This increase in current is the first of the two positive resistance sections of the characteristic curve. It is followed by a negative resistance section where the current drops to the valley current, $I_v$ (about 0.35 mA for a germanium device and about 0.45 mA for a silicon diode), when the valley voltage is $V_v$. It turns up once again (following the conventional diode characteristic with a positive resistance) to a maximum allowable peak current of $I_{fp}$ when the forward peak voltage is $V_{fp}$. In some specification sheets, $I_p$ and $I_{fp}$ coincide, resulting in a lower value of $V_{fp}$.

## Parameters of the Tunnel Diode

Peak and valley voltage and current are functions of the semiconductor materials of the device. The ratio of the currents is a more important consideration than are their absolute values. The current that is about midway between $I_p$ and $I_v$ and the voltage about midway between $V_p$ and $V_v$, are referred to as inflection current, $I_o$, and inflection voltage, respectively. Negative resistance is at its maximum at this point.

The parameters of the devices are somewhat temperature dependent. $V_p$ drops 0.1 mV and $I_p$ increases about 0.05 percent, for each °C the temperature rises. A 1°C increase in temperature produces a drop of 0.8 percent in $V_v$ and an increase of 1.5 percent in $I_v$. Silicon diodes are most frequently used as switching devices as they are not excessively temperature sensitive.

Negative resistance in the region between $I_p$ and $I_v$ is quite important when the tunnel diode is to be used as an oscillator. Like the other parameters, negative resistance varies with the material used to construct the diode as well as with $I_p$. Its value is quite temperature dependent, decreasing at the rate of about 0.5 ohm for each °C the temperature rises.

Noise at the output of a diode is directly related to the product of $I_p$ and $R_n$, the negative resistance. A noise constant is 10 $I_pR_n$. It indicates how well the diode behaves in this respect. Diodes made of gallium arsenide are the least noisy.

## Biasing the Tunnel Diode

The tunnel diode can be biased for any one of three modes of operation: monostable, astable, and bistable. The various modes of operation are illustrated using the circuit

Fig. 1-24. In a typical application a pulse is fed to the tunnel-diode which is biased by a battery.

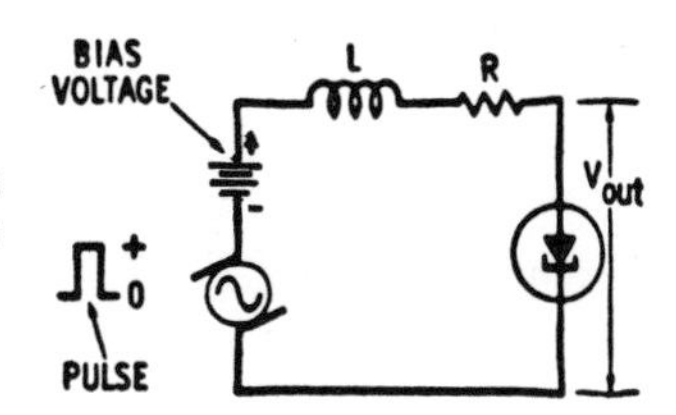


in Fig. 1-24. Here, the dc bias is obtained from a battery. A pulse is then fed to the circuit to upset the quiescent condition. If biased in the monostable mode, the load line can cross the curve at either positive resistance region. This is shown in Fig. 1-25A.

Assume first that the supply is at voltage V for load line (1) as shown in Fig. 1-25B. When a positive pulse is applied to the diode, it raises the bias voltage and the current rises along the curve to $V_p$-$I_p$. Unstable in this condition, the voltage jumps to $V_f$ while the current remains at $I_p$. Since the pulse is of short duration, the current and voltage drop to $V_v$-$I_v$, jump to $V_x$-$I_v$, and climb once again to the Q starting point. A similar analysis can be made if load line (2) in Fig. 1-25A is the monostable load line. In either case, the mode of operation can be used in a relaxation oscillator or switching circuit.

In the astable mode, as used in amplifiers or oscillators, the load line crosses the negative-resistance region, as shown in Fig. 1-26A. Here the load line starts at Q, but a pulse causes it to rise to $V_p$-$I_p$ and pursue the same course as shown in Fig. 1-25B.

The bistable circuit load line is shown in Fig. 1-26B. This bias mode is used in relaxation oscillators and switching circuits. Assume the action starts when the diode is biased at Q. If the pulse is large enough in the positive direction to raise the load line above the $V_p$-$I_p$ point, the line shifts for an instant to the location in Fig. 1-26C. The new stable bias point will be at Q1, as shown in Fig. 1-26B.

If the next pulse is in the negative direction and is of adequate amplitude to move the load line so that it is below $V_v$-$I_v$ for an instant, as in Fig. 1-26D, the quiescent point will reset itself at Q.

When the tunnel diode is used as a switch (e.g., to perform high-speed logic functions), the important parameters are the peak and valley voltages and currents. The capacitance of the diode and associated inductance are important in determining the switching speed limits.

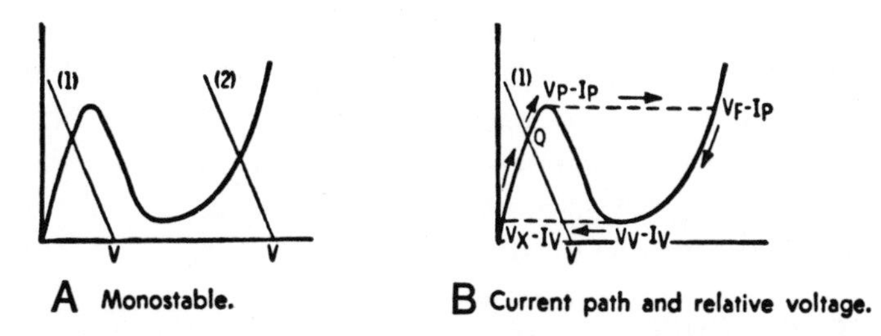


Fig. 1-25. These tunnel-diode curves include load lines.

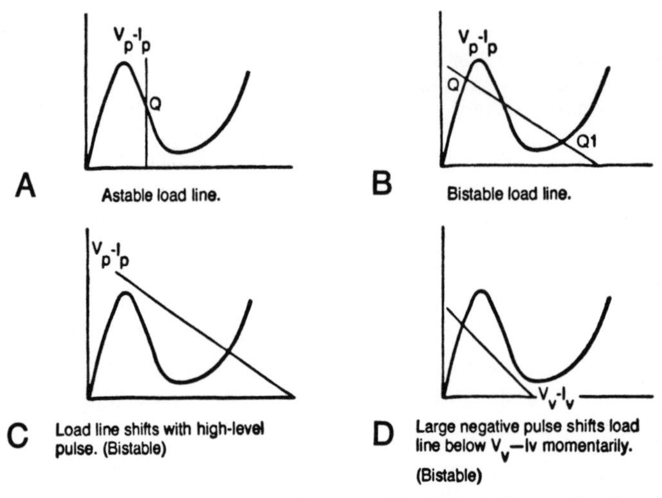


Fig. 1-26. These curves illustrate the astable, and bistable modes of operation for a tunnel-diode.

When used as an amplifier, the diode is biased in the negative resistance region and is connected to an L-C resonant tank circuit. The negative resistance cancels a portion of the positive resistance inherent in tank circuits. The noise generated by the diode should be low. It is directly proportional to the negative resistance and $I_0$, the inflection current.

As an oscillator and modulator, the tunnel diode is usually biased in the negative-resistance region of the curve. In mixer and relaxation-oscillator applications, the bias should be in the positive-resistance region nearest zero. All parameters are reasonably important in this application. The frequency of oscillation is always less than the resistive cutoff frequency $f_{ro}$ of the diode.

## Unitunnel Diode

Note the reverse characteristic of the tunnel diode in Fig. 1-23. Reverse current starts to rise at 0 volts, and continues to increase. The reverse breakdown voltage is 0 volts. Should it be required to get relatively high current near 0 volts, the diode can be biased in the reverse direction. The efficiency of rectification is very high. A diode operated in this fashion is known as a back diode or tunnel rectifier.

In the forward direction, the specially designed back diode has a slight peak current that is almost independent of voltage when small forward voltages are applied.

Back diodes perform well as mixers and detectors. They can detect the most minute signals because of their 0-volt cut-in voltage. Because of this, back diodes are useful as small-signal video detectors up to about 10 GHz.

## LIGHT SENSITIVE DIODES

The final characteristic of the junction diode to be discussed, is how some of its characteristics relate to the intensity and frequency of light impinging on the junction. The light-emitting diode or LED is also described here.

### Photodiodes

Should a diode be encased in a clear plastic container and its terminals connected to a voltage source so that the semiconductor is reverse biased, the amount of current flowing through the circuit depends upon the amount of light striking the junction. The current flowing varies almost linearly with the amount of light at the junction. This assumes that no other factors limit circuit current.

In the absence of light, the usual reverse saturation current flows through this diode just as it flows through any other junction diode. Should the temperature of the diode be raised, there will be an increase in the reverse current. Heat is energy. Light is energy. If light energy were applied to the diode rather than heat, the consequences would be similar. More reverse current would flow. Because diode current varies almost linearly with the intensity of light hitting the device, the diode is extremely useful as a movie sound track pickup device, as a light-operated switch, as a rapid counting device on production lines, and so on.

### Photovoltaic Diodes

When light strikes a photovoltaic diode, a voltage exists across its terminals. It is known as the photovoltaic emf. It is this diode that is the basis of the solar cell. Open-circuit voltage of the silicon solar cell is about 0.6 volt.

Under open-circuit conditions, no current is drawn from the cell. When the cell is shorted, the voltage across the device is zero. Obviously, power delivered by the cell is zero when the cell is shorted or when no current is being drawn. Maximum output power from the cell can be realized only when some specific intermediate-size resistor is placed across the device. Information as to the optimum size of this resistor is frequently supplied by the manufacturer of the cell.

### LEDs

When holes and electrons combine, energy is radiated in the form of heat. This energy can also be radiated in discrete bundles emitting light instead of heat. These light-emitting diodes or LEDs are operated in the low-current low-voltage regions. As a rule of thumb, 1.3 volt and 1 mA for a total of 1.3 mW input power is all that is required to turn on the device.

The LED is used primarily as an indicating light. It can also perform as the light emitting element of an optically coupled isolator. (See the section on phototransistors in the next chapter.) Because it emits small amounts of light, stray light may upset the performance of an optically coupled isolator using an LED. In order to separate the output from the LED from other sources of ambient light, the LED must be pulsed or modulated.

Pulsed light from the LED is then picked up or received by a photodiode (or phototransistor). Through the use of circuitry, the pulsed light is differentiated from the constant ambient light and amplified. These devices can now serve in burglar-alarm systems, tape recorders, and so on.

When an LED is used in an opto-isolator, it works well only when it is close to the detector element. For operation over a distance, a lens may be placed near the LED to focus the light onto the detector. Because it also directs stray light from the LED to the detector, light from the LED that would normally never reach the detector will now impinge on it, materially increasing the sensitivity of the overall system.

Most LEDs are made from a direct band-gap semiconductor material. Gallium arsenide (GaAs) is widely used.

LEDs are available in a variety of colors. Red is by far the most common. It tends to be the most visible color in many applications. Green and yellow LEDs are also fairly popular. Recent developments have made additional colors, such as blue, possible, although these units are not yet in wide use, and may be unavailable to the experimenter.

An LED is housed in an epoxy bubble. The epoxy is selected to maximize the transmission of light at the wavelength emitted by the LED. Often, a lens of some sort is built into the LED's housing. Such a lens serves two purposes, it helps enhance the light transmission from the LED, and it can also control the directional characteristics of the emitted light.

Infrared LEDs are also available. They work in fundamentally the same way as standard LEDs, but the emitted light is in the non-visible range.

A specialized variation on the basic LED is the laser diode. Laser diodes feature a very flat and uniform junction, and the emitted light is a very coherent beam. This type of laser is called the injection laser. Laser diodes are appearing in numerous low-power consumer applications, such as CD (Compact Disc) players.

In practical circuits, an LED is used with a current-limiting series resistor, as shown in Fig. 1-27. If this resistor is not used, the LED will draw an excessive amount of current and quickly burn itself out. The resistor's value will usually be somewhere in the 200 ohm to 1000 ohm (1K) range. The best value will depend on the characteristics of the specific LED used. The exact series resistance will determine how brightly the LED will glow. In most applications one of the following standard resistor values is used; 330 ohms, 390 ohms, or 470 ohms. If you're not sure what value to use for the LED you are working with, start out with one of these three values (whichever one happens to

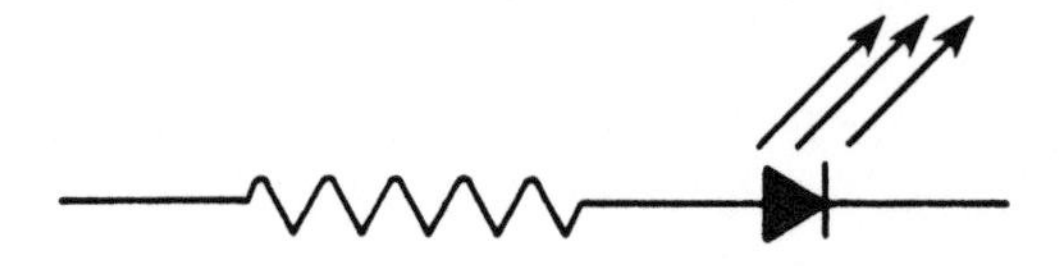

Fig. 1-27. A dropping resistor is normally used in series with an LED.

be handy). If the LED glows too brightly for your application, increase the value of the series resistor. Similarly, if the LED's glow is too dim, it can be brightened by decreasing the series resistance. But be careful. There is always a limit to how brightly an LED can glow. If you reduce the series resistance too much, you will defeat its purpose as a current limiter. Unless the manufacturer's specification sheet indicates otherwise, the series resistance should never be less than 200 ohms.

## POINT-CONTACT DIODES

Except for the brief mention of a diode made from a metal and a semiconductor when discussing hot-carrier diodes, the entire discussion until now centered around the junction device made from one n-slab and one p-slab of semiconductor material. Like the hot-carrier device, point-contact diodes are made from a semiconductor and a metal. A point metal such as tungsten makes contact with n-type semiconductor, usually silicon.

In order to forward bias the point-contact diode, the metal must be made positive with respect to the semiconductor. Electrons flow from the semiconductor to the metal. Ac resistance of this diode is usually much higher than that of the junction device. The voltage-current characteristic curve does not rise as sharply as the one drawn in Fig. 1-10, but runs somewhat lower, lying closer to the horizontal axis. Reverse leakage current is low, but seldom approaches the minimal standards set by junction devices.

Some of the prime applications of point-contact diodes are at frequencies in and above the uhf region from 300 MHz to 3 GHz, where they serve well as detectors and mixers. Also used in these situations are the Schottky-barrier diodes and the unitunnel diodes.

### Schottky-Barrier Diode

Microwave diodes can be formed with a metal point exerting pressure against a semiconductor material. It is this principle that is applied to a diode when metal particles are sprinkled onto it and made to adhere to a semiconductor slab. If the metal is aluminum and the semiconductor is n-type silicon, a Schottky diode is formed.

Several important dc characteristics of this diode should be noted. When forward biased, cut-in voltage is between 0.3 and 0.5 volt, while for a junction silicon device it is 0.6 or 0.7 volt. The reverse breakdown voltage of a Schottky diode is considerably higher than that of a point-contact device. Leakage and saturation currents are lower. The prime advantage of Schottky diodes is realized when they are used as detectors and mixers in the microwave frequency range. Schottky diodes surpass the frequency capability of point-contact devices.

### Hot-Carrier Diodes

This device can be thought of as an extension of the Schottky diode. Basically, the construction of each device is similar, with the exception that in hot-carrier diodes a pointed metal whisker, usually made from gold or silver, makes good contact with the metal dots on the semiconductor. When forward biased, electrons flow from the semiconductor to the metal. These electrons have extremely high energies and perform well at elevated temperatures. Because of this action, the diode was given the name "hot carrier."

Some of the more important characteristics of the diode are low forward ac resistance, a cut-in voltage similar to that of p-n junction devices, and lower noise than is possible from other high-frequency devices. It has the capability of dissipating relatively large amounts of power and will not be destroyed easily by instantaneous high current surges when the diode is turned on.

In its roles as detector and mixer at microwave frequencies, the hot-carrier diode is more sensitive than the other types of devices and thus performs its various functions efficiently. When reverse biased by only a fraction of a milliwatt, current ceases to flow through the diode. Hence, it exhibits excellent switching action with minimal recover time. Low reverse leakage current flows until the large avalanche or breakdown voltages are reached or exceeded.

## FOUR-LAYER DIODES

Ordinary diodes are made of two semiconductor layers (one n-type and one p-type). Since there are just two layers, there is obviously just a single junction, as illustrated in Fig. 1-28A.

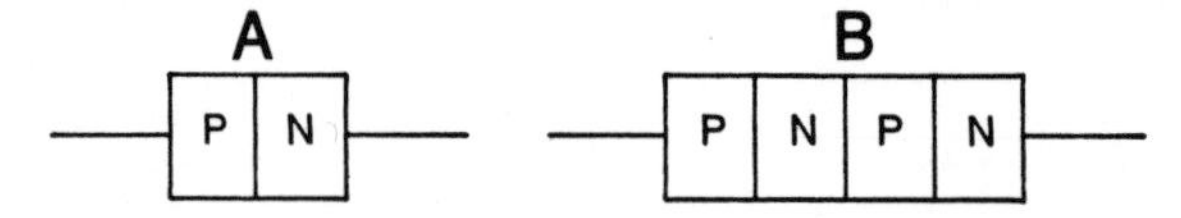


Fig. 1-28. A four layer diode is like a double-diode.

A special type of diode is the four-layer diode, which has three PN junctions, as shown in Fig. 1-28B. This type of diode is employed primarily in switching applications.

When reverse biased, a four-layer diode functions pretty much like any other diode. It exhibits a very high resistance, and virtually all current flow is blocked.

But the four-layer diode starts behaving a little differently when it is forward biased. The current remains small as the voltage is increased until a specific point is reached. Once this trigger point has been passed, the current will suddenly increase, and will continue to rise, even if the forward voltage is reduced almost to zero. A graph of the operation of a typical four-layer diode is shown in Fig. 1-29.

In most applications, a four-layer diode is used as a switch. It is switched on by a brief forward voltage pulse, called the trigger pulse. The diode will now remain in the *on* condition as long as a positive forward voltage is fed to it, no matter how small. To switch the diode back to the *off* condition, the polarity of the forward voltage must be reversed (reverse biasing the diode), or the voltage must be removed altogether.

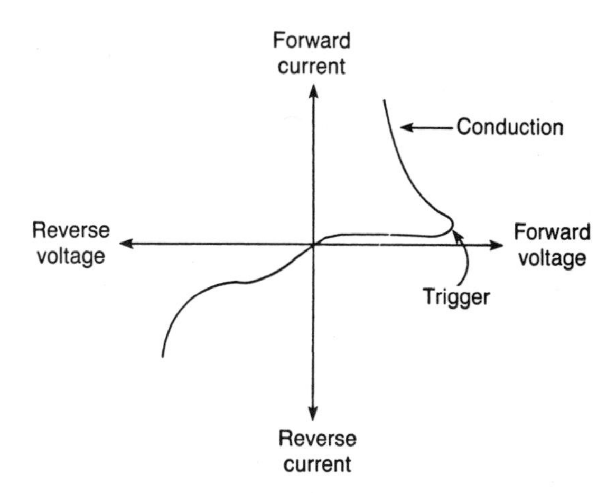


Fig. 1-29. This graph illustrates the operation of a typical four-layer diode.

Four-layer diodes can switch at very high rates. The switching action can take as little as a few nanoseconds. The switch-on time is generally shorter than the switch-off time.

The four-layer diode is closely related to the SCR (silicon controlled rectifier) and the triac, which are also four-layer devices, but with three leads instead of just two.

Sometimes four-layer diodes are known as Shockley diodes, not to be confused with the Schottky-Barrier diode discussed earlier.

## MICROWAVE DIODES

The various types of microwave diodes can be artificially split into two groups—avalanche and bulk diodes.

### Avalanche Diodes at Microwave Frequencies

Microwave avalanche devices are reverse-biased junction diodes working in the avalanche breakdown region. When high rf frequencies are applied, the maximum reverse current, $I_p$, does not occur when the reverse voltage, $V_p$, is at its peak. Because of this phase difference, this diode behaves as a negative resistance similar to the tunnel diode. Consequently, the avalanche diode is useful as an oscillator and as a negative-resistance amplifier.

Diodes that function as just described are referred to as devices operating in the *IMPATT* mode. This is also known as the *transit time* or *read-effect* mode. The term IMPATT is derived from the description of the mode, *Impa*ct ionization and *T*ransit *T*ime.

Another mode or type of operation uses harmonics from the device. This is referred to as the *TRAPATT* mode, an acronym for *tra*pped *p*lasma *a*nd *t*ransit *t*ime. In this mode of operation, current and voltage are 180° out of phase at a specific frequency, depending, of course, on the particular diode used. It is a reactance at the high frequencies.

Oscillation can be produced up to 3 GHz by simply connecting the diode to an external reactance capable of tuning out the reactance of the semiconductor device itself.

A third mode of operation boasts high efficiency despite the fact that the output from the oscillator is but 1/10 that attainable in the IMPATT mode. In this *self-pumped parametric* mode, the diode is pulsed by a voltage. A large current is established in the device because of the avalanche breakdown. Breakdown is so powerful that a plasma of holes or electrons is left in the material. A voltage swing is generated by the plasma formation. This voltage swing is reinforced when it first travels down a transmission line and then returns back to the diode. When striking the diode upon its return, the high voltage can cause an increase in avalanche current thereby forming more plasma. This process continues until the diode goes into a state of oscillation. Oscillation is relatively low in frequency, because it is limited by the time it takes for the plasma to wither.

### Bulk Diodes at Microwave Frequencies

Energy of electrons are limited to several levels or bands. Normally, electrons are in a band in which they can be mobile. In the *Gunn diode* (used in what is referred to as the transit time mode of operation), an electric field is established that causes the electrons to fall from one energy band to another. In this second band, the electrons are not as mobile as they were before. Current is reduced through the diode; hence an increase in voltage is required at the diode to produce the electric field that results in a reduction of electric current. This is the perfect situation needed to establish a negative resistance.

When biased for negative resistance, an electric field is formed in the material at the cathode. This field is known as a *field domain*. It floats through the semiconductor and disappears when it reaches the anode. A new field domain is then formed by the electric field in the negative resistance. Current flow through the device increases between the time the domain disappears and a new domain forms. The high frequency produced by this action in the Gunn diode is determined by the transit time of the domain. The frequency is related to the distance the domain must travel through the diode. It can be reduced by delaying the forming of a new domain for a short period of time. Formation of a new domain is thereby delayed in what is naturally called the *delayed domain* mode of operation.

In LSA, or *L*imited *S*pace charge *A*ccumulation mode of operation, the bulk diode is tuned to such a high frequency that there is not enough time for a full domain to form during each cycle. High frequencies can be produced regardless of the length or actual dimensions of the diode.

Gunn diodes require about 7 volts dc bias. They should be operated from a well-stabilized constant-voltage power supply. If the diode oscillates, a 0.1 $\mu$F or larger capacitor should be wired across the bias voltage supply.

Microwave diodes and related components will be discussed in greater detail in Chapter 6.

# 2

# Transistors

There are many different types of transistors. The two most common, popular classes of transistors are bipolar transistors and FETs (Field Effect Transistors). Most early transistors were of the bipolar type, and bipolar transistors still predominate, although FETs are rapidly catching up, because of their special advantages and improved technology.

We will start the discussion in this chapter with the bipolar transistor. The second half of the chapter will be concerned primarily with the various types of FET. Finally, we will look at a third, somewhat less common, but still important type of transistor—the UJT (UniJunction Transistor).

## BIPOLAR TRANSISTORS

A transistor is essentially a diode with an additional n-type or p-type slab, forming two p-n junctions. If the p-type material is sandwiched between two n-type slabs, an npn transistor is formed. In a pnp transistor, the n-type material is between two p-type slabs.

Effectively there are two diodes. The middle slab is known as the base, one end slab is referred to as the emitter, and the other as the collector. The base-emitter diode or junction is normally forward biased and the base-collector junction is reverse biased.

Current flowing through the lead connected to the collector is approximately beta ($\beta$) multiplied by the base current. Similarly, current flowing through the collector lead is approximately alpha ($\alpha$) times the current in the emitter lead. Alpha and beta are two current-gain factors to be used extensively throughout the text.

### Basic Views

The two types of transistors, npn and pnp, are drawn in Fig. 2-1. As was the case with the diode, the direction of the electron current is opposite to the direction of the arrow in the symbol.

In the npn device, the base must be made positive with respect to the emitter, if there is to be current flowing through this junction. Current will flow from the emitter to the base in the direction opposite to that depicted by the arrow in the symbol.

For the pnp device, the base-emitter junction must be biased opposite in polarity to that of the npn transistor, if there is to be current flowing from the base to the emitter. The emitter must be made positive with respect to the base. Once again, the arrow in the symbol points opposite to the direction of the electron current.

As in the case of all forward-biased diodes, a voltage is developed across the base-emitter junction. This is the usual diode voltage—about 0.2 volt for germanium transistors and 0.6 volt or somewhat more for silicon devices.

When the base-emitter junction is in the conducting mode, ac and dc resistances are in series with the emitter lead. While ac emitter resistance is referred to by the symbol $r_e$, $r_E$ represents dc emitter resistance. Ac resistance $r_e$ is equal to $26/I_E$, where $I_E$ is the emitter current expressed in milliamperes. This is identical to the resistance of the forward-biased diode. Dc emitter resistance is usually so small that it is normally ignored in circuit calculations.

Base current, multiplied by a factor of beta, is the major portion of current in the collector. Beta is merely a number indicating the ratio of collector current to base current. In a similar fashion, the major portion of the current in the collector may be related to the current in the emitter. Collector current is alpha times the emitter current. Alpha is a ratio relating the collector to the emitter current. It is always less than 1 for the junction transistor.

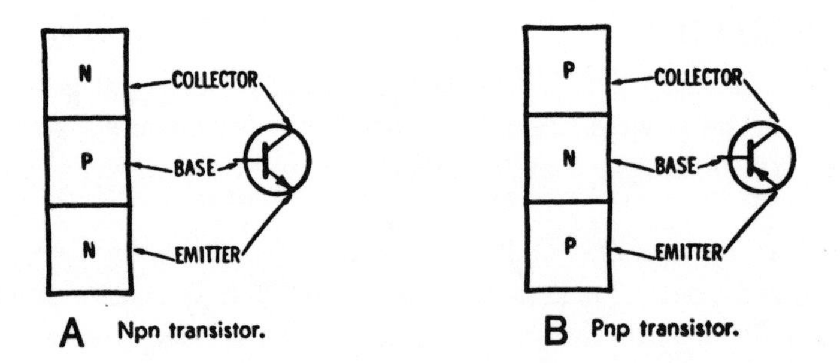


Fig. 2-1. Different schematic symbols are used for the two types of bipolar transistors.

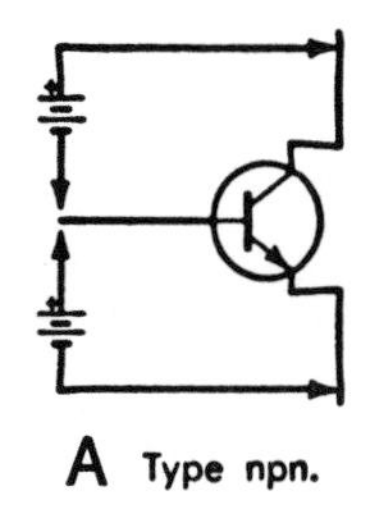

A Type npn.

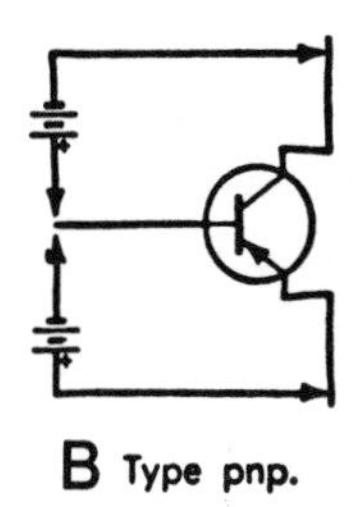

B Type pnp.

Fig. 2-2. The npn and pnp transistor differ in the polarities of the voltages to be applied to their terminals.

The total collector current is due to two major factors. One is the gain of the transistor: $\alpha I_E$ or $\beta I_B$. The second factor is due to the leakage current flowing through the reverse-biased base-collector junction.

In order to avoid confusion, from this point on, collector current due to the gain of the device will be referred to simply as collector current. The symbol is $I_C$. The collector current due to all factors will be called total collector current and will have the symbol $I_C$ (total).

The collector is usually reverse biased with respect to the base. In the npn transistor, the base is made of p-type material and the collector (as the emitter) of n-type slabs. To reverse bias the base-collector junction, the collector is made positive with respect to the base. Because the base and collector in the pnp transistor are composed of n-type and p-type slabs, respectively, the collector is reverse biased by making it negative with respect to the base.

In summary: if a transistor is to conduct, the base-emitter junction must be forward biased and the base-collector junction reverse biased. This is shown in Fig. 2-2. Should both junctions be reverse biased, there will be no conduction of current from the emitter through the collector.

The actual operation of the device can be determined using the pnp transistor in Fig. 2-3. Details apply equally to the npn transistor. In the latter instance, however, the current flows in the opposite direction.

All significant and useful transistor current, $I_E$, flows through the emitter. It is generally safe to assume that this is the maximum current that flows through any element of the transistor. Most of $I_E$ is in the collector lead as well. The portion of $I_E$ in the collector is equal to $\alpha$ (alpha). Mathematically, $\alpha$ is:

$$\alpha = \frac{I_C}{I_E} \tag{2-1}$$

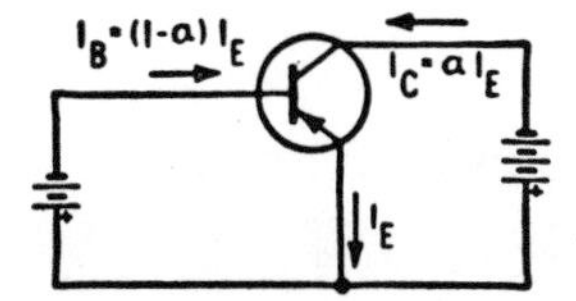


Fig. 2-3. A transistor acts as a current divider.

The collector current, $I_C$, is equal to $\alpha I_E$. $\alpha$ must be less than I for the collector current is always less than the emitter current. Usually $\alpha$ is greater than 0.95 and is frequently indistinguishable from 1.

The tiny portion of emitter current left for the base is $I_B = I_E - I_C = I_E - \alpha I_E = I_E(1 - \alpha)$.

The npn transistor operates in the same manner; only the direction of current is reversed. (Examples throughout the text will usually involve the npn transistor. Everything will apply as well to the pnp device except that the polarities of the supply voltage must be reversed.)

The transistor operates essentially as a current device. Voltage between the base and emitter must be applied in such a manner that there is a specific base current, $I_B$. Voltage between the emitter and collector must be applied in the proper polarity to allow collector current to flow through the transistor. The relative voltage between the base and collector must be such that there is no current flowing through this reverse-biased junction. (Actually, there is a leakage current flowing through this junction.)

Beta ($\beta$) has been defined as the current gain existing between the base and collector. If the base current is $I_B = (1 - \alpha) I_E$ and the collector current is $I_C = \alpha I_E$, the dc current gain of the device is:

$$\beta = \frac{I_C}{I_B} = \frac{\alpha I_E}{(1 - \alpha) I_E} = \frac{\alpha}{1 - \alpha} \qquad (2\text{-}2)$$

If $\beta$ is the known quantity, $\alpha$ can be determined from the relationship:

$$\alpha = \frac{I_C}{I_E} = \frac{\beta}{\beta + 1} \qquad (2\text{-}3)$$

The transistor is normally connected into a circuit in one of three ways: common base, common emitter, and common collector. They are also referred to as grounded base, grounded emitter, and grounded collector, respectively.

## Common Base

A drawing of the common-base circuit is shown in Fig. 2-4. Here, the input and output voltages are considered with respect to the base. The emitter-base junction is forward biased by the $V_{EE}$ supply. The collector-base junction is reverse biased by the $V_{CC}$ supply. Note the polarities.

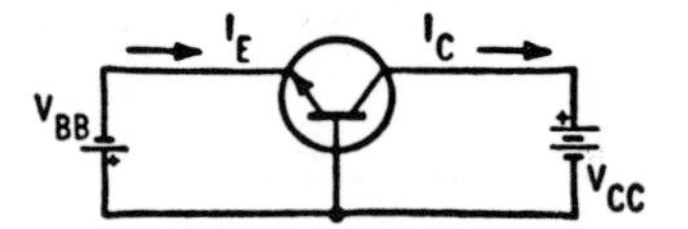


Fig. 2-4. Transistors are often used in common-base or grounded-base circuits.

The total current flowing through the collector is $\alpha$ multiplied by the emitter current or $\alpha I_E$ plus the additional leakage current flowing through the reverse-biased base-collector junction. This leakage current, $I_{CBO}$, or simply $I_{CO}$, is the current passing from the collector to the base junction with the emitter open. It is the current flowing through the collector when the emitter current is zero. The total collector current is:

$$I_C(\text{total}) = \alpha I_E + I_{CBO} \tag{2-4}$$

An example to determine the operating point of a transistor uses the circuit and information furnished in Fig. 2-5. A silicon transistor with a beta of 50 is assumed. Voltage across the base-emitter junction of a silicon transistor is approximately 0.6 volt. Current entering the emitter, by Ohm's law, is:

$$I_E = \left(\frac{4 \text{ volts} - 0.6 \text{ volt}}{10 \text{ ohms}}\right) \text{amps} = \left(\frac{3.4}{10}\right) (1000) \text{ mA} = 3.4 \times 10^2 \text{ mA}$$

The collector current due to the emitter-base current is $\alpha I_E$. From Equation 2-3, $\alpha = \beta/(1 + \beta) = 50/51 = 0.98$. The collector current is $(0.98)$ $(3.4 \times 10^2)$ mA $= 3.33 \times 10^2$ mA.

The contribution of the leakage current to the collector current is 10 $\mu$A, and is negligible in this case. The voltage drop across the 100-ohm resistor is (100 ohms) $(3.33 \times 10^2$ mA$)$ = (100 ohms) (.333 amps) = 33.3 volts. The voltage between the base and collector is the collector supply voltage less the drop in the 100-ohm collector resistor, or 50 volts − 33.3 volts = 16.7 volts. This, added to the 0.6 volt from the base to emitter, is the collector-emitter voltage. It is 16.7 + 0.6 = 17.3 volts.

Most of the power dissipated by the transistor is in the collector circuit. From the equation equating power to the product of voltage and current, the transistor dissipates 17.3 volts × 0.333 amp or 5.76 watts.

Now, superimpose an ac signal on the 4-volt dc emitter supply voltage and let the peak-to-peak voltage of the signal be 2 volts. The input will vary from 3 to 5 volts. We can successfully apply this information to the circuit when we note than the ac voltage gain is approximately equal to the ratio of the impedance in the collector circuit to the impedance in the emitter circuit. We should also note that the emitter and collector currents are just about equal to each other.

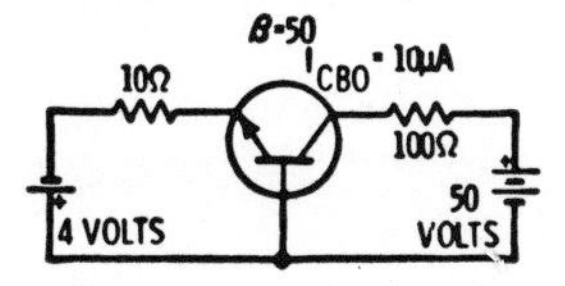


Fig. 2-5. This is an example of common-base circuit.

Emitter current in this circuit varies from (5 volts - 0.6 volt)/10 ohms, or 0.44 amps at the peak to (3 volts - 0.6 volt)/10 ohms, or 0.24 amps at the crest. The respective collector currents are alpha multiplied by these emitter currents or 0.431 and 0.235, respectively.

At the peak of the signal, the voltage across the 100-ohm resistor is $100 \times 0.431 = 43.1$ volts. At the crest, the voltage drops to $100 \times 0.235 = 23.5$ volts. The peak-to-peak voltage across the 100-ohm resistor is $43.1 - 23.5 = 19.6$ volts. Consequently, voltage gain is 19.6/2 or 9.8. This is about the same as the ratio of the resistor in the collector circuit to the resistor in the emitter circuit.

The common-base characteristics of the transistor can also be described using curves. A theoretical set of curves for the transistor used in the example is in Fig. 2-6A. The collector current, $I_C$, is plotted on the vertical axis and the collector-to-base voltage, $V_{CB}$, on the horizontal axis. Each curve shows how $I_C$ varies with $V_{CB}$ for the different currents, $I_E$, flowing in the emitter.

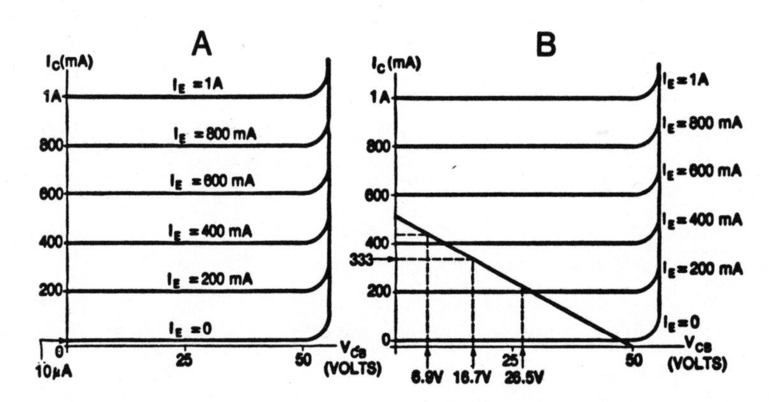


Fig. 2-6. Theoretical curves: (A) Common-base collector characteristic curves, (B) Curves with load line added.

The problem around the circuit in Fig. 2-5 can be solved graphically using the curves in Fig. 2-6A. These have been redrawn in Fig. 2-6B. Curves shown are actual plots of complex equations describing the characteristics of the transistor. A second equation must be plotted on the same graph relating voltage and current in the collector circuit. This is the load line in Fig. 2-6B determined from the equation:

$$V_{CC} = I_C R_C + V_{CB}$$

All factors or components in the collector circuit are noted in this equation. Substituting numbers from the circuit in Fig. 2-5:

$$50 = 100 I_C + V_{CB}$$

The plot of the load line can be executed by first plotting the point where $I_C = 0$. At this point $V_{CB} = 50$ volts. This is a point on the horizontal axis where $I_C = 0$ and $V_{CB} = 50$. A second point on the load line can be found by letting $V_{CB}$ in the equation be at zero. At this point, $I_C = 50/100 = 0.5$ amps or 500 mA. A second point on the load line is then at $V_{CB} = 0$ and $I_C = 500$ mA. Connect the two points just determined with a straight line. This is a plot of the dc load line in the collector circuit.

Quiescent emitter current has previously been determined to be 340 mA. Plot this point on the load line. If you extend vertical and horizontal lines from this point to the axes, you will find that one line hits the vertical axis at about 333 mA and the other line hits the horizontal axis at about 16.7 volts, the base-to-collector voltage. This agrees with calculations previously performed. The difference between the 16.7 volts and the 50-volt supply, 33.3 volts, is across the collector resistor, $R_C$.

If a 2 volt peak-to-peak ac signal is superimposed upon the load line, it will cause the emitter current to vary from 0.24 amps to 0.44 amps, as determined in the calculations. Plot these extreme points of current excursions on the load line in Fig. 2-6B. Extend a vertical line from each point to the vertical axis. $V_{CB}$ is 6.9 volts for one point and 26.5 volts for the second point. This is the voltage variation that is across $R_C$ as the voltage across $R_E$ varies from 43.1 to 23.5 volts.

Power gain is equal to the current gain multiplied by the voltage gain. Since the current gain in high beta transistors is practically equal to 1, the power gain is approximately equal to the voltage gain.

Also note that there is no phase shift between the input and output signals. They are both at their peaks and crests at the same instant of time.

### Common Emitter

A drawing of the most-used configuration, the common-emitter transistor circuit, is shown in Fig. 2-7. Of all arrangements, this circuit can supply the highest power gain. There is also current and voltage gain.

Input and output voltages are referred to the emitter. The base-emitter junction is forward biased by virtue of the polarity of $V_{BB}$, $V_{CC}$ reverse biases the collector with respect to the emitter.

Collector current is beta times the base current. There is an additional current, $I_{CBO}$, flowing through the base-collector junction because of leakage through the reverse-biased diode. Because the collector current is beta multiplied by the base current, collector current due to leakage is $(\beta)\ I_{CBO}$ and is equal to $I_{CEO}$. $I_{CEO}$ is the collector-to-emitter

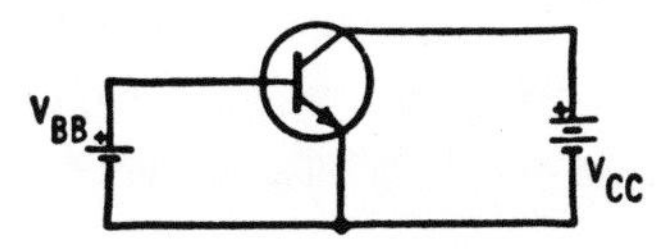


Fig. 2-7. The common-emitter transistor circuit.

current, with the base open. The total collector current due to leakage current and base current is:

$$I_C \text{ (total)} = \beta I_B + \beta I_{CBO} = \beta I_B + I_{CEO} \tag{2-5}$$

A practical circuit is shown in Fig. 2-8A. The dc bias circuit for the base-emitter diode is lifted out of Fig. 2-8A and is shown in Fig. 2-8B.

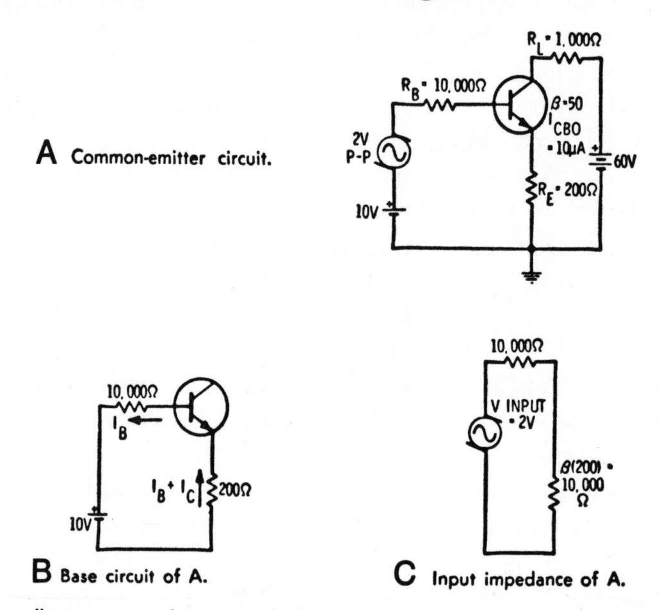


Fig. 2-8. These diagrams examine a practical common-emitter circuit.

Base current is determined by the 10-volt battery in series with the 10,000-ohm resistor, the 200-ohm resistor in the emitter circuit, the dc emitter resistance ($r_E$), the dc resistance in the base $r_B$ and the voltage drop across the junction. Let us consider each of these factors individually.

The total dc voltage supplied to the circuit is the 10 volts. The drop across the junction reduces the total available voltage. Assuming a silicon transistor, this drop is about 0.6 volt. Hence, the total voltage affecting the base current is 10 − 0.6 volts, or 9.4 volts.

The total resistance in the circuit consists of four items:

1. One item is the 10,000 ohms in the base circuit. Only base current flows through this resistor.

2. The base resistance is a resistor in the transistor that is theoretically in series with the base. $I_B$ is the only current flowing through this resistor. Only the dc resistance, $r_b$, may be considered in the base equivalent circuit. However, it is usually negligible compared to all other resistors.

3. The 200-ohm resistor in the emitter passes two currents. These are the base and the collector currents that add up to the emitter current. Written mathematically, $I_E = I_B + I_C = I_B + \beta I_B = (1 + \beta)\, I_B$. If beta is much larger than 1, as it usually is, the emitter current is $I_E = \beta I_B$. Because of this, the resistor in the emitter has a larger effect on the base current than just its 200 ohms. It behaves as if a resistor equal to $\beta \times 200$ were in the base circuit. In this case, it is $50 \times 200 = 10{,}000$ ohms.

4. The dc emitter resistance, $r_E$, is usually negligible. It should not be confused with the ac emitter resistance, $r_e$, which is equal to $26/I_E$ when $I_E$ is expressed in milliamperes. The $r_E$ is usually much smaller than $r_e$.

Base current can be found from the Ohm's law equation

$$I_B = \frac{9.4 \text{ volts}}{10{,}000 + (200)(50)}$$

$$= 470\ \mu\text{A} = 0.47 \text{ mA} = 0.47 \times 10^{-3} \text{ amperes}$$

The total collector current, $I_C$ (total), is beta multiplied by the base current or $50 \times 0.47 = 23.5$ mA plus beta multiplied by $I_{CBO}$ or $50 \times 10 \times 10^{-6} = 0.5$ mA. The sum of the two factors is 24 mA $= 24 \times 10^{-3}$ amps. Voltage across the 1,000-ohm resistor is 1000 ohms $\times\ 24 \times 10^{-3}$ amps $= 24$ volts.

The voltage across the transistor, from the emitter to the collector, is 60 volts less the sum of the voltage across the 1000-ohm resistor (24 volts) and the voltage across the 200-ohm resistor. The current flowing through the 200-ohm resistor is $I_E = I_C + I_B = 24 \times 10^{-3}$ amps $+\ 0.47 \times 10^{-3}$ amps $= 24.47 \times 10^{-3}$ amps. The voltage across the resistor is $I_E \times 200$ ohms $= 24.47 \times 10^{-3} \times 200 = 4.894$ volts.

The voltage across the transistor is $60 - 24 - 4.894 = 60 - 28.894 = 31.106$ volts. The power dissipated in the collector circuit is $I_C$ (total) $\times\ V_C = (24 \times 10^{-3}$ amps) (31.106 volts) $= 0.747$ watt.

The primary factor that must be noted from this problem is that, when seen from the base of the transistor, the resistance in the emitter appears as if it is multiplied by a factor of beta. Similarly, the resistance in the base circuit will appear divided by beta when viewed from the emitter circuit.

Common-emitter characteristic curves of the transistor can be used to find the quiescent collector current $I_C$ and collector-to-emitter voltage $V_{CE}$ of the transistor. These curves are more useful than those for the common-base circuit in Fig. 2-6. Here, the plots relate the collector current to changes in $V_{CE}$ when different values of $I_B$ are applied to the base. A plot for a transistor with $\beta = 50$ is in Fig. 2-9.

The load line can be determined from the equation:

$$E_{CC} = I_C R_C + I_E R_E + V_{CE}$$

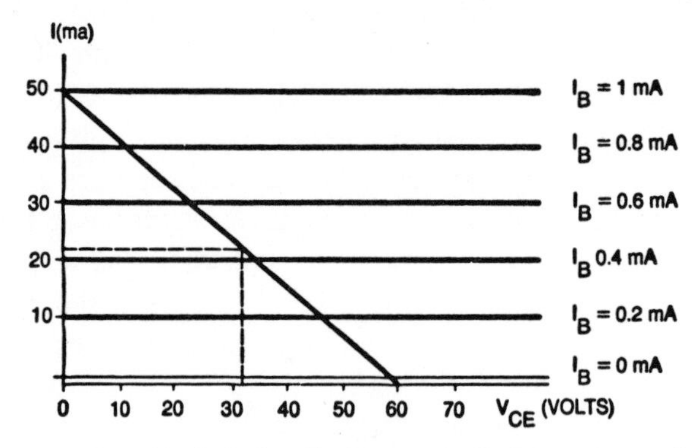


Fig. 2-9. Collector curves are used in designing circuits arranged in the common-collector, and common-emitter configurations. A 1200 ohm load-line has been drawn.

Using the approximation that $I_C = I_E$ and substituting numbers from Fig. 2-8 into the equation, the load line for that circuit can be drawn from the following relationship:

$$60 = I_C(1000 + 200) + V_{CE} = 1200I_C + V_{CE}$$

Two points on the graph are used to determine the load line. One point is at $I_C = 0$, $V_{CE} = 60$. This point is on the horizontal axis at $V_{CE} - 60$ volts. The second point is determined from the equation when $V_{CE} = 0$. Now, $I_C = 60/1200 = 0.05$ amperes $= 50$ mA. Plot this point on the vertical axis as $I_C = 50$. Connect the two points with a straight line.

It was determined that for this circuit, $I_B = 0.47$ mA. Plot this point on the load line as shown and draw lines from this point to the horizontal and vertical axes. One line crosses the horizontal axis at about $I_C = 24$ mA and the second line crosses the vertical axis at about 31.1 volts. This is approximately equal to the transistor current and voltage as calculated above.

The approximate ac voltage gain of the transistor itself for the circuit in Fig. 2-8 is the ratio of the resistor in the collector circuit to the resistor in the emitter of 1000/200 = 5. The voltage gain of the transistor may also be defined as the ratio of the ac voltage across the collector load resistor or between the collector and ground, to the voltage between the base and ground. Since the current gain of the transistor is $\beta$, the power gain is $\beta$ multiplied by the voltage gain or $\beta R_L/R_E$. There is a 180° phase shift between the input and output voltages. When the voltage at the base is at a peak in the cycle, it is at its crest at the collector.

## Common Collector

In the common-collector circuit in Fig. 2-10, the input and output voltages are stated with respect to the collector. The transistor is in the conducting mode when the supplies are connected as shown in Fig. 2-10A or 2-10B.

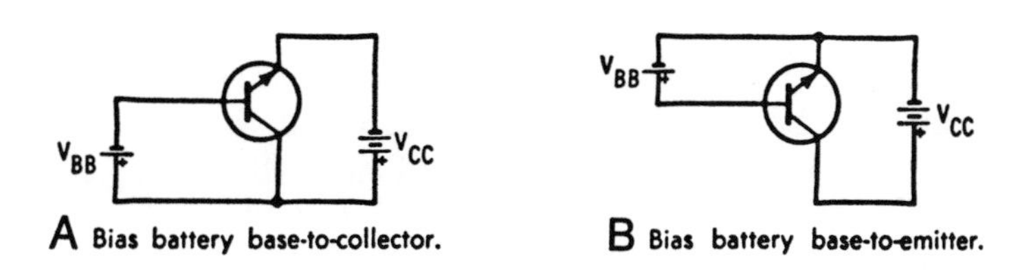


Fig. 2-10. Here are two versions of the basic common-collector circuit.

The ratio of the emitter current, $I_E$, to the base current, $I_B$, is $I_E/I_B = 1/(1 - \alpha)$, from Fig. 2-3. If $\alpha$ is large, approximately equal to 1, the current ratio is approximately equal to the current gain of the common-emitter circuit, or $\beta$.

The total emitter current is $(\beta + 1)$ multiplied by the total base current. If there is no base-to-collector leakage, it is simply $(\beta + 1)I_B$. Should there be leakage (and it is assumed that all the leakage current flows through the base-emitter junction), the total emitter current is:

$$\begin{aligned} I_E \text{ (total)} &= (\beta + 1)I_B + (\beta + 1)I_{CBO} \\ &= I_C = I_B + I_{CEO} + I_{CBO} \end{aligned} \qquad (2\text{-}6)$$

A practical circuit is shown in Fig. 2-11. Note the similarity between this and the one in Fig. 2-8A.

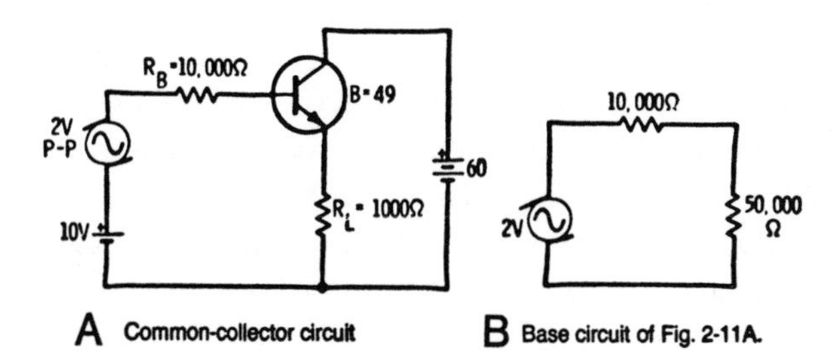


Fig. 2-11. This is a practical common-collector circuit.

The dc base current is primarily due to the 10-volt supply minus the 0.6-volt drop across the base-emitter junction, the 10,000-ohm resistor in the base circuit, and, finally, the 1000-ohm load resistor reflected into the base circuit as $(\beta + 1)$ (1000) = 50,000 ohms.

The total resistance in the base circuit is 10,000 ohms + 50,000 ohms = 60,000 ohms. The dc base current is (10 − 0.6) volts/60,000 ohms = $1.57 \times 10^{-4}$ amps. Assuming zero leakage current, the emitter current, from Equation 2-6, is equal to $(\beta + 1)\, I_B = 50\,(1.57 \times 10^{-4}$ amps$) = 7.85 \times 10^{-3}$ amps. The voltage across the load resistor is $I_E R_L = (7.85 \times 10^{-3}$ amps) (1000 ohms) = 7.85 volts.

The collector supply is 60 volts. The voltage across the transistor is 60 − 7.85 = 52.15 volts. With an emitter (and hence collector) current of about $7.85 \times 10^{-3}$ amps, the power dissipated by the transistor is $52.15 \times 7.85 \times 10^{-3} = 0.41$ watt.

The 2 volts peak-to-peak ac signal superimposed on the 10-volt supply will force the supply to vary from 9 to 11 volts. When the supply is at 9 volts, the current flowing through the emitter circuit is approximately $[50(9 - .6)]/60{,}000 = 7.00 \times 10^{-3}$ amps. The voltage across $R_L$, by Ohm's law, is $(7 \times 10^{-3})(1000) = 7$ volts.

During the portion of the cycle when the supply is at 11 volts, the emitter current is $[50(11 - .6)]/60{,}000 = 8.67 \times 10^{-3}$ amps. The voltage across $R_L$ is $(8.67 \times 10^{-3})(1000) = 8.67$ volts. The peak-to-peak ac voltage across $R_L$ is 8.67 volts − 7.00 volts = 1.67 volts.

With a 2-volt ac signal input, the ac output is 1.67 volts. The voltage gain of the *overall circuit* is less than 1. In this case the voltage gain is 1.67/2 or 0.835.

The ac voltage across the 50,000-ohm impedance in the voltage divider formed by $R_B$ and the reflected impedance of $R_L$ into the base circuit, $\beta R_L$, (see Fig. 2-11B) is $[50{,}000/(50{,}000 + 10{,}000)]2$ volts = 1.67 volts. Hence 1.67 volts is the actual voltage applied to the transistor proper. This is the same as the output voltage previously calculated as appearing across $R_L$. The gain of the *transistor circuit itself* is 1. In actual common collector circuits, the ac voltage gain of the transistor circuit is slightly less than 1. This does not include the losses due to the resistance of the voltage or signal source.

The input impedance of the transistor itself is $\beta R_L$. As for the output impedance, an external load looking back into the emitter circuit sees the resistor in the emitter circuit as being connected in parallel with the series combination composed of the impedance in the base circuit divided by beta and the emitter resistance, $r_e$. In the example in Fig. 2-11, the emitter current was calculated at 7.85 mA, so the ac emitter resistance is 26/7.85 = 3.3 ohms. The resistor in the base circuit divided by beta is 10,000/49 = 204.1 ohms. The sum of this and the emitter resistance is 204.1 + 3.3 = 207.4 ohms. This resistance is in parallel with the 1,000-ohm resistor in the emitter circuit, $R_L$, so that the resistance seen by an external load looking back into this emitter circuit is $(207.4 \times 1000)/(207.4 + 1{,}000) = 172$ ohms. This is the parallel resistance of $R_L$ with the sum $r_e$ + (resistance in the base circuit)/$\beta$.

## Voltage Limits

Regardless of the circuit or mode of operation, the size of voltages that may be applied across the transistor, are limited. Excess voltage can result in either an avalanche or punch-through breakdown. You know that the voltage breakdown point has been reached when the reverse current begins to increase rapidly above $I_{CBO}$. Collector-base avalanche breakdown voltage has been assigned the symbol $BV_{CBO}$.

Should the base be left disconnected and the emitter is connected to the voltage supply in its stead, there can be a breakdown between the collector and the emitter.

This breakdown voltage, $BV_{CEO}$, is about one-half of $BV_{CBO}$. Collector-to-emitter breakdown voltage can be increased by connecting resistors or other dc loads between the base and emitter rather than leaving the base floating. Collector-to-emitter breakdown voltage increases as the size of the dc resistance between the base and emitter is reduced.

Punch-through mechanism differs from that of avalanche breakdown. In both types of breakdowns, collector and emitter current increase tremendously when an excessively large collector-base junction voltage is applied. An excessively large collector-base voltage increases the size of the depletion region of that junction. The width of the base is narrowed. It may be narrowed sufficiently for the depletion region at the collector-base junction to reach through to the base-emitter junction. High current will flow from the collector to the emitter. Without the intervening base, the device no longer acts as a transistor. If avalanche breakdown will not limit the voltage that may be applied between the collector and base, punch-through will set the boundary of operation.

## HIGH-POWER TRANSISTORS

Semiconductor crystals are fairly delicate devices. They can be damaged in a number of ways. Two of the biggest dangers to semiconductor junctions are heat and over-voltage. Either can destroy a semiconductor device very quickly.

Most transistors are low-power devices. The power ratings are generally in the milliwatt range. For most applications this is fine. But, some applications (especially those involving large amounts of amplification) require an active device that can handle a much larger amount of power.

In the early days of transistors the circuit designer was forced to rely on tubes for high power applications. There has been considerable development in the area of high-power transistors. Transistors with power ratings as high as 200 or 300 watts are available. Some of these units offer voltage ratings up to 2 $kV_{CBO}$.

A high-power transistor tends to be physically larger than the low-power units. While low-power transistors are packaged in either plastic or metal housings, high-power transistors are almost always in a metallic container for maximum heat transfer. Adequate heat-sinking is an absolute must for all high-power applications.

High-power transistors tend to be a little more expensive, and harder to find, than ordinary low-power transistors. However, if it is needed, it is worth a little extra effort to seek out a high-power unit.

## PHOTOTRANSISTORS

Like the photodiode, current flowing through the phototransistor increases with the amount of light hitting a junction in the device. Although conventional transistors are mounted in light-proof cases, elements of the phototransistors are exposed to light. They are usually used in the common-emitter configuration. $I_C - V_{CE}$ characteristics look very much like those in Fig. 2-9, with the exception that units of light intensity are substituted for base current at each curve. Collector current increases with light intensity.

As is the case with the conventional transistor, a dc power supply is connected between the collector and emitter. The base of the phototransistor is usually not connected to a source of current. If the transistor is an npn device, the collector is invariably made positive with respect to the emitter. The collector-base junction is exposed to light. When there is no light impinging on this junction, ordinary leakage current, $I_{CBO}$, flows through it. Collector and emitter currents are $\beta I_{CBO}$ or $I_{CEO}$. As energy in the form of light hits this junction, $I_{CBO}$ increases, as do the collector and emitter currents.

Phototransistors are frequently mounted in a single package with an LED to form an optoisolator. There is no electrical contact between the two devices. Because of the isolation of the LED from the phototransistor, the optoisolator can function as a relay. In this application, an applied current turns the LED on in one circuit. Light from the LED initiates an increase in the amount of current flowing through the phototransistor connected in a second circuit.

Besides being used as a relay switch, the optoisolator can be used to control quantities of current flowing through the phototransistor. At low collector voltage, the impedance of the transistor is a function of the collector current flowing through it. Wired properly in the circuit, the optoisolator can be used to optically couple audio signals from one circuit to another. Using this very same variable resistance characteristic, the device can also be used as a remote optically coupled level control.

Factors such as rise time, delay time, storage time, and fall time (time for the phototransistor's collector current to drop from 90 percent to 10 percent of its maximum level) were described in the discussion of switching diodes. These factors also describe the time it takes for the phototransistor collector current to react to switching commands. Reaction time is in microseconds.

## TRANSISTOR SWITCHES

The transistor can be used as a switch. Consider the common-emitter circuit in Fig. 2-12. Assume that the voltage drop across the base-emitter junction is negligible. Then forty volts or more peak-to-peak signal is required to switch the transistor in this example from a maximum conducting to a nonconducting state. This can be shown arithmetically as well as through the use of the $V_{CE}I_C$ curves in Fig. 2-13. Draw a load line for the collector circuit in Fig. 2-12. This is accomplished using steps similar to those described above with reference to Fig. 2-9.

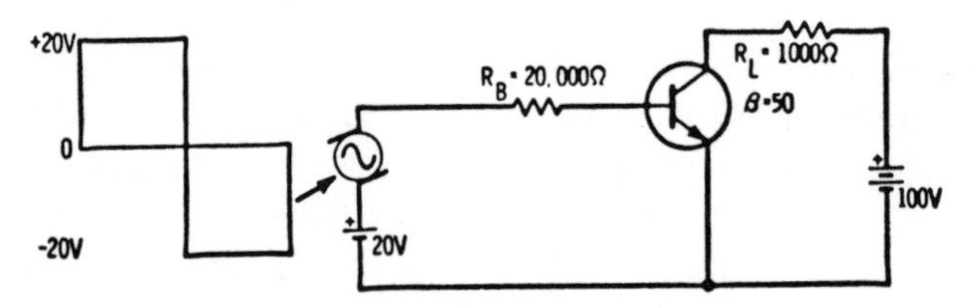


Fig. 2-12. Bipolar transistors are frequently used in switching applications.

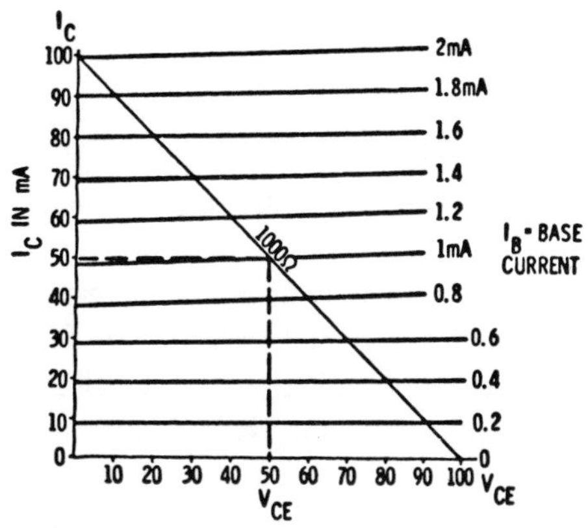

Fig. 2-13. Graph plots collector voltage versus collector current.

Should the base current swing to 2 mA, the collector current rises to 100 mA and the collector voltage drops to zero. This is the condition when the +20 volts of the square wave is superimposed on the 20-volt supply. Any further increase in base current does not change the collector voltage or current.

Should the base current swing to 0 mA, the collector voltage (or voltage from emitter to collector) rises to the supply voltage of 100 volts and the collector current drops to zero. This is the condition when the superimposed square wave drops to −20 volts or less.

It is instructive to follow the transistor current and voltage excursions as the input signal varies. The voltage and current variations are tied to the load line. Any "collector voltage-collector current-base current" condition must be a point on the load line. All input signal variations are considered as points traveling up and down the load line.

## FIELD-EFFECT TRANSISTORS

The input impedance of bipolar transistors is relatively low. This characteristic presents obstacles to the designer whenever a circuit or transducer at the input to the device must see a high impedance. Here is where the field-effect (FET) excels. Its input impedance may be measured in the megohms range.

There are several different variations on the FET. The most basic type is referred to as the *Junction Field-Effect Transistor*, and is abbreviated as JFET. Generally, when the generic term FET is used, a JFET is implied. To avoid confusion in this book we will use the more specific JFET designation.

The second important type of FET is the IGFET, or *Insulated Gate Field-Effect Transistor*. This type of device is often referred to as a MOSFET, or MOST device, the latter initials being an abbreviation of *Metal Oxide Semiconductor Transistor*. The terms IGFET and MOSFET are essentially interchangeable.

The primary differences between the JFET and the IGFET lie in the input impedances, and required bias voltages. There are also several secondary differences that will be discussed.

A recent variation on the MOSFET is the V-MOSFET, or VFET. The V stands for "Vertical." This will be explained shortly.

## Junction Field-Effect Transistors

A drawing of the JFET and the schematic representations are shown in Fig. 2-14. The semiconductor slab connecting two JFET terminals, the source and drain, is known as the channel. The channel may be an n- or p-type material. The n-channel device and its schematic drawing are shown in Figs. 2-14A and 2-14B, respectively. Similarly, the p-channel device and representation are in Figs. 2-14C and 2-14D. A semiconductor material opposite in type to that used for the channel is placed against the channel. This material, or transistor terminal, is called the gate. Polarities of voltages to be applied to the various terminals are usually as shown. All voltages are considered with respect to the source terminal, which in this drawing is at ground.

Although the n-channel JFET will be the center of the following discussion, all factors refer to the p-channel device as well, with only the voltage polarities reversed. Electron current flow for the n-channel device is from the terminal marked "source" to the terminal marked "drain."

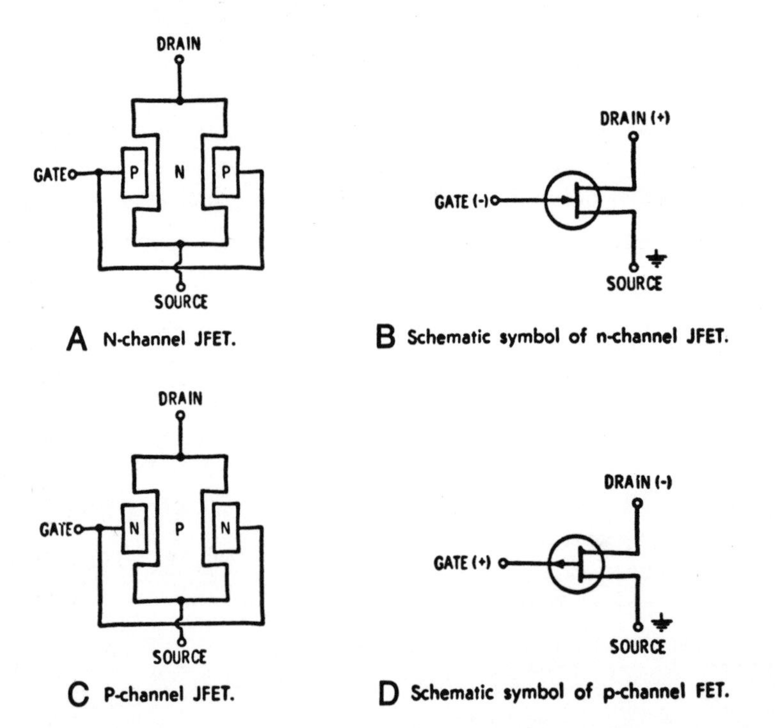


Fig. 2-14. The JFET has a different structure than bipolar transistors, different schematic symbols are used.

The amount of current flowing between the source and drain is dependent on several factors. Up to a specific drain-to-source voltage, $V_{DS}$, the channel acts as a resistance. This resistance increases with an increase in temperature. Source-to-drain current is also a function of the size of the reverse voltage applied between the gate and source, $V_{GS}$. This can be clarified with the help of Fig. 2-15 which shows a typical set of n-channel JFET curves.

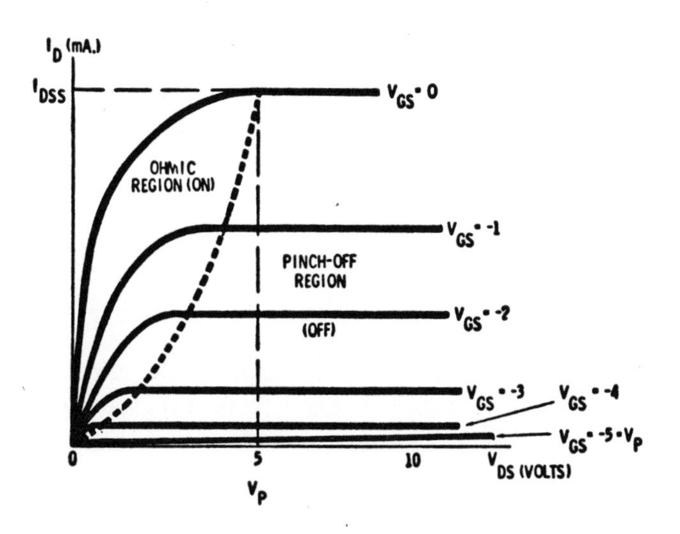


Fig. 2-15. This curve illustrates the output (or drain) characteristic curves of JFETs.

The curves shown indicate how the drain current, $I_D$, varies with the voltage, $V_{DS}$, across the drain and source terminals when different negative or reverse-bias voltages, $-V_{GS}$, are applied between the gate and the source. When $V_{GS} = 0$, or the gate is connected to the source, the drain current is $I_{DSS}$. The drain-to-source voltage at which $I_{DSS}$ reaches its maximum on the $V_{GS} = 0$ curve is referred to as the pinch-off voltage, $V_P$. If $V_{GS}$ were made equal to $V_P$ instead of to zero, the lowest curve in the drawing applies. There is an infinitesimal amount of current flowing through the channel when $V_{GS} = V_P$.

Following the curve $V_{GS} = 0$, the drain current, $I_D$, rises relatively linearly to $I_{DSS}$. In this "ohmic region" (or on region), the current is dependent on the resistance of the channel material. This dependence ceases at $V_P$ volts. $V_{DS}$ may be increased after $V_P$ is reached, but the drain current remains relatively constant at $I_{DSS}$. In this "pinch-off region" (or off region), the output impedance of the transistor is very high.

As the gate is made more negative with respect to the drain, there is less current flowing through the channel. Finally, a $V_{GS}$ is reached where the current ceases. This is the pinch-off voltage. It is identical in value to the pinch-off voltage previously defined. Hence, the pinch-off voltage may be determined by finding the value of $V_{GS}$ that halts the drain current, or it is the value of $V_{DS}$ at which the drain current reaches saturation when $V_{GS}$ is 0. The saturation current in the latter case is $I_{DSS}$.

Resistance of the JFET in the ohmic region is relatively low. At these low drain voltages, each curve has a different dynamic or ac resistance. Although not directly applicable to linear amplification, changes in drain circuit resistance can be used in various situations. One such application is the speech compressor, where the transistor shunts a portion of the applied signal. As larger signal voltages are reduced more by the transistor resistance shunt than smaller signals, the overall circuit behaves as a compressor.

Curves of the transistor are practically horizontal in the pinch-off region. This indicates a high drain-to-source impedance. It is this linear portion of the characteristic that is normally used for audio and rf amplification purposes.

The two portions of the characteristic are shown separated by a broken line. The equation of this line is:

$$V_{DS} = V_P + V_{GS} \tag{2-7}$$

Drain current is related to the pinch-off voltage, $V_P$, the current in the pinch-off region when the gate-to-source voltage is zero, $I_{DSS}$, and the gate-to-source voltage, $V_{GS}$. This relationship is defined by the equation:

$$I_D = I_{DSS}\left(1 - \frac{|V_{GS}|}{|V_P|}\right)^2 \tag{2-8}$$

where,

$|V_{GS}|$ is the absolute value of $V_{GS}$, disregarding polarity,

$|V_P|$ is the absolute value of $V_P$, disregarding polarity.

Once the pinch-off voltage and $I_{DSS}$ for any transistor has been determined from data or measurements, the relatively horizontal curves in the pinch-off region may be derived using Equation 2-8. Note that pinch-off voltages increase 2.2 mV for every degree Celsius the temperature rises while $I_{DSS}$ drops by about 35 percent from its 25°C figure as the temperature increases to 125°C, and rises by about 45 percent above its 25°C value as the temperature drops to −55°C.

Equation 2-8 can also be used to provide information for plotting the transconductance curve shown in Fig. 2-9. It is a plot of the drain current against the gate-to-source voltage in the pinch-off region of the drain characteristic. Transconductance curves can be generated by simply determining the drain current at the various gate-to-source voltages from Equation 2-8 or Fig. 2-8 and plotting these relationships on an $I_D$ versus $V_{GS}$ graph. The transconductance curve as derived is very important to the designer.

*Transconductance* is a measure of the gain of the transistor. With $I_D$ in amps and $V_{GS}$ in volts, transconductance can be stated mathematically by the equation:

$$g_m = \frac{\Delta I_D}{\Delta V_{GS}} \text{ mhos} \tag{2-9}$$

$g_m$ is represented by the slope of the curve. $g_m$ is not constant, but changes with the collector current and the applied gate-to-source voltage.

The curve hits the $I_D$ axis at a point where $I_D = I_{DSS}$, for it is here that $V_{GS} = 0$. Transconductance at this point is assigned the symbol $g_{mo}$. The slope of the curve, and hence the transconductance at this point, can be determined by drawing a line tangent to the curve (touching the curve at one point without intersecting it) at the $V_{GS} = 0$, $I_D = I_{DSS}$ point. This line hits the $V_{GS}$ axis at a point equal to half the pinch-off voltage, $V_P/2$. Transconductance, $g_{mo}$, can be determined by calculating the slope of this tangent line. It is equal to:

$$g_{mo} = \frac{I_{DSS} - 0}{|V_P|/2 - 0} = \frac{2I_{DSS}}{|V_P|} \tag{2-10}$$

Transconductance, $g_m$, at any other point can be determined from the slope of the tangent to the curve at that point, or more directly from Equations 2-8 and 2-9.

Transconductance at any point is:

$$g_m = g_{mo}\left(1 - \frac{|V_{GS}|}{|V_P|}\right) \tag{2-11}$$

Different relationships exist between transconductance and the other circuit factors. These are summarized in Equation 2-12. Note in the equations that other symbols besides $g_m$ are used in the literature to represent transconductance.

$$\begin{aligned} g_m &= \frac{\Delta I_D}{\Delta V_{GS}} = g_{fs} = y_{fs} = \frac{2I_{DSS}}{V_P}\left(1 - \frac{|V_{GS}|}{|V_P|}\right) \\ &= g_{mo}\left(1 - \frac{|V_{GS}|}{|V_P|}\right) = \frac{1}{r_{ds}}\left(1 - \frac{|V_{GS}|}{|V_P|}\right) \\ &= g_{mo}\left(\frac{I_D}{I_{DSS}}\right)^{1/2} \end{aligned} \tag{2-12}$$

where,

$\Delta V_{GS}$ is the change in gate-to-source voltage,

$I_D$ is the quiescent drain current,

$\Delta I_D$ is the change in drain current due to $\Delta V_{GS}$,

$I_{DSS}$ is the drain current in the pinch-off (or off) region when $V_{GS} = 0$,

$V_P$ is the pinch-off voltage,

$|V_{GS}|$ is the absolute value of the gate-to-source voltage ignoring the + or − signs,

$|V_P|$ is the absolute value of the pinch-off voltage ignoring the + or − signs,

$g_{fs}$ is another symbol appearing in the literature to represent the same relationship as $g_m$,

$g_{mo}$ is $g_m$ when $V_{GS} = 0$ at the $I_D = I_{DSS}$ curve. ($g_{mo} = 2I_{DSS}/V_P$)

$r_{ds}$ is the resistance of the channel (from source to drain) when the transistor is operated in the ohmic (or on) region. Using Fig. 6-2, it is $\Delta V_{DS}/\Delta I_D = I/g_{os}$. $g_{os}$ is the output conductance.

$y_{fs}$ is the transadmittance and is equal to $g_m$ at low frequencies. Use of $g_m$ at high frequencies instead of $y_{fs}$ is misleading, as $g_m$ does not include the effects of the capacities of the JFET. $y_{fs}$ does include these capacity effects. It should be noted that $g_m$ and $I_{DSS}$ drop 0.5 percent for every 1°C rise in temperature.

While $g_m$ was defined as an ac transconductance, there is likewise a dc $g_m$ or $g_{fs}$. It is the ratio of the drain current to the gate-to-source voltage at a particular point on the output curve. It differs from point to point on the curve.

You may use the $g_m$ at the particular quiescent point in your design procedure. A more accurate approach is to determine an average transconductance, $\overline{g}_m$, for the range over which the input signal swings $V_{GS}$. In Fig. 2-16 assume the signal swings $V_{GS}$ from

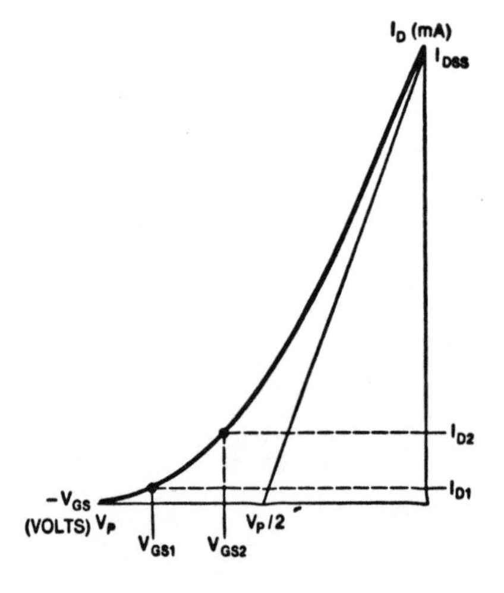


Fig. 2-16. This is an example of a transconductance-curve graph.

$V_{GS1}$ to $V_{GS2}$. $I_D$ will vary from $I_{D1}$ to $I_{D2}$. Average $g_m$ is the slope of the line joining these two points on the curve.

$$\overline{g_m} = \frac{I_{D2} - I_{D1}}{V_{GS1} - V_{GS2}} \tag{2-13}$$

If $I_{D2}$ is at $I_{DSS}$ so that $V_{GS2} = 0$, the equation becomes:

$$\overline{g_m} = \frac{I_{DSS} - I_{D1}}{V_{GS1}} \text{ or } I_{D1} = I_{DSS} - g_m V_{GS1} \tag{2-14}$$

## Insulated Gate FETs

The IGFET is made in two basic arrangements, each not radically different from the JFET just discussed. Each arrangement may use an n-channel or a p-channel semiconductor. Hence, four types of IGFETs are common. All types are illustrated in Fig. 2-17. Once again, only the n-channel will be considered; p-channel transistors use similar mechanisms, but the terminal voltages must be reversed.

In Fig. 2-17A, the substrate (foundation) is a highly resistive p-slab. Two n-slabs are diffused (joined) into the p-material. An oxide insulator covers this combination. A metal gate placed over the insulator, in conjunction with the p-substrate, forms a capacitor. A positive voltage on the gate causes negative charges to be induced into the p-substrate. These charges allow conduction between the source and drain n-slabs. The greater the charge (or the more positive the voltage placed on the gate with respect to the source) the lower the resistance between the source and drain. Drain current is increased or enhanced by an increase in gate potential. Hence, the name enhancement-type transistor.

Depletion-type IGFETs behave very much like the JFETs. The drawings of these are in Fig. 2-17C and 2-17D with the schematic representation in Figs. 2-17E and 2-17F. The depletion-type device is similar in structure to the enhancement-type IGFET, with the addition of a medium resistance n-channel connecting the source and drain. There is current between the source and drain even with low values of $V_{DS}$ voltages. Negative gate voltages will cause the current to decrease. As drain current rises when $V_{GS}$ is decreased and drops as $V_{GS}$ is increased, this device is most accurately described as both a depletion and enhancement type of IGFET.

.The three FET devices are classified by type. Type A is a depletion-type transistor only, and is represented by the JFET. Type-B devices, such as the last one discussed, exhibit both depletion and enhancement characteristics. Approximate curves are shown in Fig. 2-18A. Type-C devices are strictly enhancement-type. Figure 2-18B shows theoretical drain characteristic curves for this type of device.

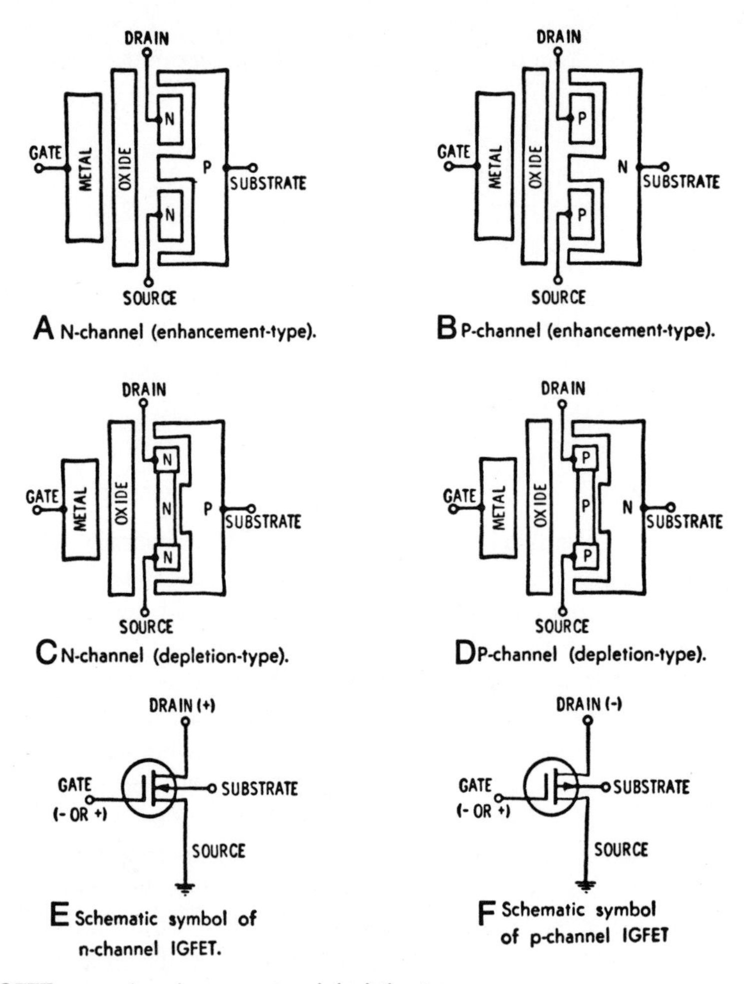


Fig. 2-17. IGFETs come in enhancement and depletion types.

Drain resistance, $r_d$, of IGFETs ranges up to about 50,000 ohms, while that of the JFET can be as high as 1 MΩ. As drain resistance is determined from the slope of one of the $V_{GS}$ curves, it is equal to the change of drain-to-source voltage, $\Delta V_{DS}$, divided by the change of drain current, $\Delta I_D$, or $\Delta V_{DS}/\Delta I_D$.

Another difference between JFETs and IGFETs is in the term *pinch-off voltage*. It is generally not applied to the enhancement-type device as it does not conduct drain current until a threshold gate-to-source voltage is exceeded. It is arbitrarily assumed than conduction starts only after $I_D$ exceeds 10 $\mu$A. Thus the threshold gate-to-source voltage, $V_{GST}$ (symbols $V_T$ and $V_{GS(th)}$) are used by some manufacturers), is at the point when $I_D = 10\ \mu$A. The gate is connected to the drain when determining the $V_{GS}$ at which $I_D = 10\ \mu$A. As for enhancement/depletion types of devices, the symbol $V_{GS(off)}$ is used to indicate the gate-to-source voltage when the drain current is very low.

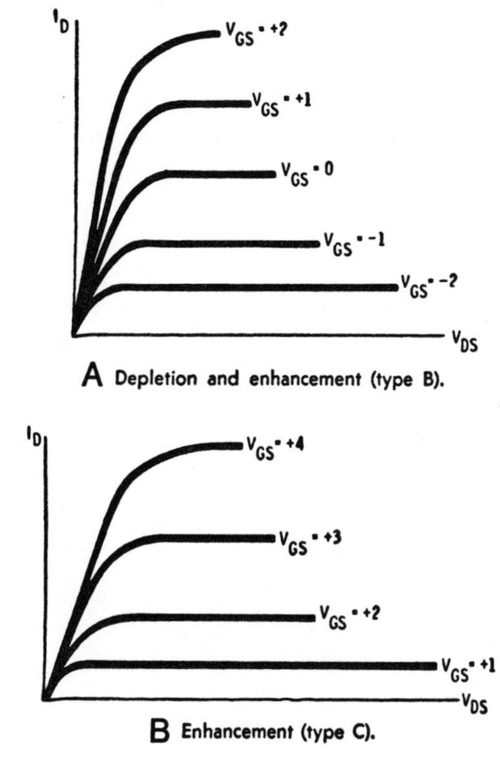


A Depletion and enhancement (type B).

B Enhancement (type C).

Fig. 2-18. Graphs illustrating depletion and enhancement curves.

$I_{DSS}$ retains the definition it has for the JFET when it is applied to the IGFET. For enhancement devices, it is about 10 $\mu$A, the same as the drain current that flows when $V_{GS} = V_{GST}$. $I_{DSS}$ of depletion devices is similar in magnitude to that of the JFET current.

Transconductance curves can best be generated using the information present in the drain characteristic curves in Fig. 2-18. Such curves are shown in Fig. 2-19 for a

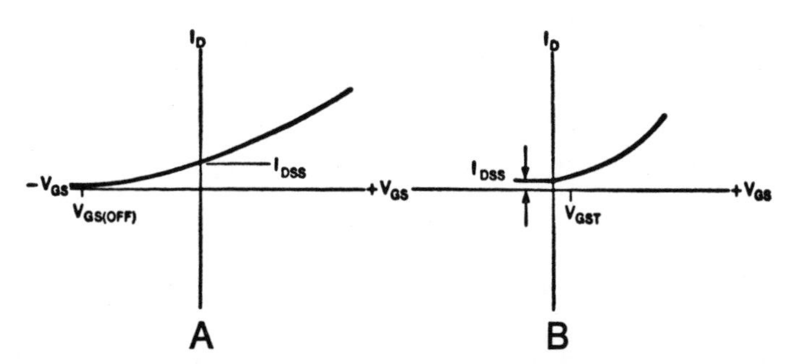


Fig. 2-19. The transconductance-curves for typical IGFETs. (A) Depletion-type device. (B) Enhancement-type device.

hypothetical n-channel device. Actual curves differ somewhat from those shown because of temperature variations. Drain current and $g_m$ drop about 25 percent for an 80°C rise in temperature, while $I_{DSS}$ doubles for every 10°C the temperature increases over the 25°C ambient. Curve variation may also be due to deviations of a transistor's characteristics from the average parameters assigned to a specific type number. Changes of the curves are minimized by using a large resistor, possibly in the order of 1,000 ohms, in the emitter circuit of the transistor.

In the n-channel device, there is a substrate made of p-type material. Caution must be observed not to make the substrate positive with respect to the channel, or there will be conduction between the substrate and various n-slabs in the IGFET. Frequently, the manufacturer connects the substrate to the source inside the IGFET and no separate lead is brought out of the case. If a lead from the substrate is available, it should usually be connected to the source.

### The V-MOSFET

A fairly recent variant of the basic IGFET is the V-MOSFET. Figure 2-20 shows the construction of a standard IGFET device. All of the contact electrodes are on one side of the semiconductor chip. This type of construction results in several design constraints. The most significant of these is that the maximum power handling capability is limited because there is no easy way to extract waste heat generated within the chip.

By redesigning the FET's internal structure with a vertical groove as shown in Fig. 2-21, performance can be substantially improved. The required size of the FET for a given power handling capability is greatly reduced.

Essentially, the V shaped cut-out effectively increases the chip area without increasing its actual size.

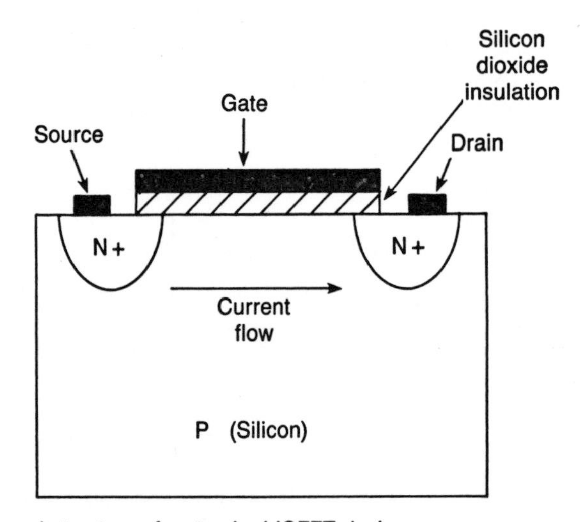


Fig. 2-20. The internal structure of a standard IGFET device.

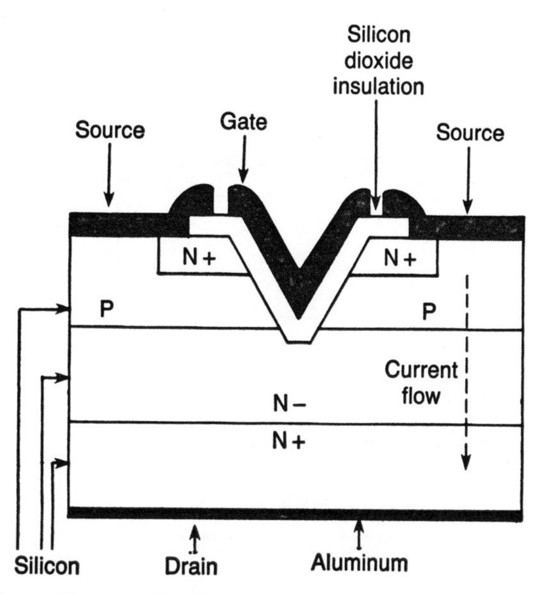


Fig. 2-21. Power FETs usually use a V-shaped groove to increase power handling capability.

This type of device is known as a V-MOSFET. The V can be interpreted in two ways. The most obvious is the shape of the groove. It looks like a letter "V." A more precise interpretation is that the V stands for "Vertical." In standard IGFETs, the current flow is in a lateral direction. A V-MOSFET features vertical current flow. This shortens the current path, therefore, the component's switching speed is much faster. In addition, the channel resistance in the "on" state is lower than for conventional IGFETs.

Unfortunately, the vertical groove structure illustrated in Fig. 2-21 is also limited in its power handling capability. Although, the limit is higher than for standard IGFETs. Physically cutting a groove into the chip introduces both mechanical and electrical stresses.

Other physical configurations can also offer the advantages of vertical current flow. One such arrangement is shown in Fig. 2-22. This device may also be labelled a V-MOSFET, or it may be called a D-MOSFET (*D*ouble—diffused *M*etal *O*xide *S*emiconductor *F*ield *E*ffect *T*ransistor). In this transistor, the current flows vertically from the source connection to the drain connection.

Another feature of V-MOSFETs is a relatively high gate capacitance. This capacitance is typically in the 500 to 800 pF range. Thanks to this high input capacitance, the gate will tend to draw fairly large currents at moderately high frequencies (the upper audio range, and well into the ultra-audio and rf range).

V-MOSFETs offer a number of important advantages. The most important of these are as follows;

- Very high switching speeds are possible (in the nanosecond range)
- Extremely low "on" resistance (as low as a few tenths of an ohm)

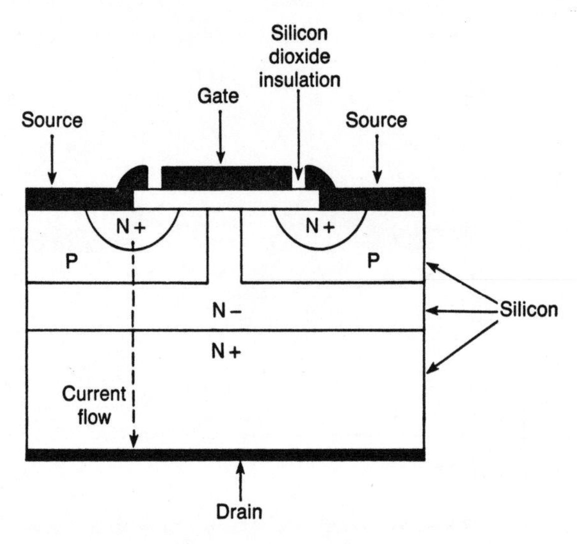


Fig. 2-22. A variation on the V-MOSFET is the D-MOSFET.

- Can switch heavy loads (up to hundreds of volts, and tens of amperes)
- Very high input impedance
- Low voltage drive requirements
- Better thermal stability than bipolar transistors
- Can be operated as switches or linear amplifiers
- Can be easily connected in parallel (no special design procedures or constraints)

Of course nothing is perfect, and that certainly includes electronic components. There are disadvantages to V-MOSFETs. The most important disadvantage is that they are quite susceptible to electrostatic discharge, requiring special handling. Secondly, they are more expensive than conventional semiconductor devices. While the price may come down somewhat, V-MOSFETs will probably always be relatively expensive, because their manufacture requires several extra processing steps. Because of lower drive requirements, some circuits can work out to be less expensive overall when V-MOSFET design is used.

There are many potential applications for V-MOSFETs. In fact, they can be used in almost any circuit requiring an active device. We will briefly consider just a few general examples.

In some power supply circuits, the series-pass stage includes two or more bipolar transistors wired in parallel with at least two power resistors for each individual transistor. These resistors force the transistors to share the load current more or less equally. If the resistors were not in the circuit, and the temperature of the transistors varies for any reason, the current will be split unevenly among the transistors. One could end up drawing an excessive amount of current and destroying itself, as well as a few other components.

V-MOSFETs are much more thermally stable than bipolar transistors, or even IGFETs. The FETs feature a negative temperature coefficient. The current drawn by the device decreases as its temperature rises. This prevents the problem of thermal runaway. No power resistors are required, so the circuit is simplified.

V-MOSFETs are also used in audio amplifiers, especially in the output stage. They offer a number of advantages in this type of application. Thermal runaway is often a problem for bipolar transistors in an output stage. V-MOSFETs are not subject to this type of failure.

Since the input impedance of a V-MOSFET is so high, it requires less drive power for the same output level in a bipolar transistor circuit. The driver stage can use less expensive components rated for lower power levels.

Finally, a V-MOSFET has more linear operating characteristics than bipolar transistors, so there is less distortion of the amplified signal.

V-MOSFETs offer similar advantages in rf amplifiers and power switching applications.

### The COMFET

A recent variant of the basic IGFET is the COMFET. The letters of the name for this device stand for *CO*nductivity-*M*odulated *F*ield *E*ffect *T*ransistor.

To a large degree the COMFET can be considered a combination of IGFETs, bipolar transistors, and thyristors (see Chapter 12). It shares many features with these semiconductor devices.

The COMFETs resistance is considerably lower than an IGFET of the same size. This resistance is typically just 0.1 ohm or less, with a maximum drain current of approximately 20 amps passing through the COMFET.

Though no larger than a typical IGFET, a COMFET can stand up to much more power. A typical COMFET can block up to 400 volts in the forward direction, and up to 100 volts when reverse-biased.

Both the high power capability and the low on resistance make the COMFET very useful for high-voltage, high-power applications. The higher on resistance of a standard IGFET would cause increased power losses and voltage drops.

A COMFET might not be ideal for all applications. Its switching speed is slower than that of an IGFET. This might not be a major limitation. In many cases a COMFET's switching speed is in the same range as a bipolar transistor. A COMFET can switch fast enough for use in the majority of low to medium frequency applications.

Functionally, a COMFET behaves as if it was a heavy duty IGFET feeding into a pair of direct-coupled bipolar transistors. The equivalent circuit for a COMFET is illustrated in Fig. 2-23.

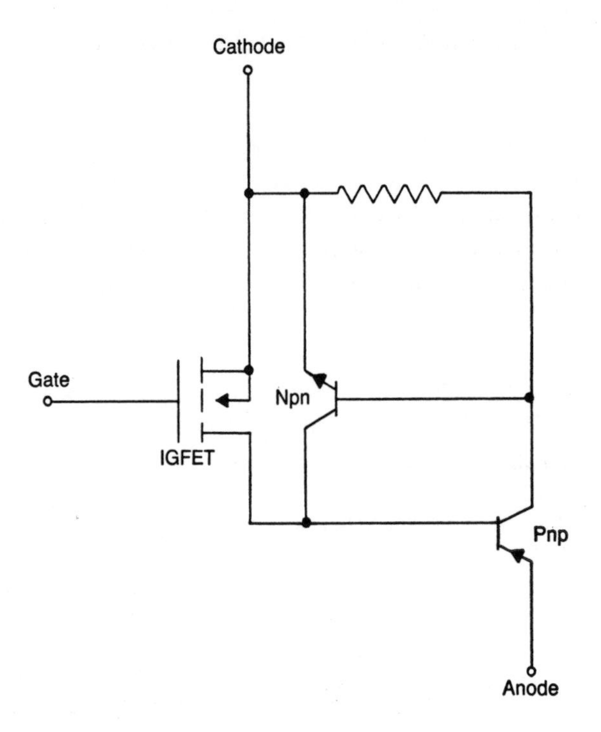


Fig. 2-23. The equivalent circuit for a COMFET.

Like standard IGFETs, the COMFET features high input impedance, reducing its drive requirements, and often simplifying its supporting circuitry.

## PHOTOFETS

Much as the phototransistors consisting of bipolar devices, light impinging upon the gate-channel junction of the JFET increases the flow of gate current. Ordinarily, only leakage current, $I_{GSS}$, flows through the reverse-biased gate junction. $I_{GSS}$ flows not only through the gate and drain leads, but also flows through the resistor in the gate circuit. A voltage equal to $I_{GSS}R_G$ is developed across $R_G$ by the leakage current. In the absence of light, this is the dark-gate current. Current increases with the intensity of the light impinging on the junction.

The polarity of the voltage developed across the gate resistor due to the increase in current, is such as to increase the dark-gate current, $I_{GSS}$, and in turn increase the drain current. Because of the high impedance in the gate circuit, the current gain, or the ratio of the change in drain current to the change in gate current, can be as high as several million.

The quantity of drain current flowing is a function of the intensity of the light striking the junction. A voltage is developed across the load in the drain circuit because of the current flowing through it. Thus the voltage developed in this circuit (as well as the current) is related to the quantity of light at the junction.

## FET CIRCUITS

Just as was the case with bipolar device, the FETs can be used in three basic circuit arrangements. Each arrangement exhibits a specific set of characteristics.

### Common Source Circuit

By far, the common source circuit is used more than any of the other FET configurations. It is shown in Fig. 2-24. All drawings in this figure and all equations stated here, are for the IGFET. They apply as well to the JFETs.

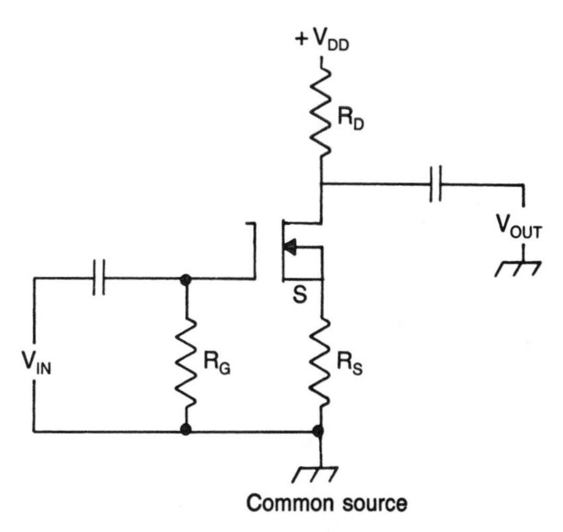


Fig. 2-24. FETs are sometimes used in common-source circuits like this one.

The input signal, $V_{in}$, sees an impedance equal to $R_G$. At the output, the impedance presented by the transistor is equal to $r_{ds}$, the drain-to-source resistance of the device when $R_S$ is equal to zero. If $R_S$ is greater than zero, the output impedance of the device is increased to:

$$R_{out} = r_{ds} + (1 + g_m r_{ds})R_s \tag{2-15}$$

If there is a load at $V_{out}$, it sees an impedance equal to $R_D$ in parallel with $R_{out}$.

Drain and source currents flow through $R_S$. A dc voltage is developed across this resistor equal to the product of the source current and $R_S$. This voltage is also between the gate and source, biasing the gate negative with respect to the source.

There is negative ac feedback caused by the presence of $R_S$ in the circuit. In the absence of $R_S$, the ac input signal, $V_{in}$, is applied between the gate and the source, so that the entire amplified output is across $R_D$. Should $R_S$ be in the circuit, a portion of the amplified output signal is developed across this resistor. Refer to this signal as $V_R$. The input circuit of the FET sees the signal across $R_S$ as being in series with $V_{in}$. This signal is added to $V_{in}$. The combination of $V_{in} + V_R$ is applied across the gate-to-source input terminals of the FET. The relative polarity of the two signals is such that when added together, $V_{in} + V_R$ is less than $V_{in}$ by itself. Because the actual signal voltage at the gate-to-source input of the transistor is reduced to $V_{in} + V_R$ by the presence of voltage across $R_S$, the output voltage, $V_{out}$, across $R_D$ is also reduced. Thus, for a fixed input signal $V_{in}$, the output signal $V_{out}$ is reduced because of the presence of $V_R$. This indicates a reduction in the overall voltage gain of the circuit, or negative feedback.

As far as the ac or signal voltage is concerned, the effects of the source resistor can be eliminated substantially be wiring a large capacitor across it. The capacitor is a short circuit for ac signals, while still permitting the dc bias voltage to remain untouched.

The voltage gain of the circuit as drawn is:

$$A_{VS} = \frac{g_m r_{ds} R_D}{r_{ds} + R_D + (1 + g_m r_{ds}) R_S} = \frac{g_m (R_D || r_{ds})}{1 + g_m R_S}$$

Should $R_S$ be set equal to zero, the equation reduces to:

$$A_{VS} = g_m (r_{ds} || R_D) \tag{2-17}$$

where $r_{ds} || R_D$ is the symbol for the resistance of $r_{ds}$ in parallel with $R_D$.

### Common Drain Circuit

This circuit arrangement is often referred to as a *source follower*. Two common drain circuits are shown in Fig. 2-25. In Fig. 2-25A, the input impedance is simply $R_G$, while in Fig. 2-25B it increases to:

$$R_{in} = \frac{R_G}{1 - A_{VD}} \tag{2-18}$$

where $A_{VD}$ is the voltage gain of the circuit. Voltage gain is equal to:

$$A_{VD} = \frac{g_m R_S}{1 + g_m R_S} = \frac{A_{VS}}{1 + A_{VS}} \approx 1 \tag{2-19}$$

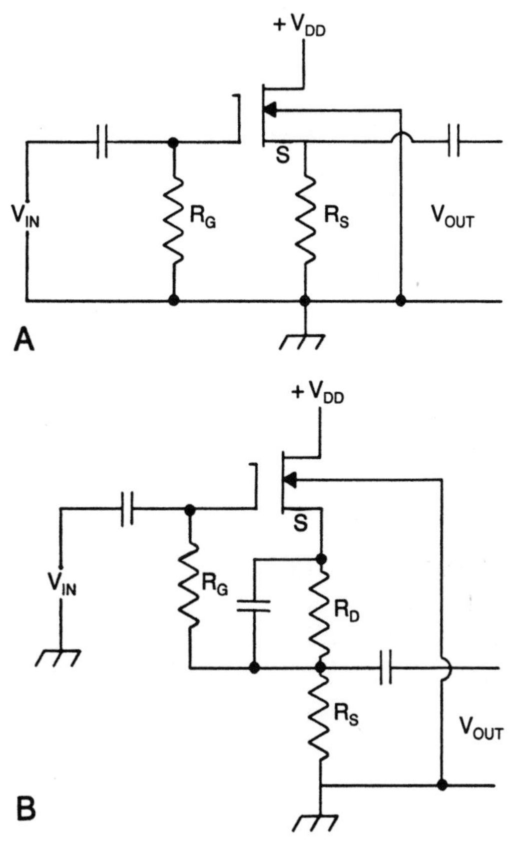

Fig. 2-25. Common-drain circuits are also widely used with FETs.

The impedance seen by $V_{out}$ looking back into the circuit is approximately equal to:

$$R_{out} = \frac{R_S}{1 + g_m R_S} \approx \frac{1}{g_m} \tag{2-20}$$

If there is a load resistor across $R_S$, $R_S$ in the equation should be modified to equal the resistance of $R_S$ in parallel with the load resistor.

### Common Gate Circuit

The third basic FET circuit is, of course, the common gate circuit, which is illustrated in Fig. 2-26. The voltage gain of this circuit can be approximated as:

$$A_{VG} = \frac{g_m R_D}{1 + g_m R_S} \tag{2-21}$$

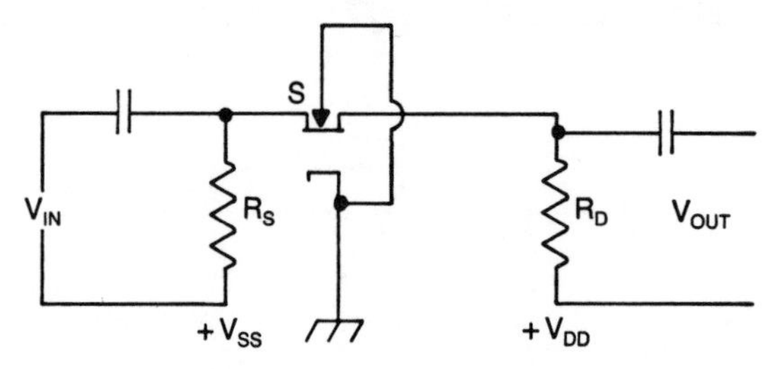


Fig. 2-26. The third basic FET configuration is the Common-gate circuit.

While the output resistance is about equal to $R_D$, the input impedance approaches:

$$R_{in} = \frac{g_m R_S + 1}{g_m} \tag{2-22}$$

## UJTS

The vast majority of transistor devices are of the bipolar or field effect types, or some variation of one of these. There is another distinct type of transistor that should be considered here. This is the *UniJunction Transistor*, or UJT. The name is a significant clue to the structure of this device. A bipolar transistor has two PN junctions, but a UniJunction Transistor has only one. However, it is far removed from the simple diode. The basic structure of a UJT is illustrated in Fig. 2-27, and the schematic symbol for this component is shown in Fig. 2-28.

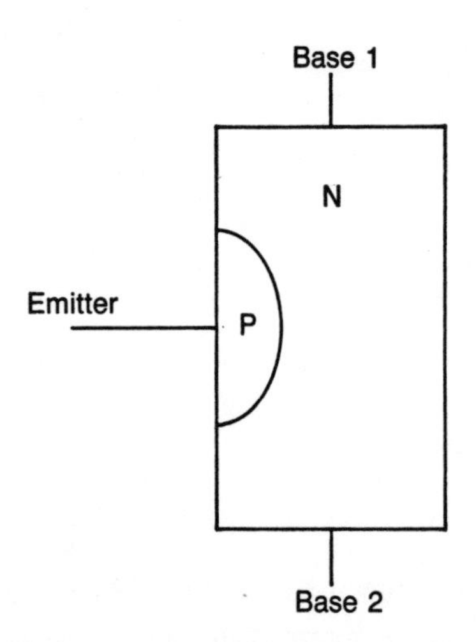


Fig. 2-27. A UJT has an emitter and two base connections.

Notice how the three leads of this device are labeled. There is an emitter and *two* bases, but no collector. Electrically, the main body of the semiconductor (the n-type section) acts like a voltage divider with a diode (the PN junction) connected to the common

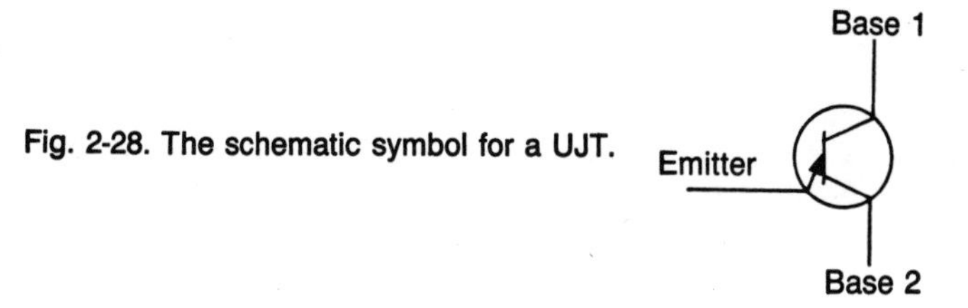


Fig. 2-28. The schematic symbol for a UJT.

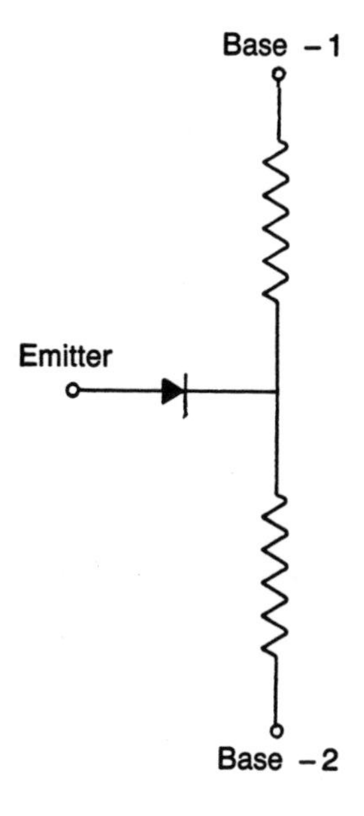


Fig. 2-29. The equivalent circuit for a UJT.

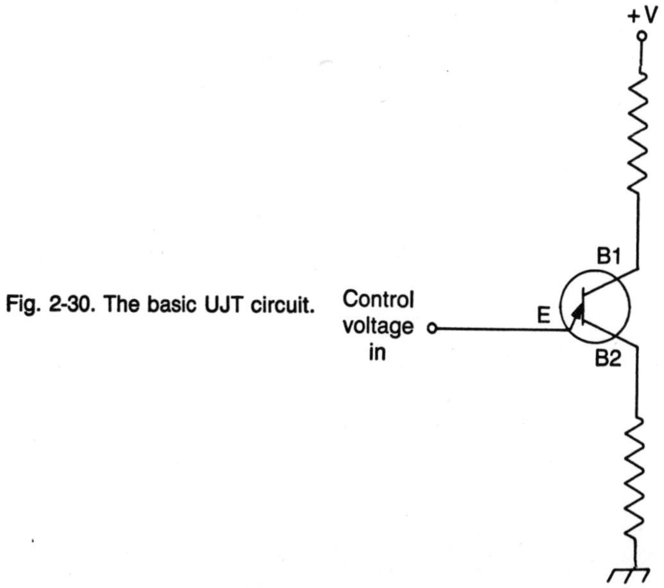


Fig. 2-30. The basic UJT circuit.

ends of the two resistive elements. Figure 2-29 shows a simplified equivalent circuit for a UJT.

In operation, a voltage is applied between Base-1 and Base-2. This voltage puts a reverse bias on the diode. As a result, no current will flow from the emitter to either base. A second variable voltage source is connected between the emitter and Base-1, as shown in Fig. 2-30. As this control voltage is increased, a point will be reached when the diode becomes forward biased. From then on, current can flow between the emitter and the bases.

Unijunction transistors are most frequently used in oscillator circuits and timing applications.

# 3

# Biasing Transistors

In order to function properly, any transistor must be correctly biased. This simply means that a suitable voltage of the proper polarity must be presented to each of the device's terminals. Two factors should be considered here. The first is that a transistor is a complex device. It is temperature-sensitive and has many voltage, current, and power limitations. The frequency range in which a particular device can perform effectively is severely limited. The second factor is the economic impracticality of designing haphazardly and using only the basic circuit configurations shown in Chapter 2.

This chapter details methods used to bias transistors. Bias is used to set the specific quiescent operating point, Q, for the transistor. The base-emitter junction of bipolar transistors is forward biased and the base-collector junction is reverse biased. Base current is adjusted to yield a specific collector current. The particular voltage polarities required for specific devices is also applied to properly bias the different types of FETs.

## BIAS CIRCUITS USED FOR BIPOLAR DEVICES

There are essentially nine circuits used although each one is a variation on the other. Each circuit has its own specific characteristics and limits.

### Bias Circuit I

The first arrangement to be discussed is depicted in Fig. 3-1A. Voltage across the base-emitter junction is $V_{BE}$. Leakage current flowing through the base-collector junction is $I_{CBO}$.

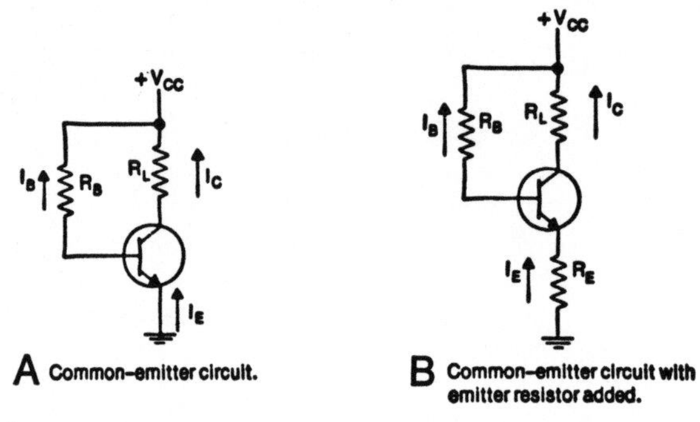


Fig. 3-1. Bias circuit I.

Voltage supplying power to the collector circuit is also used as the source for base current. The only component connecting the supply voltage, $V_{CC}$, to the base is resistor $R_B$.

The resistance in the base circuit is $R_B$, the dc base resistance $r_B$, and the dc emitter resistance $r_E$. When viewed from the base circuit, the last resistance must be multiplied by $\beta$. Stated mathematically, the resistance in the base circuit is $R_B + r_B + \beta r_E$. The total voltage across this combination of resistors is $V_{CC}$ minus the voltage drop from the base to the emitter, or $V_{CC} - V_{BE}$.

Assuming that $V_{BE}$, $r_B$, and $\beta r_E$ are all negligible, the base current, $I_B$, from Ohm's law, is $V_{CC}/R_B$. Some or all of the negligible factors may be added to the equation if more accuracy is required.

Figure 3-1A differs from Fig. 3-1B in that an emitter resistor is added to the latter. In this case, the emitter resistor must be multiplied by $\beta$ when reflected into the base circuit. Now, the base current becomes:

$$I_B = \frac{V_{CC}}{R_B + \beta R_E}$$

The total quiescent collector current can be determined from Equation 2-5.

### Bias Circuit II

To help maintain temperature stability, a resistor, $R_X$ is added to the circuit in Fig. 3-1. This is shown in Fig. 3-2A. Current flowing through $R_B$ is divided between the base and $R_X$. The Thevenin equivalent of the network feeding the base circuit, is used to determine the base current. The network is shown in Fig. 3-2B. Now, determine components of the Thevenin equivalent of the network.

First consider that the equivalent circuit of a power supply is a voltage in series with a resistor. The Thevenin theory is used to reduce complex networks to this form. The network in Fig. 3-2B is converted using the following procedure.

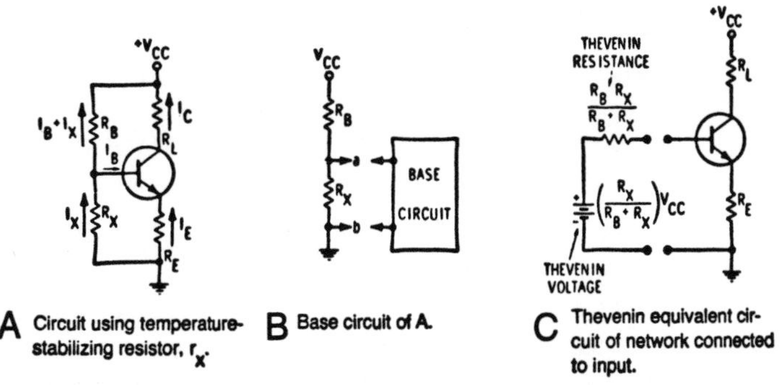


A Circuit using temperature-stabilizing resistor, $r_X$.

B Base circuit of A.

C Thevenin equivalent circuit of network connected to input.

Fig. 3-2. Bias circuit II.

With the base circuit opened the voltage across $R_X$ is $[R_X/(R_B + R_X)]V_{CC}$. This is the Thevenin equivalent voltage. Short the supply voltage. $V_{CC}$ is at ground. $R_B$ is across $R_X$. The two resistors in parallel, or $R_BR_X/(R_B + R_X)$, is the Thevenin equivalent resistance. The complete Thevenin equivalent circuit of the network, connected to the input of the transistor, is drawn in Fig. 3-2C.

The base current can be determined as before. It is the Thevenin voltage minus the base-emitter voltage (if it is of significant magnitude), divided by the sum of the Thevenin equivalent resistance, $R_BR_X/(R_B + R_X)$ and $\beta R_E$. (Ignore $r_B$ and $r_E$ in this particular example.) The collector current can be determined as in the discussion of Bias Circuit I, using Equation 2-5.

## Bias Circuit III

Some specialized circuits require a positive and negative voltage, each with respect to ground, to control the base current. In this case, the bias circuit in Fig. 3-3A can be used.

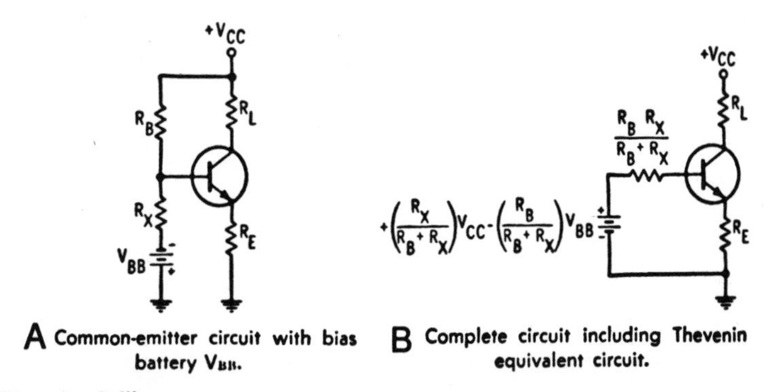


A Common-emitter circuit with bias battery $V_{BB}$.

B Complete circuit including Thevenin equivalent circuit.

Fig. 3-3. Bias circuit III.

To calculate the base current, first separate the components in the base-emitter circuit from the transistor. Now, use the Thevenin theorem to obtain the equivalent circuit for the bias network.

The superposition theorem should be used along with the Thevenized network. The theorem states that if there is more than one voltage or current source acting on a circuit, consider the effects on the circuit of one source at a time and then add the cumulative effects of all the individual sources.

In order to accomplish this, all but one source must be disabled for each calculation. A short circuit across a voltage source disables it. As for the current source, at least one of the leads must be opened to defeat it. Using the superposition theorem, the Thevenin equivalent voltage is found to be:

$$V_{TH} = +\left(\frac{R_X}{R_B + R_X}\right) V_{CC} - \left(\frac{R_B}{R_B + R_X}\right) V_{BB} \tag{3-1}$$

The equivalent resistance between *a* and *b* is the calculated value of $R_B$ in parallel with $R_X$.

$$R_{TH} = \frac{R_B R_X}{R_B + R_X} \tag{3-2}$$

The complete circuit, including the Thevenin equivalent, is shown in Fig. 3-3B. The analysis of this is identical to that of bias circuit II.

### Bias Circuit IV

In Fig. 3-4A, the emitter resistor is returned to a fixed voltage rather than to ground. The base-emitter circuit has been redrawn in Fig. 3-4B. The supply is shown as a battery.

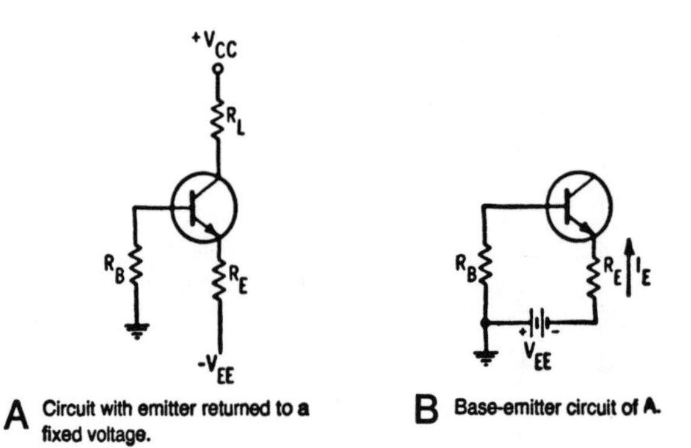


A Circuit with emitter returned to a fixed voltage.

B Base-emitter circuit of A.

Fig. 3-4. Bias circuit IV.

Emitter current, $I_E$, passes through the supply and $R_E$. Emitter current has no effect on the voltage source, $-V_{EE}$, because as an ideal voltage source, it has zero internal impedance. The total base circuit resistance, ignoring $r_B$ and $r_E$, is $R_B + \beta R_E$. Supply voltage is $V_{EE}$ and base current is approximately $V_{EE}/(R_B + \beta R_E)$. As usual, the collector and emitter current is approximately equal to $\beta$ multiplied by the base current. The voltage across $R_L$ and $R_E$ are the collector or emitter current multiplied by the respective resistor.

### Bias Circuit V

The bias circuit in Fig. 3-5 is unique in that the collector voltage, rather than the power-supply voltage, is used to determine the base current. $R_B$ completes a feedback circuit to be discussed in a later chapter.

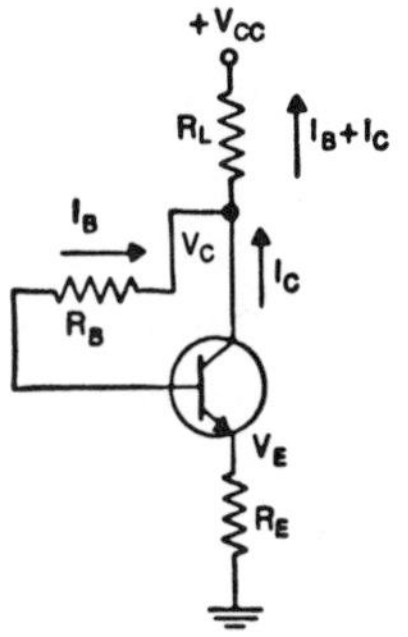

Fig. 3-5. Bias circuit V.

Circuit where collector voltage is used to determine base current.

Voltage across the load resistor, $R_L$, is $R_L$ multiplied by the total current flowing through this resistor. In a perfect transistor, the total current flowing through $R_L$ would be the sum of the base and collector currents, $I_B + I_C$. Since $I_B$ is much less than $I_C$, the voltage across $R_L$ is about $I_C R_L$. The voltage $V_C$ at the collector with respect to ground is $V_{CC} - I_C R_L$.

Base current is $I_B = V_C/(R_B + \beta R_E)$. Collector current is the base current multiplied by beta.

### Bias Circuits VI, VII, VIII and IX

Many of the circuits discussed above are designed with the primary goal of maintaining a constant bias despite variations of the ambient temperature around the transistor. Besides the methods used here to stabilize the bias, circuits have been designed to compensate for various factors that cause instability. In Fig. 3-6, a diode is used for temperature stabilization. This is essentially bias circuit II with an additional forward-biased diode in the base circuit.

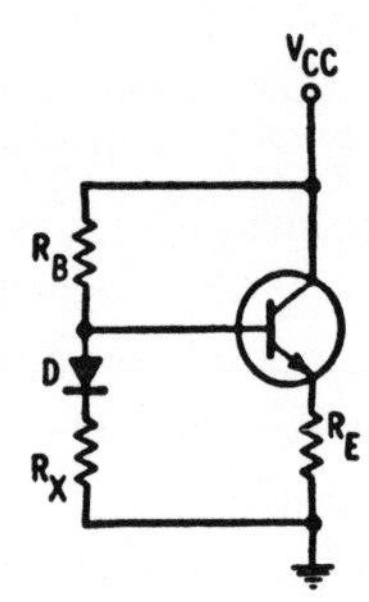


Fig. 3-6. Bias circuit VI uses a diode for temperature stabilization.

As will be pointed out later in this chapter, two characteristics of the transistor are primary factors in causing the bias point to shift with temperature changes. This is so because these two factors are themselves temperature dependent. They are $I_{CBO}$ and $V_{BE}$. $I_{CBO}$ almost doubles with every 10° C rise in temperature. $V_{BE}$ decreases about 2.5 mV for every degree Celsius the temperature increases. A diode placed in a circuit as in Fig. 3-6, helps to compensate for changes in $V_{BE}$. This is particularly important where silicon transistors are involved, because in these devices, the variation of $V_{BE}$ with temperature is the prime cause of thermal instability.

In Fig. 3-6, the voltage across the forward-biased diode is equal to the voltage between the base and emitter if both devices are maintained at the same temperature and constructed in a similar fashion of the same material. Because $R_E$ and the base-emitter junction are in parallel with $R_X$ and the diode, the voltages across the parallel connected components are identical. Since the voltages across the diodes are equal, the voltages across $R_X$ and $R_E$ are equal.

The analysis of this circuit is identical to that of bias circuit II. Ignore the diode and base-emitter voltages and proceed with the mathematical methods previously employed.

Three circuits using semiconductor temperature compensating devices, are drawn in Fig. 3-7. This is in addition to the one in Fig. 3-6. In each case, it is assumed that the device, be it a diode or thermistor, is at the same temperature as the transistor being stabilized.

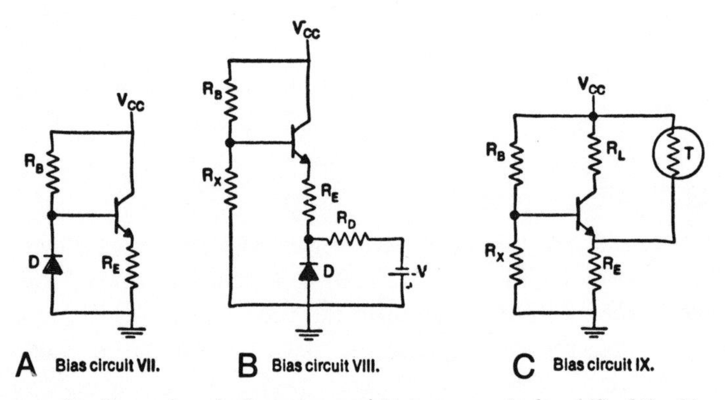


Fig. 3-7. In some circuits semiconductors are used to compensate for shift of the bias point.

Bias circuit VII is used to compensate for variations of $I_{CBO}$ with temperature—the most important factor causing thermal instability in germanium transistors. Leakage current variations with temperature of the diode should be similar to the $I_{CBO}$ variations of the transistor. A thermistor can be used in place of the diode. The resistance of the thermistor should be relatively large so as not to load the input circuit of the transistor. Accuracy of the compensation can be increased by altering the thermistor's temperature characteristics through use of one of the circuits in Fig. 1-1.

Bias circuit VIII is used in an effort to compensate for the change of $V_{BE}$ with temperature. The diode is forward biased by battery voltage V. Assuming the semiconductor material used to form both the diode and transistor are identical, voltages across the forward-biased junctions of both devices change identically with variations of temperature. As the change of voltage across one device compensates for the equivalent change of the $V_{BE}$ voltage of the other device, collector current remains relatively constant.

Resistance of the thermistor in bias circuit IX decreases as the temperature climbs, while collector current increases as the temperature rises. The circuit is arranged so that current flowing through the thermistor because of the reduction of its resistance with rising temperature limits the increase in collector current in the transistor. As the temperature rises and the resistance of the thermistor drops, more current is allowed to flow through $R_E$. The amount of current flowing through the base-emitter junction is determined by the sum of the voltages across $R_E$ and $R_X$, as these components are connected in series across the junction. The size of this current is a determining factor in the size of the collector current, $I_C$. As the heated thermistor allows more current to flow through $R_E$, the voltage across that resistor increases. In turn, there is a reduction of the voltage across, and consequently the current flowing through, the base-emitter junction. Collector current is thereby reduced. It is prevented from increasing endlessly as the temperature rises.

Positive temperature coefficient thermistors can be used instead of $R_E$ or in parallel with $R_B$. If one of these alternate devices is used, the thermistor in bias circuit IX should be omitted. The posistor can be chosen experimentally. The following procedure can be used. Substitute a variable resistor for the posistor. Let the circuit operate at a normal room temperature of approximately 25° C. Adjust the variable resistor so that the collector current is at its desired value. Now increase the ambient temperature. Readjust the variable resistor to a value that will provide the same collector current as was flowing when the device was operating at 25° C. Repeat this at several temperatures. Select a posistor that will more or less provide these resistances at the various temperatures.

### Effects of $r_d$ and $r_c$ on the Bias

Collector-to-base and collector-to-emitter resistances have been ignored. These factors are shown in Fig. 3-8 for the common-base and common-emitter circuits.

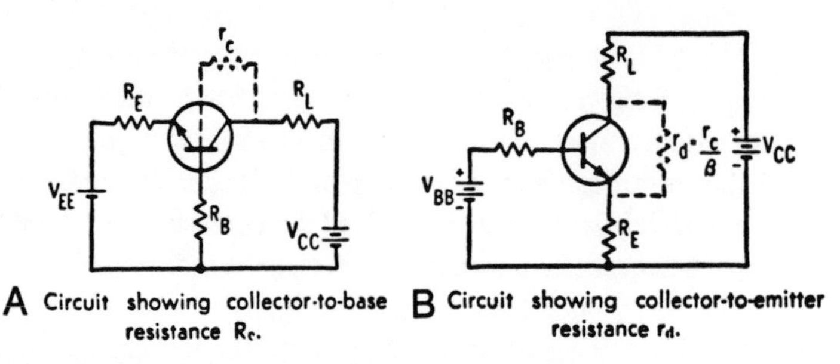


Fig. 3-8. Diagrams illustrating collector resistance considerations.

There is a resistance across the elements in the output circuit of the transistor. In Fig. 3-8A, the collector resistance, $r_c$, is between the collector and base, while in Fig. 3-8B, the resistance $r_d$ is between collector and emitter. Usually, $r_d$ ranges from 10,000 to 100,000 ohms and $r_c$ is $\beta$ times as large. The sizes of these resistances can be determined from the characteristic curves of the transistor. Because it is difficult to determine $r_c$, the method for determining $r_d$ will be illustrated in Fig. 3-9; $r_c$ can be determined from $r_d$ by simply multiplying it by $\beta$.

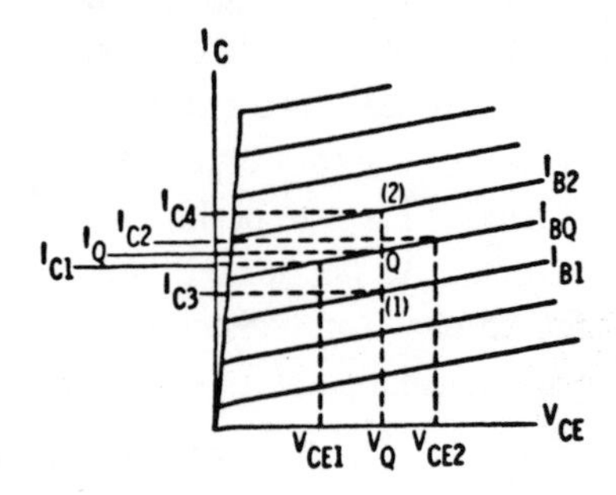


Fig. 3-9. A set of theoretical transistor curves.

First, determine the approximate quiescent base current. Also approximate the quiescent collector current and collector-emitter voltage from the considerations discussed above.

Now, mark two points on the straight portion of the $I_{BQ}$ curve, approximately equidistant from Q. Draw horizontal and vertical lines to the axes and note the currents and voltages. Voltages and currents determined from the points where the lines intersect the vertical and horizontal axes are used to determine $r_d$. It is an ac resistance equal to:

$$r_d = \frac{V_{CE2} - V_{CE1}}{I_{C2} - I_{C1}} \tag{3-3}$$

Use the same curves to determine $\beta$. Extend the vertical line from $V_Q$ to $I_{B2}$. Mark points (1) and (2) where this line intersects $I_{B1}$ and $I_{B2}$, respectively. Draw horizontal lines to the $I_C$ axis. Mark points $I_{C3}$ and $I_{C4}$ on the axis. Dc beta is the collector current divided by the base current at a specific point on the output curve. At this point, ac $\beta$ is the ratio of a change in collector current caused by a change of base current. These are $I_{C4}-I_{C3}$ and $I_{B2}-I_{B1}$ respectively, around the Q point for the curves in Fig. 3-9:

$$\beta = \frac{I_{C4}-I_{C3}}{I_{B2}-I_{B1}} \tag{3-4}$$

and

$$r_c = \beta_{r_d} \tag{3-5}$$

For circuits similar to those in Fig. 3-8B, the collector current has been shown equal to $\beta I_B + I_{CEO}$. There is an additional collector current due to $r_d$. This current is $V_{CE}/r_d$. If $r_d$ is not much larger than $R_L$, $r_d$ cannot be ignored. In this case, the current due to $r_d$ must be added to $\beta I_B$ and $I_{CEO}$ to determine the total collector current, $I_C$. The collector-to-emitter voltage can easily be derived from:

$$V_{CE} = V_{CC} - I_C(R_L + R_E) \tag{3-6}$$

where $I_C$ is the total collector current.

The collector current due to $r_d$ is:

$$I = V_{CE}/r_d = (V_{CC}-I_CR_L-I_CR_E)/r_d \tag{3-7}$$

The effects of $r_d$ are more far-reaching than just to set the bias conditions. Where $r_d$ is a factor in the circuit, it alters the effective value of $\beta$. When determining ac gain, the effective beta, $\beta_{eff}$ becomes $\beta r_d/(r_d + R_L)$. The input impedance due to $R_E$ is more accurately, though not precisely, determined from $(\beta_{eff})\ R_E$. The ac current gain is $\beta_{eff}$ and the voltage gain is $\beta_{eff}(R_L/R_{BR})$, where $R_{BR}$ is all the impedance reflected into the base circuit. In all cases where $r_d$ is negligible compared to $R_L$, the gain and impedance revert to the originally stated simpler forms. Usually, $r_c$ is so large as to be negligible.

## FACTORS AFFECTING THE BIAS POINT OF BIPOLAR TRANSISTORS

It does no good to perform sophisticated calculations to set an optimum quiescent collector current if it will shift radically with temperature changes. Methods must be devised to maintain $I_C$ at or near its calculated value despite temperature and other variations. $I_{CBO}$, $V_{BE}$, $\alpha$, and $\beta$ all vary with temperature and are the four major contributors to the drift of the quiescent collector current.

Quiescent operating conditions at two temperatures can differ from each other. Changes due to temperature are further complicated by the variation of several parameters with the quiescent collector current. Parameters that vary with collector current are $\alpha$, $\beta$, $V_{BE}$, $V_{ce}$, $I_E$, $r_e$, $r_c$, $r_b$ and $I_{CBO}$. In addition to these factors, bias is affected by supply voltage changes and manufacturer's tolerances. As to the latter, it should be recalled that a particular transistor type is assigned a specific $\beta$. But $\beta$ is not a constant. The $\beta$ within a specific group may vary by more than a factor of 1:3. Equipment must be designed so that these variations are not significant.

Quantities $\alpha$ and $\beta$ usually decrease at extremely low and extremely high values of collector current. At high currents, this phenomenon is known as "alpha crowding." Some manufacturers use curves to specify the variation of $\beta$ with $I_C$. Specified parameters are relatively constant over the major portion of the operating range.

The base-to-collector voltage, $V_{CB}$, and the emitter-to-collector voltage, $V_{CE}$, are interrelated and are almost equal. They differ only by the forward voltage drop from the base to the emitter, $V_{BE}$. As a general rule, $\alpha$,$\beta$,$I_{CBO}$,$r_b$,$r_c$, and $r_d$ increase with $V_{CB}$, and $r_c$ and $r_d$ decrease as the base-emitter diode enters the reverse-biased region, $-V_{BE}$. Normally, the collector current affects $V_{CE}$, but the effect of $V_{CE}$ on the collector is negligible.

Since $I_C$, $I_E$, and $I_B$ are closely interrelated, an increase in $I_E$ usually signals a comparable increase in the other two factors. Since $r_e$ is inversely related to $I_E$, $r_e$ will decrease as $I_E$ rises.

Evidently collector current has a marked effect on the various transistor parameters. But those parameters that vary with temperature have a marked effect on the collector current. For example, both $\alpha$ and $\beta$ vary with temperature. They usually increase with temperature up to a peak, at which time they may start to decrease.

$V_{BE}$ decreases with temperature rise at the rate of $-2.5$ mV/°C. Since the base current is dependent on this voltage, $I_B$ will also be affected by the temperature. In turn, $I_C$ will change.

$I_{CBO}$ approximately doubles for every 10° C rise in temperature, evidently affecting $I_C$. Thus if the temperature increases 20° C, the $I_{CBO}$ stated in specification for 25° C should be multiplied by four; if it increases 30° C, $I_{CBO}$ should be multiplied by eight. Another rule-of-thumb is to multiply the 25° C specified $I_{CBO}$ by 10 for every 35° C increase in temperature. There are, of course, other variations, but these are the generally applicable values.

## BIAS AND STABILITY OF BIPOLAR TRANSISTORS

An important factor affecting collector current when the temperature changes is $I_{CBO}$. A stability factor relating the change in collector current to the change in reverse leakage current due to all factors is:

$$S = \frac{\Delta I_C}{\Delta I_{CBO}} ; \Delta I_C = S(\Delta I_{CBO}) \quad (3\text{-}8)$$

$\Delta$ indicates a change in the quantity or a difference in size of a quantity under different conditions.

Two other major factors that affect the collector current are the base current, $I_B$ (as a function of the base-emitter voltage, $V_{BE}$), and the $\alpha$ or $\beta$ of the transistor. Compared to $I_{CBO}$, the effects of these factors are minor for germanium devices. The other factors are the prime cause of the instability when silicon transistors are used.

$I_B$ is a function of the base circuit supply voltage, $V_{BB}$. A stability factor, $S_E$, can be defined to relate the change in collector current to the change in $V_{BB}$.

$$S_E = \frac{\Delta I_C}{\Delta V_{BB}}; \Delta I_C = S_E(\Delta V_{BB}) \quad (3\text{-}9)$$

This stability factor is quite important as $V_{BB}$ can change with the varying power-line voltages as they are converted to dc. $V_{BB}$ also changes with signal amplitude, temperature, and the cyclic variations of the applied intelligence. In turn, changes in $V_{BB}$ affect $V_{BB}$ significantly.

Since $I_C$ also depends on $\alpha$ and $\beta$, a stability factor, $S_\beta$, can be defined to relate the change in collector current to the change in $\beta$ (or $\alpha$):

$$S_\beta = \frac{\Delta I_C}{\Delta \beta} ;\Delta I_C = S_\beta(\Delta \beta) \quad (3\text{-}10)$$

Stable circuits can be designed readily by using the extreme-points procedure, as follows. First determine the components necessary to establish the desired quiescent requirements for a circuit similar to that in Fig. 3-1B. Next, apply Thevenin's theorem or the equations stated below to the bias circuit you wish to use, relating resistors to be used in that circuit with those necessary for the circuit in Fig. 3-1B.

The actual circuit we will use in the extreme-points design procedure is shown in Fig. 3-10. It is a slight variation on the circuit in Fig. 3-1B. Separate voltages, $V_{BB}'$ and $V_{CC}$, are used as base and collector supplies. A prime (′) symbol has been added to those of the resistors and to the base supply voltage. This is to indicate that these values are only for the circuit as shown in Fig. 3-11, and not for the final circuit derived from the one in the figure.

Establish two basic equations for the circuit. One equation relates all factors adding up to $V_{BB}'$ when the collector current, $\beta$, and the temperature of the device, are at a maximum. These are $I_C$(max), $\beta$(max) and T(max), respectively. When this condition

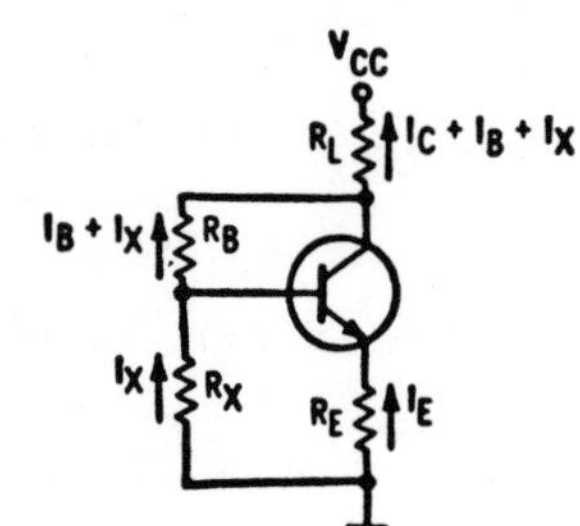


Fig. 3-10. This circuit is used to determine components when applying the extreme points design procedure.

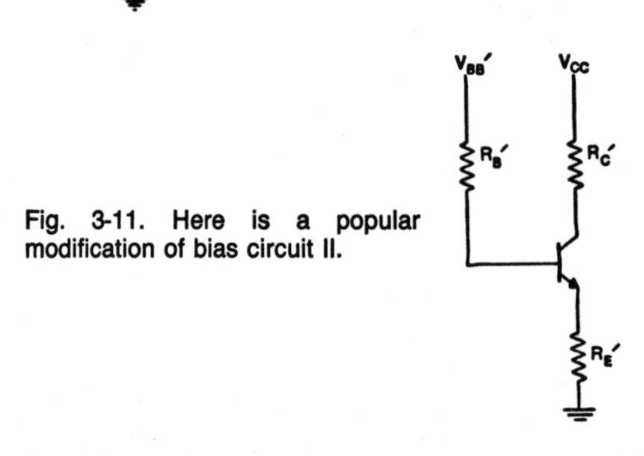


Fig. 3-11. Here is a popular modification of bias circuit II.

exists, the base-emitter voltage is at a minimum, $V_{BE}$(min). The second equation relates all factors adding to $V_{BB}'$ in the same circuit when $I_C$, $\beta$, and temperature are at a minimum and $V_{BE}$ is at a maximum. Now the symbols are $I_C$(min), $\beta$(min), T(min), and $V_{BE}$(max), respectively. All factors but temperature are used in the equations.

$$V_{BB}' = I_C(\text{min}) \left(R_E' + \frac{R_B'}{\beta(\text{min})}\right) + V_{BE}(\text{max}) - I_{CBO}(\text{min})R_B' \tag{3-11}$$

$$V_{BE}' = I_C(\text{max})\ R_E' + \left(\frac{R_B'}{\beta(\text{max})}\right) + V_{BE}(\text{min}) - I_{CBO}(\text{max})R_B' \tag{3-12}$$

Subtracting one equation from the other yields:

$$0 = I_C(\text{min}) \left(R_E' + \frac{R_B'}{\beta(\text{min})}\right) - I_C(\text{max}) \left(R_E' + \frac{R_B'}{\beta(\text{max})}\right) + V_{BE}(\text{max}) - V_{BE}(\text{min}) + I_{CBO}(\text{max})R_B' - I_{CBO}(\text{min})R_B' \tag{3-13}$$

Simplify Equation 3-13. First note that the allowable limits of the quiescent collector current, $I_C$(min) and $I_C$(max), for the proper reproduction of the input signal, can be determined from the required current swing in the output circuit. Bias limits should be set so that peaks in the signal should not push the collector current into saturation or cutoff. Once the quiescent values of $I_C$(max) and $I_C$(min) have been determined substitute them into Equation 3-13.

$\beta$(max) and $\beta$(min) are limits of $\beta$ as provided on most specification sheets. $I_{CBO}$(max) and $I_{CBO}$(min) can be determined from data provided by the manufacturer of the transistor in question, along with calculations at the chosen temperature extremes. $V_{BE}$(max) and $V_{BE}$(min) need not be calculated individually because the difference of these two voltages is $2.5 \times 10^{-3}$ multiplied by the difference in the desired operating temperature extremes, when these are measured in degrees Celsius. Substitute all this information into Equation 3-13.

Various factors besides stability dictate the size of either $R_E'$ or $R_B'$. If minimum voltage gain, $A_V$, is the limiting consideration, $R_E'$ can be determined from the relationship that $A_V$ is approximately equal to $R_C'/R_E'$. As $R_C'$ must be known in order to specify $I_C$(max) and $I_C$(min) intelligently, the minimum $R_E'$ can be determined easily from the ratio. If distortion is an important consideration, $R_E'$ must be as large as possible, consistent with the gain requirements. Input impedance is limited by the sizes of $R_E'$ and $R_B'$, being approximately equal to the resistance of $R_B'$ in parallel with $\beta R_E'$. Once the ohmic size of one of the resistances has been established, the other resistance can be determined from calculations performed after this and the other factors have been substituted into Equation 3-13.

$V_{BB}'$ does not appear in Equation 3-13, but it can be calculated by substituting numbers into Equations 3-11 or 3-12.

Now all component values and voltages have been determined for use in the circuit in Fig. 3-10. Do not use the circuit as is. It should serve only as a Thevenin equivalent for one of the bias circuits detailed above. For good stability, the bias circuit in Fig. 3-2A should be used. The relationships in Equations 3-14 through 3-16 apply:

$$R_B = \frac{V_{CC} R_B'}{V_{BB}'} \tag{3-14}$$

$$R_X = \frac{V_{CC} V_{BB}'}{V_{CC} - V_{BB}'} R_B' \tag{3-15}$$

$$R_E = R_E' \tag{3-16}$$

Better stability can be achieved using the circuit in Fig. 3-11. All equations except 3-16 also apply to this circuit. However Equation 3-16 now becomes:

$$R_E = R_E' - \frac{E_{BB}' R_C}{E_{CC}} \tag{3-17}$$

After the stable stage has been designed, it can be capacitor or direct coupled to a second and even to a third stage. There are many schemes of direct coupling one stage to another. Temperature stability of each stage is affected by the leakage current in the previous stage. Stability equations for dc-coupled amplifiers are long and complex. Instead of using these complex relationships, design each stage for its best stability characteristics. Then check the overall design in the laboratory at extreme operating temperatures.

## FET BIAS CIRCUITS

A major consideration when designing bias circuits for the various types of FETs is whether voltage should be negative, positive or zero at the gate with respect to the source. Positive or zero bias may be used on enhancement types of IGFETs. The IGFET never draws gate current regardless of the polarity of the bias voltage. The JFET does draw current when the n-channel device gate bias runs more than 0.5 volt positive. Although negative or positive bias can be applied, negative bias is usually employed when using depletion-type IGFETs.

Two methods of achieving negative bias are shown in Fig. 3-12. In Fig. 3-12A, the gate is made negative by virtue of the bias supply voltage, $V_{GG}$. The voltage from the gate to the source is equal to $V_{GG} - R_G I_{GSS}$, where $I_{GSS}$ is the leakage current of the

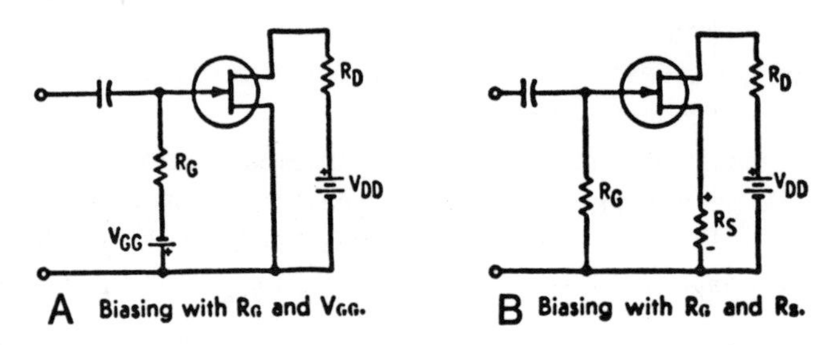


Fig. 3-12. There are several methods for biasing FETs.

reverse-biased junction. The $R_G I_{GSS}$ voltage drop can be ignored if it is much less than $V_{GG}$.

$I_{GSS}$ is similar to the $I_{CBO}$ characteristic of the bipolar-type transistor. It is current flowing from the gate to the channel when the source is shorted to the drain. In the JFET, it doubles for every 10° C rise in temperature. As for the IGFET, the $I_{GSS}$ leakage current is unaffected by temperature and is much lower than the leakage current of the JFET.

In Fig. 3-12B, a voltage developed across $R_S$ is equal to that resistor multiplied by $I_S$, the quiescent source current. Since the source and drain currents are equal, this voltage is $I_D R_S$. The top or source end of the resistor, $R_S$, is positive with respect to ground. The gate is connected to ground through $R_G$. Hence, the source end of the resistor, or source, is positive with respect to the gate. This is the proper polarity for putting a negative bias on the FET. If the voltage due to leakage, $R_G I_{GSS}$, is negligible, the bias voltage from gate to source is equal to the voltage across $R_S$.

Should an input ac signal voltage be fed to the gate circuit of the transistor, an amplified version will appear in the drain and source circuits across $R_D$ and $R_S$. The ac voltage across $R_S$ is in series with the input. Because of the phase relationship, the signal between the gate and source is reduced by the size of the ac voltage across $R_S$. In effect, this is a reduction in the overall gain of the stage, for now less signal appears between the gate and source than would be present when $R_S$ is not in the circuit. For the same input signal at the gates of Figs. 3-12A and 3-12B, less signal is developed across $R_D$ when $R_S$ is finite than when it is set equal to 0 ohms. The effective loss of gain can be overcome by shunting $R_S$ with a large capacitor. The impedance of the capacitor at the lowest frequency to be amplified should be equal to one-tenth of the ohmic value of the resistor, or less.

The method used for biasing the transistor shown in Fig. 3-12B is more desirable than that in Fig. 3-12A because of the better dc stability of the circuit with the source resistor than the stability of the circuit without it. Stability can be further improved by making $R_S$ very large. In this case, the bias voltage would be very high. This drawback can be overcome through use of the circuit in Fig. 3-13. Here, the bias voltage from the gate to the source is:

$$V_{GS} = I_D R_S - V_{GG} - I_{GSS} R_G \tag{3-18}$$

Fig. 3-13. This circuit illustrates a scheme to increase $R_S$.

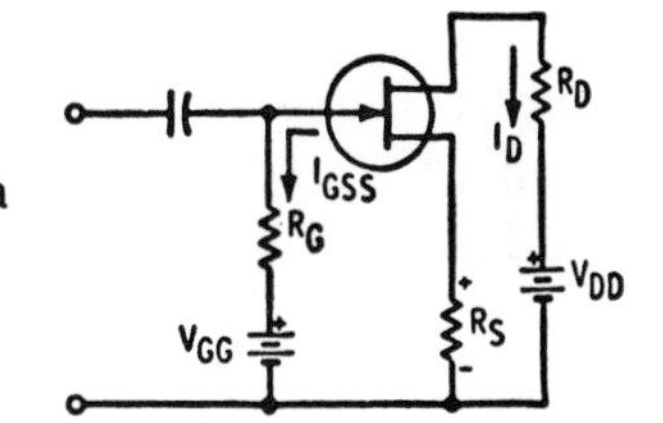


Each transistor parameter is specified to be within a specific range rather than at one absolute value. This cannot be ignored when biasing the device. If $\Delta I_D$ is the variation of the drain current and $\Delta V_{GS}$ is the variation of the gate-to-source voltage, Equation 3-18 shows that:

$$\Delta V_{GS} = \Delta I_D R_S - \Delta I_{GSS} R_G \tag{3-19}$$

where $\Delta I_{GSS}$ is the change in gate current over the temperature range. To determine $R_G$, permissible drain current variations are substituted into the equation along with the allowable gate-to-source voltage change and the $\Delta I_{GSS}$ over the temperature range of operation.

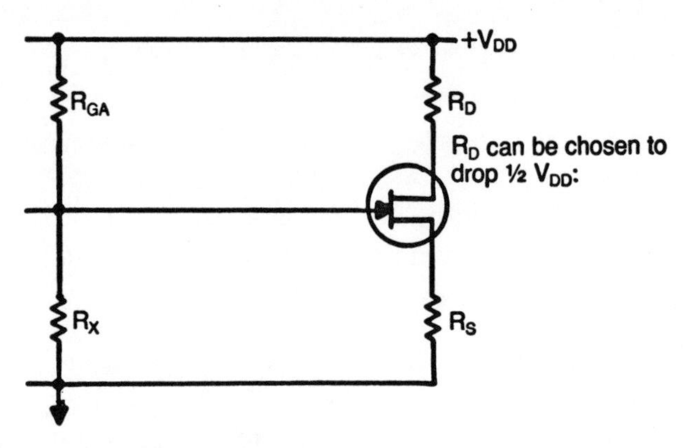


Fig. 3-14. Here is a practical FET circuit.

A practical version of this circuit without use of an extra $V_{GG}$ supply, is shown in Fig. 3-14. Let $V_{DD}$ = 20 volts. Using information derived for the circuit in Fig. 3-13, a 1000-ohm resistor can be used for $R_S$. Resistors for the bias circuit are determined here by solving two simultaneous equations. One equation considers the parallel combination of $R_{GA}$ and $R_X$, which must be equal to the $R_G$ determined from Equation 3-19. Hence:

$$R_G = \frac{R_{GA}R_X}{R_{GA} + R_X} \tag{3-20}$$

The second equation (Equation 3-21) uses voltage-divider methods to get the required voltage at the gate, $V_{GG}$. (The required $V_{GG}$ was calculated by substituting information into Equation 3-18).

$$V_{GG} = \left(\frac{R_X}{R_{GA} + R_X}\right)V_{DD} \tag{3-21}$$

Circuit components will be determined by solving these two equations. Solutions are noted as Equations 3-22 and 3-23.

$$R_{GA} = \frac{R_G V_{DD}}{V_{GS}} \tag{3-22}$$

$$R_X = \frac{R_G V_{DD}}{V_{DD} - V_{GS}} \tag{3-23}$$

The procedure just described is sufficient once circuit components such as $R_S$ and $R_D$ have been determined from other factors, along with the drain current. Other procedures should be followed if the circuit must be designed "from scratch." Two procedures are noted here. One uses transconductance curves, while the other applies transistor data directly. A little algebra must be used in the second procedure. Each method has its own advantages.

## USING TRANSCONDUCTANCE CURVES TO DESIGN JFET BIAS CIRCUITS

The transconductance curve for any JFET is defined by Equation 2-8. The design procedure should commence by using the equation and the extremes of $I_{DSS}$ and $V_P$ specified for a particular transistor. Here, let's use transistor type 2N5558 where $I_{DSS}$ is specified as ranging from 4 mA to 10 mA and $V_p$ ranges from 1.5 volts to 6 volts. Substitute this information into Equation 2-8 to derive the upper and lower curves in Fig. 3-15 when the temperature is at 25° C.

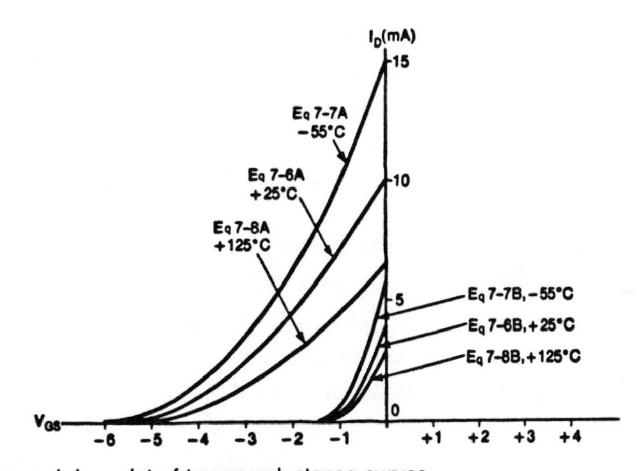


Fig. 3-15. This graph is a plot of transconductance curves.

Equations should also be written for the conditions when the temperature is at its extremes—namely at −55° C and +125° C. It was noted in Chapter 2 that $V_p$ increases at the rate of 2.2 mV ($2.2 \times 10^{-3}$ volt) for each degree the temperature rises and $I_{DSS}$ increases by 45 percent at −55° C and drops by 35 percent +125° C. These changes are, of course, from the $V_p$ and $I_{DSS}$ values stated at 25° C. Using these values as the base, the $I_{DSS}$ that is 10 mA at 25° C becomes 14.5 mA at −55° C and 6.5 mA at +125° C. The $I_{DSS}$ that is 4 mA at 25° C becomes 5.8 mA at −55° C and 2.6 mA at +125° C.

Substitute this newly derived data into Equation 2-8 and plot the curves at the temperature extremes and in the high and low $I_{DSS}$ categories. These are shown in Fig. 3-15 along with the curves when the temperature is 25° C.

The upper three transconductance curves in Fig. 3-15 are at the various temperatures when a device with a maximum $I_{DSS}$ is used and the lower three curves are for these temperatures when the $I_{DSS}$ is at the lower limits for the transistor type. Some manufacturers provide them on specification sheets. This eliminates the need to pursue all the mathematical manipulations.

If the transistor is to be used within some narrow temperature limits, the equations can still be used, but now $I_{DSS}$ and $V_P$ change by only a portion of the amounts shown.

In order to design a circuit properly, we need the highest and lowest curves in Fig. 3-15 that represent the limits of the temperature range we are interested in. If we are interested only in operating at 25° C, a bias point must be chosen between the curves derived from Equation 2-8. Should operation be required between 25° C and 125° C, we must work within the limits set by the upper +25° C curve and the lower +125° C curve. The +125° C on the upper group of curves and the +25° C on the lower group of curves, are between the two curves set as the required limits of operation.

As a design example using these curves, start with the circuit in Fig. 3-13. Transistor 2N5558 is to be used. Should only temperatures between 25° C and 125° C be of concern in this application, the bias must be between the curves just noted. The voltage gain of the transistor can be determined from Equation 2-16. By feeding a specific signal voltage to the input of the transistors, after amplification a signal is developed across $R_D$. The amplitude of the output signal is dependent upon the size of the signal at the input. To determine the gain of the circuit, first assume that the output voltage across $R_D$ is 5 volts peak-to-peak and that the $R_D$ in the circuit is 6800 ohms. Hence the maximum $I_D$ can swing because of the 5 volts is $\Delta I_D = 5/6800 = 0.74$ mA $= \pm\ 0.37$ mA.

Now separate the limiting curves from all curves in Fig. 3-15, and plot them as Fig. 3-16. Set bias points on the two curves so that the $I_D = \pm\ 0.37$ mA swing can be executed easily. Stay away from the pinch-off region. A reasonable bias point on the lower curve is at $I_D = 1.4$ mA. $I_D$ can swing the required $\pm\ 0.37$ mA easily around this point. As for the upper curve, choose a bias point at about 1 mA above that chosen on the lower curve, or at $I_D = 2.5$ mA. Draw a line connecting these two quiescent bias points. The line crosses the $V_{GS}$ axis at $V_{GS} = +\ 3$ volts. $V_{GS}$ is the dc voltage required between the gate and ground.

Source resistor $R_S$ can be determined from the slope of the line just drawn. Noting the two quiescent points, one gate voltage is 3 and the other is 0.3, for a difference of $\Delta V_{GS} = 3 - 0.3 = 2.7$ volts. Drain current at these two points was arbitrarily chosen as 2.5 and 1.4 so that $\Delta I_D = 1.1$ mA $= 1.1 \times 10^{-3}$A. From Ohm's law, $R_S$ is equal to $\Delta V_{GS}/\Delta I_D = 2.7/1.1 \times 10^{-3} \approx 2{,}500$ ohms.

If $I_{GSS}$ is negligible, $V_{GG}$ in our example is equal to the gate-to-ground voltage, −3 volts, as determined from the intercept of the line in Fig. 3-16 with the $V_{GS}$ axis. Should

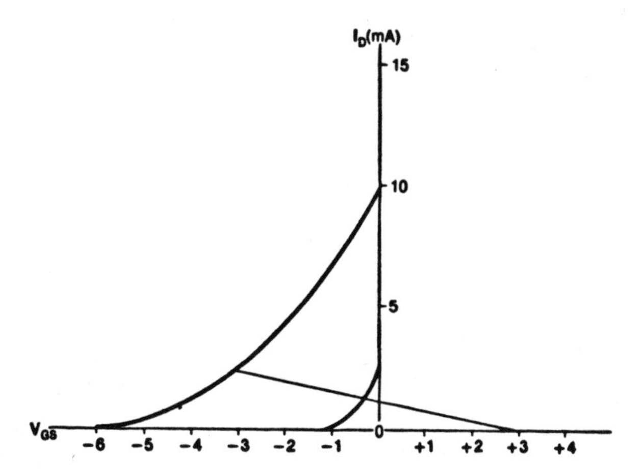


Fig. 3-16. Transconductance curves can be used to design the bias circuit.

$I_{GSS}$ be high, $V_{GG}$ must be sufficiently large to compensate for any voltage, $I_{GSS}R_G$, developed across the gate resistor because of the leakage current. The resistance of $R_G$ must be sufficiently large so as not to excessively load $V_{in}$, the signal source. The minimum size of this resistor can be selected by assuming the voltage developed across this resistor should never exceed 1/10 of $V_{GG}$ or 0.3 volt. So if $I_{GSS}$ is specified at a maximum of 0.1 nA at 25° C, its maximum at 125° C is 100 nA, because $I_{GSS}$ doubles for every 10° C that the temperature rises. Using Ohm's law, the maximum $R_G$ is equal to 0.3 volt/100 nA = 0.3 volt/$10^{-7}$A = 3 MΩ.

In order to establish a minimum value for $V_{DD}$, several items must be considered. First, drain current flows through $R_D$ and $R_S$. A voltage is developed across these resistors because of the drain current. As the maximum quiescent drain current is 2.5 mA, the voltage developed across the two resistors is $(2.5 \times 10^{-3})(R_S + R_D) = 2.5 \times 10^{-3}(2{,}500 + 6{,}800) = 23.25$ volts. Next, voltage must always be present across the transistor if it is to perform its function. It is normally more than 1½ times the size of the maximum specified pinch-off voltage. For the 2N5558, it is (1½)(6 volts) = 9 volts. As $V_{DD}$ must be equal to or greater than these two factors, the minimum $V_{DD} = 23.25 + 9 = 32.25$ volts.

The circuit shown in Fig. 3-13 is not truly practical, because it uses more than one power supply. A more practical circuit arrangement is shown in Fig. 3-14. $R_{GA}$ and $R_X$ in this circuit can be determined from two simultaneous equations. Numbers derived above should be substituted into Equations 3-22 and 3-23. The resistors are then calculated as follows.

$$R_{GA} = \frac{(3 \times 10^6)(35)}{3} = 35\ \text{M}\Omega$$

$$R_X = \frac{(3 \times 10^6)(35)}{35-3} = 3.3\ M\Omega$$

## DESIGNING JFET BIAS CIRCUITS USING THE AVERAGING METHOD

Instead of using the transconductance curves, average values for the various parameter can be determined and used in the design.

Average $I_{DSS}$:$\overline{I}_{DSS}$ is a current midway between the minimum and maximum specified values of $I_{DSS}$. For the 2N5558 transistor, $\overline{I}_{DSS} = (10 + 4)/2 = 7$ mA.

Average pinch-off voltage: $\overline{V}_P$ is a voltage midway between the minimum and maximum specified values of $V_P$. Here it is $\overline{V}_P = (6 + 1.5)/2 = 3.75$ volts.

Average $V_{GS}$ for Class-A operation, $\overline{V}_{GS}$, is usually chosen to be about 40 percent of the pinch-off voltage so that the transconductance will be substantial and $I_D$ will not be readily cut off by a negative swing of the signal. Here, $\overline{V}_{GS} = 0.4(3.75) = 1.5$ volts.

Putting all this information into Equation 2-8, average drain current, $I_D$, can be calculated to be:

$$\overline{I}_D = 7 \times 10^{-3}\left(1 + \frac{1.5}{3.75}\right) = 2.52\ \text{mA}$$

Equation 3-18 defines the gate circuit in Fig. 3-13. If the voltage across $R_G$ is negligible (it usually is compared to the effect of $I_D$ on the bias voltage), a relationship can be established between $V_{GG}$ and $R_S$.

$$\overline{V}_{GS} = \overline{I}_D R_S - V_{GG}$$
$$1.5 = 2.52 \times 10^{-3} R_S - V_{GG}$$

If $V_{GG}$ is set to zero, $R_S = 1.5/2.52 \times 10^{-3} = 595$ ohms. A design using this resistor in the emitter circuit can be fairly stable. Stability can be improved considerably by doubling, tripling and even quadrupling the size of $R_S$.

Procedures and considerations for determining $R_G$ in Fig. 3-13 as well as $R_{GA}$ and $R_X$ in Fig. 3-14, are identical to those pursued using the design procedure involving transconductance curves.

## BIASING IGFETS

Although circuits in Figs. 3-12B and 3-14 are shown using JFETs, identical arrangements are also used for biasing IGFETs. If there is an external lead originating at the substrate of the IGFET, it is to be connected to the source.

The first step in designing the bias circuit for the IGFET is to determine the desired quiescent drain current, $I_D$. This can be derived from the $g_m$ required to fulfill the circuit

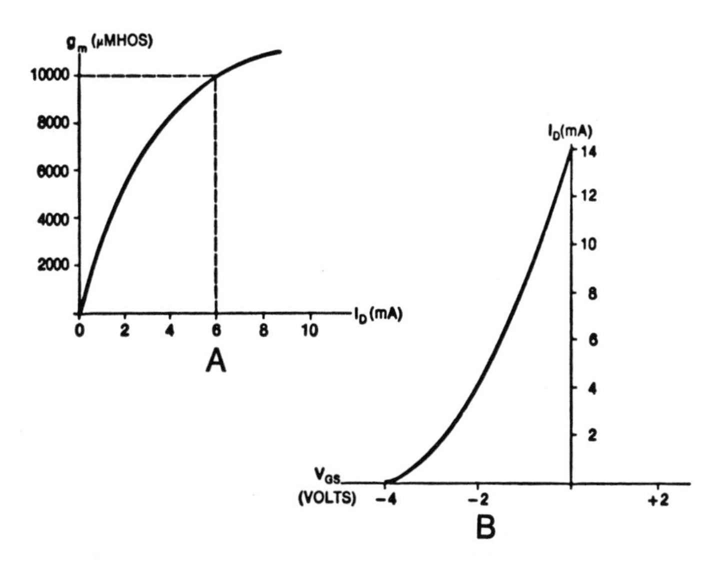


Fig. 3-17. These graphs illustrate typical IGFET characteristics used in the design of the bias circuit.

functions. A curve relating these two factors is in Fig. 3-17A. Once the value of $I_D$ to be used as the quiescent drain current has been decided upon, the gate-to-source voltage, $V_{GS}$, which will provide the required drain current, is to be determined from the curve in Fig. 3-17B. It should be noted that the conditions shown in the curves exist at a specific drain-to-source voltage when the ambient temperature is 25° C. Alternate curves must be used for different ambient temperatures and different drain-to-source voltages. For an intermediate drain-to-source voltage, $V_{DS}$, appropriate curves do not exist. The various factors should be interpolated from the curves made available to you.

We now have data for $V_{DS}$, $I_D$, and $V_{GS}$. For the circuit in Fig. 3-12B, the voltage across $R_S$ is $V_{GS}$. This infers that $R_S = V_{GS}/I_D$, through the use of Ohm's law. The drain supply voltage, $V_{DD}$, must then be equal to the sum of the voltage across the IGFET (namely $V_{DS}$, and the voltage across $R_D$ and $R_S$ or $I_D(R_D + R_S)$.

As for the circuit in Fig. 3-14, when $R_S = 0$, $V_{GS}$ must be supplied by dividing $V_{DD}$ between $R_{GA}$ and $R_X$. Instead of using $+V_{DD}$ as the gate supply, $R_{GA}$ should be connected to a $-V_{GG}$ voltage. In this case:

$$V_{GS} = \left(\frac{R_X}{R_X + R_{GA}}\right)(-V_{GG}) \tag{3-24}$$

Should $R_S$ be finite, $V_{GS}$ determined above is equal to the sum of the voltages across $R_X$ and $R_S$. Considerations as to the size of $R_S$ remain as described above for the JFET, as do the sizes of $R_X$ and $R_{GA}$, although $I_{GSS}$ is seldom large enough in the IGFET to be of interest. Also stability or thermal runaway are not factors because the $I_D$ of the IGFET drops as the temperature increases.

In most circuits, the substrate is connected to the source. It has been noted earlier that a separate voltage can be applied to the substrate. This voltage is used to set the drain current as the gate voltage is varied. In this mode, the transistor operates in the ohmic region as a variable resistance controlled by voltage applied through a potentiometer to the gate.

# 4

# Audio Voltage Amplifiers

Up to now, we have been primarily concerned with dc-biasing the transistor and maintaining stability within reasonable bounds. In this chapter, the discussion centers on ac gain and ac impedance. The three different circuit arrangements for bipolar and FET transistors will be detailed in this respect. The ac equivalents are shown for the circuit arrangements. As in the case with all ac signal equivalent circuits, dc components have been omitted. However, it must be remembered that dc bias currents do affect the values of ac parameters.

In the interest of flexibility, several different equivalent circuits have been devised for each arrangement when it is applied to the bipolar transistor. The equivalent "T" representation is popular since it looks very much like the actual transistor. The "h" or hybrid equivalent of the transistor circuit has been in widespread use for many years. Some manufacturers specify their transistors using h-parameters.

As for the FET, its equivalent circuit and ac gain equations for the various circuit arrangements are presented. Because of the high impedances associated with the device, capacitances between the various elements of the device become important at relatively low frequencies. These capacitances affect the gain and relative phase of the input and output signals.

It should be remembered that most parameters of the bipolar transistor vary with temperature, collector current, and collector-emitter voltage. Correction factors are frequently presented in the specification sheets detailing the parameter variations. When the correction factors are available, they should be applied to the specified parameter before numbers are substituted into the equations.

## EQUIVALENT CIRCUITS OF BIPOLAR TRANSISTORS

The common-base circuit and its equivalent T and equivalent h representations are shown in Fig. 4-1. Similar representations are presented in Fig. 4-2 and Fig. 4-3 for the common-emitter and common-collector circuits, respectively. In each case, the equivalent T circuit, shown in B of the figures, is self-explanatory. It resembles the transistor in the particular arrangement. The parameters remain the same for all types of circuits.

The dynamic emitter resistance, $r_e$, is equal to $26/I_E$ at 25° C, where $I_E$ is expressed in milliamps. The collector resistance in the common-base configuration is $r_c$; $r_d$ is the collector resistance in the common-emitter configuration. Should the impedance between the base and emitter leads be small, the collector resistor approaches $r_c$, a large value. If the impedance between the base and emitter is large, the collector resistance approaches $r_d$, a small number. In either case, $r_c$ and $r_d$ are related by the equation $r_c = \beta r_d$.

The ac and dc values of $\beta$ can be derived from the collector characteristic curves such as those shown in Fig. 4-4. Mark the collector-emitter voltage, $V_{CEQ}$, on the horizontal axis and draw a vertical line from this point. Next, draw horizontal lines from the points at which the $V_{CEQ}$ line just drawn crosses the $I_B$ base current plots. The dc $\beta$, $\beta_{dc}$, is the ratio of the collector current to the base current at any point on the $V_{CE}$

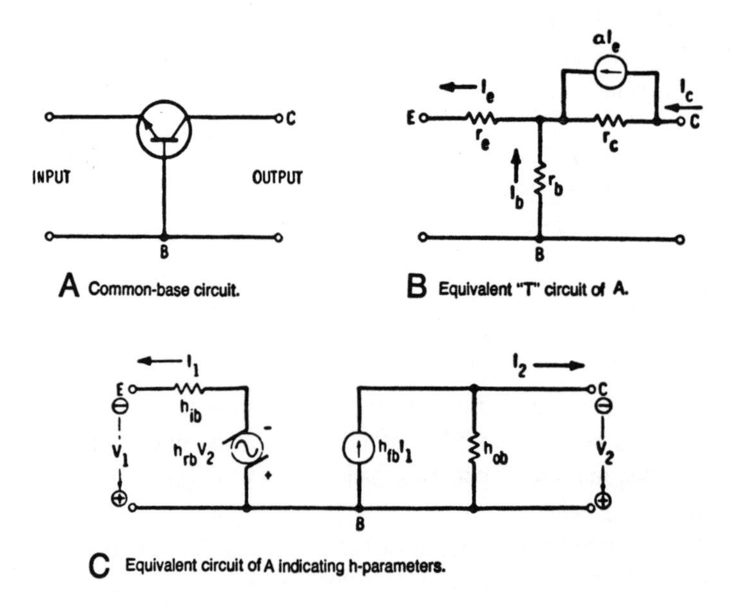


Fig. 4-1. Diagrams illustrate the common-base connection and its equivalent circuits.

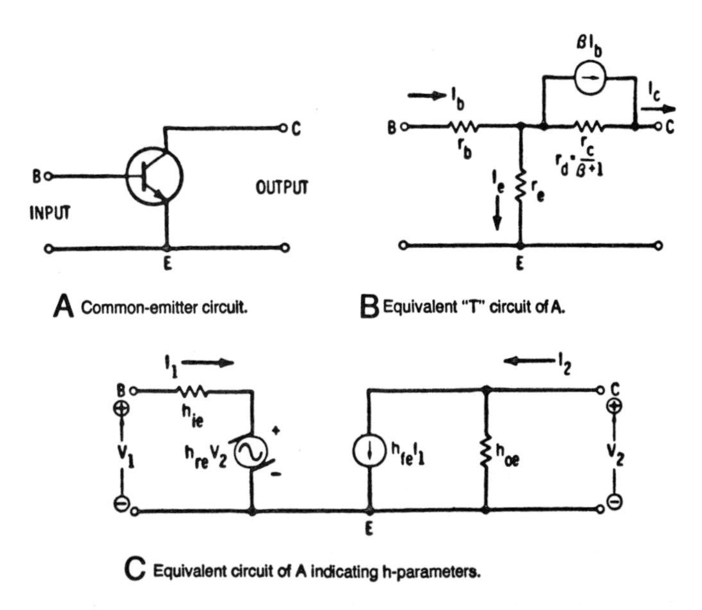


Fig. 4-2. The diagrams illustrate the common-emitter connection and its equivalent circuits.

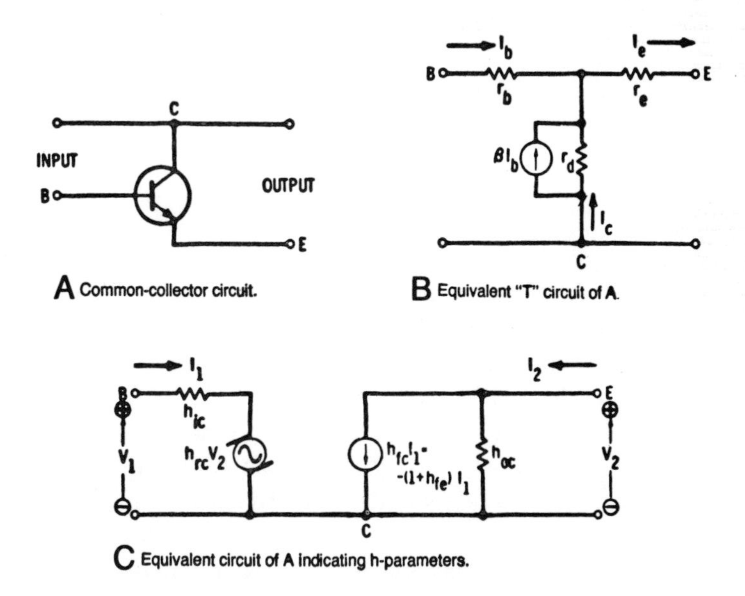


Fig. 4-3. The diagrams illustrate the common-collector connection and its equivalent circuits.

line. It will vary with base current. When the collector current is $I_{C3}$, for example, $\beta_{dc}$ is $I_{C3}/I_{B3}$.

The ac beta is the current gain for an ac signal. It is approximately equal to the ratio of the change of collector current to the change in base current causing it. It is different

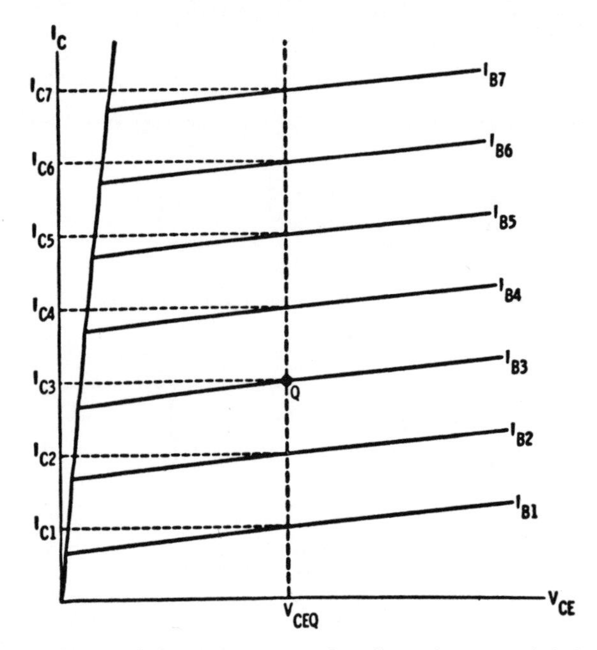

Fig. 4-4. The collector characteristic curves are used to determine ac and dc beta.

at the various portions of the display. Around $I_{B3}$, the ac $\beta$ is:

$$\beta_{ac} = \frac{I_{C4} - I_{C2}}{I_{B4} - I_{B2}}$$

The ac and dc beta frequently do not differ by a large factor. The relative ac and dc alphas can be derived from $\beta$ using the equation $\alpha = \beta/(\beta + 1)$.

The equivalent h or hybrid equivalent circuits all appear in C of Figs. 4-1 through 4-3. The letter h indicates that a parameter refers to the hybrid circuit. The first subscript letter next to the h indicates the type of parameter. The second subscript letter describes the circuit to which the parameter refers. If the second subscript is b, the parameter is for the common-base circuit, e refers to the common-emitter circuit; and c refers to the common-collector circuit.

The type of hybrid parameters has also been defined by numerical subscripts. The following is a list of comparative symbols and definitions of hybrid parameters. Refer to any of the figures to see where the symbols are used.

$h_{11} = h_i = V_1/I_1$ = The input impedance when the output is short-circuited, as when $V_2 = 0$.

$h_{22} = h_0 = I_2/V_2$ = The output admittance when the input is open-circuited, as when $I_1 = 0$. The output impedance is $1/h_0$.

$h_{21} = h_f = I_2/I_1 = -\alpha$ for the common-base circuit. In the common-emitter circuit, it is equal to $\beta$. The ratio of the output current to the input current is $h_f$ when the output is short-circuited, as when $V_2 = 0$. It is the current gain from the input to the output.

$h_{12} = h_r = V_1/V_2 =$ The ratio of the voltage at the input circuit fed back through the transistor from the output circuit, to the voltage at the output. Here, the input is open-circuited, as when $I_1 = 0$.

## COMMON-BASE CIRCUIT

A common-base circuit is shown in Fig. 4-5. The input resistance of the transistor, or the resistance presented to the input source by the emitter-base circuit, is:

$$R_{in} = r_e + \frac{r_b + R_B}{\beta} \quad (4\text{-}1)$$

The output resistance looking back into the collector-base circuit is:

$$R_{out} = \frac{r_c[r_b + R_B + \beta(r_e + R_G)]}{\beta(R_B + r_b + r_e + R_G)} \quad (4\text{-}2)$$

The load resistor, $R_L$, looks into the transistor and sees $R_{out}$. Any load connected across $R_L$ sees $R_L$ in parallel with $R_{out}$ or $R_LR_{out}/(R_L + R_{out})$.

The current gain is:

$$A_i = \frac{I_C}{I_E} = \alpha = \frac{\beta}{\beta + 1} \text{ (less than 1)} \quad (4\text{-}3)$$

while the voltage gain is the ratio of the output load impedance to the input impedance, or:

$$A_v = \frac{V_{out}}{V_{in}} = \frac{R_L}{(R_B + r_b)/\beta + r_e + R_G} \approx \frac{R_L}{r_e + R_G} \quad (4\text{-}4)$$

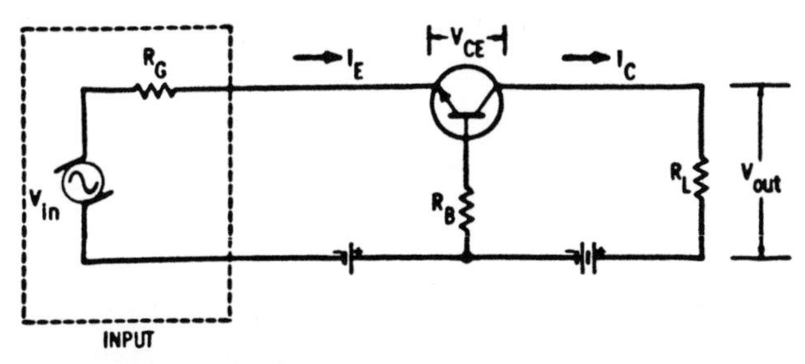


Fig. 4-5. A typical common-base circuit.

If another load is across $R_L$, the $R_L$ in the voltage gain equation no longer refers to the resistor in the circuit only, but to the resistance of the parallel combination of the load across $R_L$ and $R_L$.

The power gain, G, is the voltage gain multiplied by the current gain. Numerically, it is slightly less than the voltage gain.

When h-parameters are listed in transistor specification sheets, they may be converted to the T-parameters. The h-parameters are stated at a specific quiescent collector current and collector-emitter voltage. Should the transistor be used under other quiescent conditions, the h-parameters must be changed so that the parameters will be correct for the actual $I_E$ and $V_{CE}$ at which the device is operating. Curves defining these correction factors are available for the different transistors. Curves for one particular device are shown in Figs. 4-6 and 4-7.

In Fig. 4-6, the emitter current is plotted on the horizontal axis. The ratio of the h-parameters at the actual quiescent emitter current to the h-parameters for 1 mA emitter current are shown on the vertical axis. The curves describe how the ratios change for different values of emitter current.

For example, if the actual emitter current is 2.4 mA, mark this point on the horizontal axis, and draw a vertical line from this axis through the various curves. The ratios of

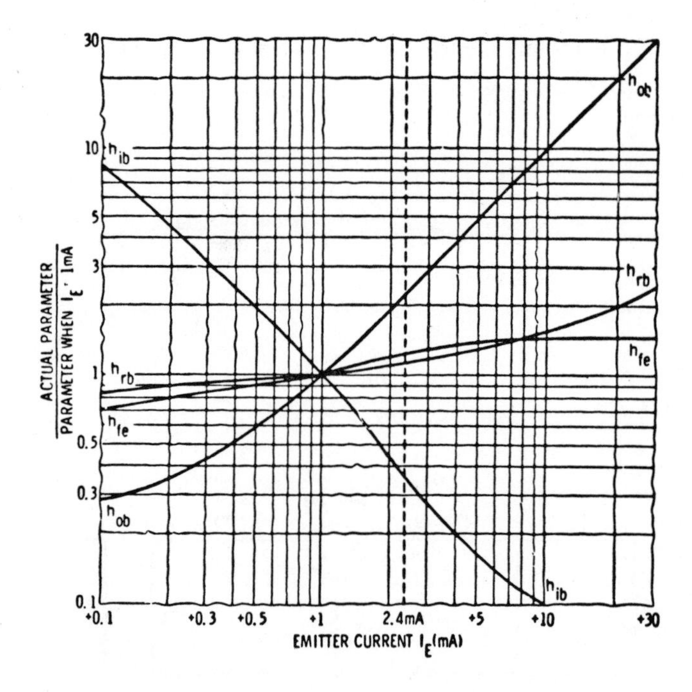


Fig. 4-6. Graph illustrating the variation of h-parameters relating to emitter current.

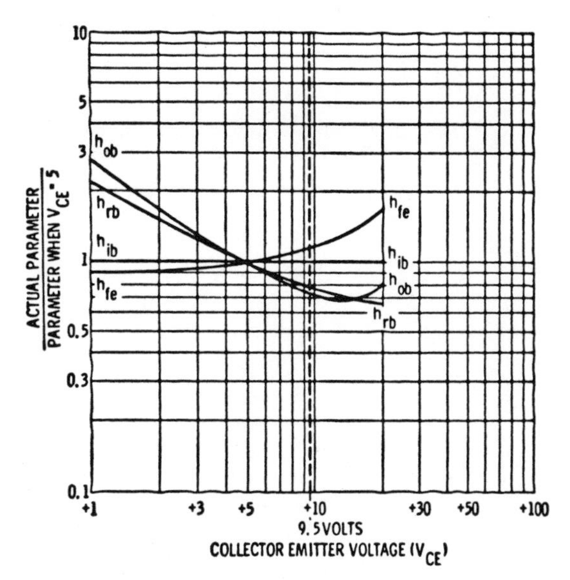


Fig. 4-7. This graph illustrates the variation of h-parameters with collector-emitter voltage.

the various h-parameters for the actual operating condition when $I_E$ = 2.4 mA, to the h-parameter when $I_E$ = 1 mA, are as follows:

For $h_{ib}$ the ratio is 0.37:1.
It is 1.2:1 for $h_{rb}$, 1.3:1 for $h_{fe}$, and 2.2:1 for $h_{ob}$

These are the emitter current correction factors.

In Fig. 4-7, the collector-to-emitter voltage is shown along the horizontal axis. The ratios of the h-parameters at the actual quiescent $V_{CE}$ to the h-parameters for $V_{CE}$ = 5 volts are shown along the vertical axis. The curves are a plot of the changes in this ratio with the variation in $V_{CE}$.

If the actual $V_{CE}$ were 9.5 volts, mark this point on the horizontal axis. Draw a vertical line from this point. The correction factors for the various h-parameters due to this $V_{CE}$ are as follows: the ratio for $h_{ib}$ is 1:1. It is 0.79:1 for $h_{rb}$, 1.2:1 for $h_{fe}$ and 0.72:1 for $h_{ob}$.

The corrected parameters are equal to the particular parameter value at $I_E$ = 1 mA and $V_{CE}$ = 5 volts, multiplied by both correction factors just determined. If $h_{ib}$ is equal to 35 ohms at $I_E$ = mA and $V_{CE}$ = 5 volts, and if $h_{rb} = 10^{-3}$, $h_{ob} = 10^{-6}$ siemens and $h_{fe}$ = 50 at these same currents and voltages, the corrected parameters at $V_{CE}$ = 9.5 volts and $I_E$ = 2.4 mA, are:

$h_{ib} = 35 \times 0.37 \times 1 = 13$ ohms
$h_{rb} = 10^{-3} \times 1.2 \times 0.79 = 9.5 \times 10^{-4}$

$h_{ob} = 10^{-6} \times 2.2 \times 0.72 = 1.58 \times 10^{-6}$ siemens, and
$h_{fe} = 50 \times 1.3 \times 1.2 = 78$

Use Equations 4-5 through 4-9 to convert the h-parameters to T-parameters. Then use Equations 4-1 through 4-4 to determine gain and impedance.

$$r_e = h_{ib} - \frac{h_{rb}(1 + h_{fb})}{h_{ob}} = h_{ib} - \frac{h_{rb}}{h_{ob}(h_{fe} + 1)} \tag{4-5}$$

$$r_b = \frac{h_{rb}}{h_{ob}} \tag{4-6}$$

$$r_c = \frac{1}{h_{ob}} \tag{4-7}$$

$$r_d = \frac{r_c}{h_{fe}} = \frac{1}{h_{ob}h_{fe}} \tag{4-8}$$

$$\alpha = -h_{fb} = \frac{h_{fe}}{1 + h_{fe}} \tag{4-9}$$

$$\beta = -\frac{h_{fb}}{1 + h_{fb}} = \frac{\alpha}{1-\alpha} = h_{fe} \tag{4-10}$$

## COMMON-EMITTER CIRCUIT

The conventional common-emitter circuit is shown in Fig. 4-8. The equations describing this circuit are:

$$R_{in} = r_b + \beta(R_E + r_e) \tag{4-11}$$

This is the resistance in the base circuit added to $\beta$ multiplied by the resistors in the emitter circuit.

$$R_{out} = r_d \frac{[(R_G + r_b) + (r_e + R_E)\beta]}{(R_G + r_b + r_e + R_E)} \tag{4-12}$$

$$A_i = \beta = \frac{\alpha}{1-\alpha} \tag{4-13}$$

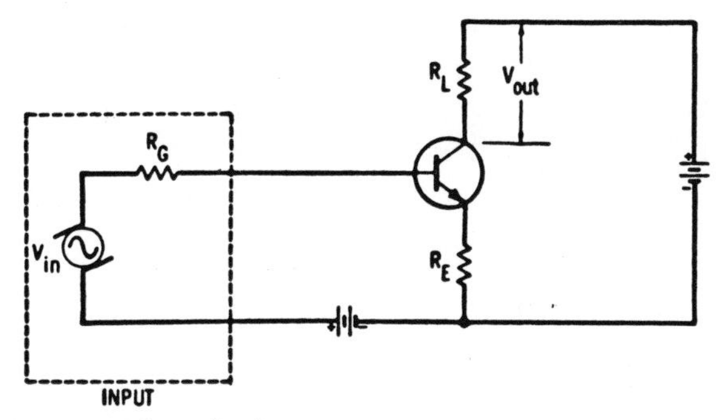

Fig. 4-8. A typical common-emitter circuit.

$$A_v = \frac{\beta R_L}{\beta(r_e + R_E) + r_b} \approx \frac{R_L}{R_E} \qquad (4\text{-}14)$$

$$G = \frac{\beta^2 R_L}{\beta(r_e + R_E) + r_b} \qquad (4\text{-}15)$$

G is the power gain. If another load is in parallel with $R_L$, $R_{out}$ and $R_L$ are treated as described in the common-base section.

If the equivalent h-parameters are specified by a manufacturer for the common-emitter mode of operation, the h-parameters should be converted to the common-base forms using the relationships:

$$h_{ib} = \frac{h_{ie}}{h_{fe} + 1} = r_e + (1-\alpha)r_b \qquad (4\text{-}16)$$

$$h_{rb} = \frac{h_{ie}h_{oe}}{h_{fe} + 1} - h_{re} = \frac{r_b}{r_c} \qquad (4\text{-}17)$$

$$h_{fb} = -\frac{h_{fe}}{h_{fe} + 1} = -\frac{\beta}{\beta + 1} = -\alpha \qquad (4\text{-}18)$$

$$h_{ob} = \frac{h_{oe}}{h_{fe} + 1} = \frac{1}{r_c} \qquad (4\text{-}19)$$

Next, use Equations 4-5 through 4-10 to derive the equivalent T parameters for substitution into the common-emitter equivalent T equations. Finally, substitute the T parameters into Equations 4-11 through 4-15.

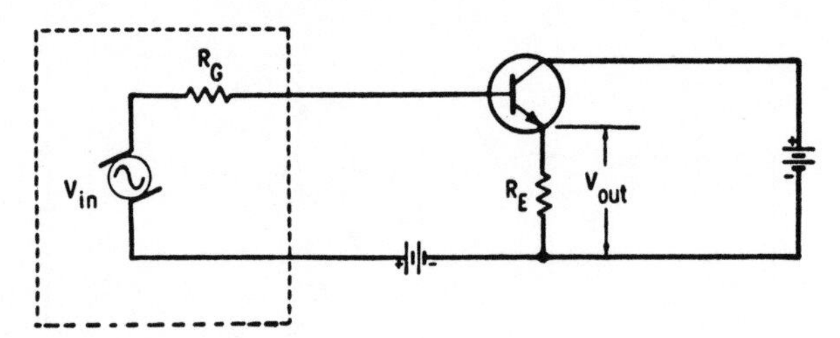


Fig. 4-9. A typical common-collector circuit.

## COMMON-COLLECTOR CIRCUIT

A common-collector circuit is shown in Fig. 4-9. The equations for this circuit are:

$$R_{in} = r_b + \beta(R_E + r_e) \tag{4-20}$$

This is the resistance in the base in addition to $\beta$ multiplied be the sum of the resistances in the emitter circuit. Should another load resistor shunt $R_E$, the $R_E$ in the equation changes to the resistance of the shunt resistor in parallel with $R_E$

$$R_{out} = \frac{R_G + r_b}{\beta} + r_e \tag{4-21}$$

The impedance presented to the output load is the emitter resistance, $r_e$, plus $\beta$ divided into the impedance in the base circuit. Impedances from the emitter circuit, when transferred to the base circuit, appear as if they were multiplied by $\beta$. In the reverse direction, impedances in the base circuit, when transferred to the emitter circuit, appear as if they were divided by $\beta$. Looking from an external load that may be connected across $R_E$, the load will see an output impedance equal to $R_{out}$ in parallel with $R_E$.

$$A_v \approx 1 \tag{4-22}$$

$$A_i = \frac{1}{1-\alpha} \approx \beta \tag{4-23}$$

$$G = \frac{1}{1-\alpha} \approx \beta \tag{4-24}$$

If thė equivalent h-parameters are specified in the common collector mode, convert them to the common-base equivalent h-parameters using the equations:

$$h_{ib} = -\frac{h_{ic}}{h_{fc}} r_e + r_b\,(1-\alpha) \tag{4-25}$$

$$h_{rc} = h_{rc} - 1 - \frac{h_{ic}h_{oc}}{h_{fc}} = \frac{r_b}{r_c} \quad (4\text{-}26)$$

$$h_{fb} = -\frac{h_{fc} + 1}{h_{fc}} = -\alpha \quad (4\text{-}27)$$

$$h_{ob} = -\frac{h_{oc}}{h_{fc}} = \frac{1}{r_c} \quad (4\text{-}28)$$

Now substitute these newly derived h-parameters into Equations 4-5 through 4-10 to determine the equivalent T-parameters for substitution into Equations 4-20 through 4-24.

## FET AMPLIFIERS

The major differences between using a JFET or using an IGFET are the bias voltages and polarities. In the equations used to calculate the amplifier gain and the various impedances, both types of transistors are treated identically.

A small-signal equivalent circuit of the FET is shown in Fig. 4-10. The solid lines depict the JFET. The IGFET equivalent circuit is identical to the one shown, with the addition of a capacitor, $C_{ds}$, at the output. It is drawn in broken lines in the figure. $C_{gs}$ and $C_{gd}$ (or $C_{rss}$) are the gate-to-source and gate-to-drain capacities, respectively. The sum of $C_{gs}$ and $C_{gd}$ is $C_{iss}$, the common source input capacity with the output shorted. While the output impedance of the JFET is resistive, that of the IGFET is primarily capacitive. It is equal to $1/g_{os} = 1/y_{os}$, the reciprocal of the output admittance. The output resistance from source to drain at the point of operation of the transistor is $r_{ds}$. When the transistor is operated in the on region, $r_{ds}$ increases with temperature. The notation for the drain-to-source resistance is $r_{ds(ON)}$ when the transistor is fully on, and $V_{GS}$ and $V_{DS}$ are both zero.

Low-frequency amplifier characteristics of the various arrangements was discussed in Chapter 2. There, the effects of capacities shown in the equivalent circuit have been

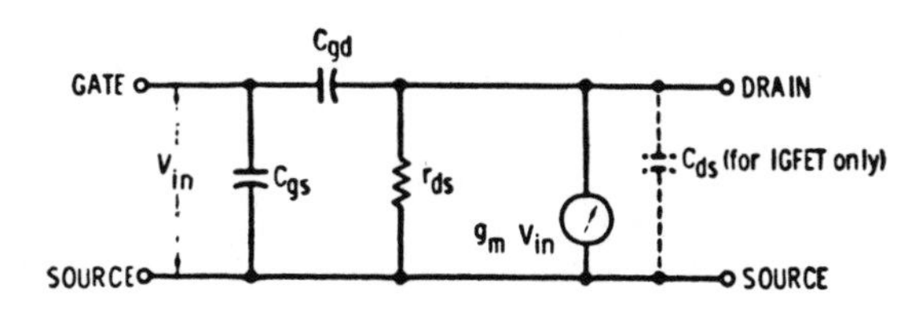


Fig. 4-10. This equivalent circuit illustrates the action of a FET.

ignored. The effect of the capacitances is most notable at higher frequencies where they are instrumental in limiting the bandwidth.

The FET is usually used in any one of the circuit arrangements shown in Figs. 4-4 and 4-12 although the common gate circuit (not drawn here) is quite common where rf amplification is involved. In the common-source and common-drain circuits, the input resistance is essentially equal to $R_G$. The input capacity is:

$$C_{in} = C_{gs} + C_{gd}(1 + g_m R_D) = C_{iss} + A_v C_{gd} \tag{4-29}$$

The upper end of the frequency spectrum is limited in gain by the presence of $C_{ds}$ and $C_{gd}$. If the generator feeding the amplifier is a voltage source, the very low resistance of the generator does not affect the frequency response. The gain of the common-source circuit is reduced 3 dB from that calculated with Equation 2-16 at:

$$f_o = \frac{1}{2\pi R_D(C_{ds} + C_{gd})} \tag{4-30}$$

$f_o$ is considered the highest frequency of interest before the gain rolls off drastically.

If the resistance of the generator is comparable to the input resistance of the amplifier, the gain of the common-source circuit is 3 dB below that predicted in Equation 2-16 at:

$$f_o \quad \frac{1}{2\pi R_{GA}[C_{gs} + C_{gd}(1 + g_m R_D)]} = \frac{1}{2\pi R_{GA} C_{in}} \tag{4-31}$$

where

$R_{GA}$ is the resistance of the generator and is in parallel with any resistor across the input of the FET.

As an example, assume that in Fig. 4-11, $R_G = 1\ M\Omega$, $R_S$ is not bypassed and is equal to 620 ohms, and $R_D = 10{,}000$ ohms. The supply is 20 volts. A voltage source feeds this circuit so that $R_{GA} = 0$. In the transistor to be used, $V_P = 2$ volts, $I_{DSS} = 1$ mA, $r_{ds} = 10^5$ ohms, $C_{gs} = 5$ pF, $C_{gd} = 1.5$ pF, and $C_{ds} = 0$. Çalculate the drain

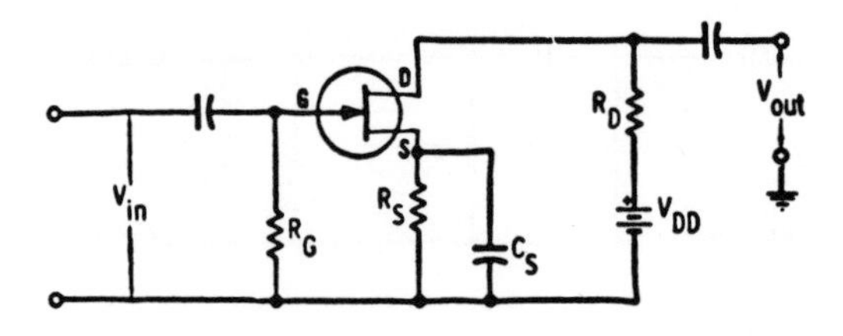


Fig. 4-11. This is a typical common-source circuit.

current, voltage gain, input capacitance, output resistance seen by the load, and frequency at which the gain dropped 3 dB.

According to Equation 2-8, the drain current is:

$$I_D = 10^{-3}\left(1 - \frac{|V_{GS}|}{|2|}\right)^2$$

A second equation, derived from bias considerations, is:

$$V_{GS} = I_D R_S = 620 I_D$$

Solving the two equations, $I_D = 0.62$ mA and $V_{GS} \approx 0.4$ volt.

In order to determine the voltage gain, $g_m$ should be calculated from Equation 2-12:

$$g_m = \frac{2I_{DSS}}{V_P}\left(1 - \frac{|V_{GS}|}{|V_P|}\right) = \frac{2 \times 10^{-3}}{2}\left(1 - \frac{0.4}{2}\right) = 8 \times 10^{-4}$$

The voltage gain, from Equation 2-16, is found to be:

$$A_V = \frac{8 \times 10^{-4}(10^4||10^5)}{1 + (8 \times 10^{-4})620} = 4.9$$

If $R_S$ had been bypassed by a large capacitor, the voltage gain, using Equation 2-17, would have been $8 \times 10^{-4}(10^4||10^5) = 72.7$. The input capacitor, according to Equation 4-29, is:

$$C_{in} = 5 \times 10^{-12} + 1.5 \times 10^{-12}[1 + 8 \times 10^{-4}(10^4)] = 18.5 \text{ pF}$$

and the resistance looking back from the load, $R_D$, to the transistors is, by Equation 2-15:

$$R_{out} = 10^5 + [1 + (8 \times 10^{-4})10^5]620 = 150{,}220 \text{ ohms}$$

while the frequency where the gain is reduced by 3 dB is, from Equation 4-30:

$$f_o = \frac{1}{2\pi(10^4)(15 \times 10^{-12})} = 10.6 \text{ MHz}$$

As for the common-drain (source follower) circuits in Fig. 4-12, the input capacitance is lower than it is for the common-source arrangement. It is:

$$C_{in} = C_{gd} + (1 - A_{VD})C_{gs} = C_{iss}(I - A_{VD}) + C_{rss}A_{VD} \qquad (4\text{-}32)$$

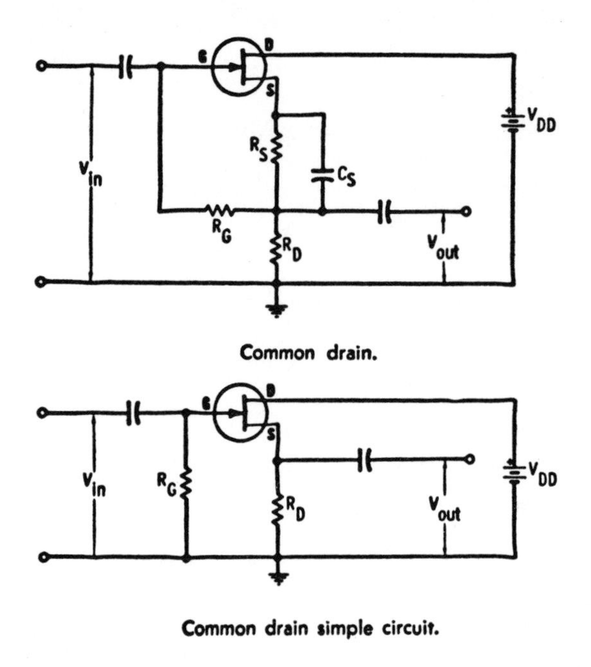


Fig. 4-12. Here are two common-drain amplifier circuits.

## NOISE

Small signal amplifiers are plagued by noise induced from environmental sources, generated in the circuit components at the input to a transistor, and from the transistor itself. Signal-to-noise ratios stated in terms of voltage or power are indications of the relative amount of signal overriding the noise.

### Signal-to-Noise Ratio

The usual method of determining the relative noise levels, involves feeding the amplifier a signal with a frequency at the geometric mean of the baud the amplifier has to reproduce. Thus if the lowest frequency of interest is $f_L$ and the highest is $f_H$, the geometric mean frequency is $\sqrt{f_L f_H}$. Advance the controls on the signal generator feeding the amplifier until the point where the maximum desirable, usable, or undistorted signal across an output load, is observed on a 'scope. Measure this signal voltage. It is $V_{SO}$.

Remove the generator and in its stead substitute a resistor load, $R_G$, equal to the impedance presented at the input of the circuit under test by a transducer or antenna. With all controls set in a normal operating condition, or at maximum, measure the voltage at the output when the appropriate resistor or capacitor load is at the input. This is $V_{no}$, the noise voltage.

Signal-to-noise ratio describes how much more signal there is at the output of the equipment than noise. Basically, the ratio is $V_{so}/V_{no}$ when measured across the same or equal load resistors at the output. It is usually expressed as a ratio in decibels (dB):

$$dB = 20 \log \frac{V_{so}}{V_{no}} \tag{4-33}$$

Considering the signal and noise powers, the dB ratio becomes:

$$dB = 10 \log \frac{P}{P_{no}} \tag{4-34}$$

where,

$P_{so}$ is the signal power at the output of an amplifier.
$P_{no}$ is the noise power at the output of the same amplifier.

Table 4-1 is a chart showing the relationship between the noise (or a signal) ratio as expressed in dB and the noise (or a signal) ratio when expressed as a simple voltage or power relationship. Using the chart, the dB figure can be determined readily from the voltage or power ratio. A voltage ratio of 10:1 and a power ratio of 100:1 both refer to 20 dB. Numbers or ratios not shown on the chart can be derived from those that are listed. Just follow the simple rule of adding the numbers when expressed in dB and multiplying the corresponding voltage or power ratios. Thus 26 dB is the sum of 6 dB and 20 dB. The corresponding voltage ratios are 2:1 and 10:1, so the signal-to-noise voltage ratio represented by 26 dB is 2:1 × 10:1 = 20:1. The corresponding power ratios are 4:1 and 100:1, so the signal-to-noise ratio of the powers for a 26 dB figure

Table 4-1. dB Notation for Voltage and Power Ratios.

| dB | Voltage ratio | Power ratio |
|---|---|---|
| 0 | 1:1 | 1:1 |
| 1 | 1.1:1 | 1.3:1 |
| 2 | 1.3:1 | 1.6:1 |
| 3 | 1.4:1 | 2:1 |
| 6 | 2:1 | 4:1 |
| 7 | 2.2:1 | 5:1 |
| 10 | 3.2:1 | 10:1 |
| 12 | 4:1 | 15.9:1 |
| 13 | 4.5:1 | 20:1 |
| 14 | 5:1 | 25:1 |
| 17 | 7:1 | 50.1:1 |
| 20 | 10:1 | 100:1 |
| 40 | 100:1 | $10^4$:1 |
| 60 | $10^3$:1 | $10^6$:1 |

is 4:1 × 100:1 = 400:1. It should also be noted that for the same numeric ratio of voltage and power, the number in dB for the voltage ratio is double the number in dB for the power ratio. This should be obvious by comparing Equation 4-33 with Equation 4-34.

### Noise Factor and Noise Figure

Manufacturers indicate the quality of a transistor by using the *noise factor* ratio, F, or the *noise figure* ratio, NF. These are defined by Equations 4-35 and 4-36 respectively.

$$F = \frac{P_{si}/P_{ni}}{P_{so}/P_{no}} = \frac{P_{no}}{GP_{ni}} = \frac{GP_{nit}}{GP_{ni}} \qquad (4\text{-}35)$$

$$NF = \log_{10}F \qquad (4\text{-}36)$$

where,

$P_{si}$ is the signal power at the input,

$P_{ni}$ is the noise power at the input and generated by the circuit components at the input,

$P_{nit}$ is the total noise power at the input due to all sources,

G is the power gain of the transistor,

$P_{so}$ and $P_{no}$ were defined earlier.

Lower noise figures indicate that under proper loading conditions, the particular device will deliver less noise to the output than will be delivered by devices with higher noise figures.

### Noise Analysis of Bipolar Devices

Different types of noise are generated by the bipolar transistor and its associated circuits. Noise varies with bandwidth and with the audio frequencies involved. The noise at the output of a transistor circuit can be divided into three frequencies categories:

1. One type, known as *white noise*, covers the entire frequency spectrum. Here equal noise power is present at all frequencies. This noise can be due to thermal phenomenon in the components, to *shot noise* generated by the random motion of charge in the semiconductor, and to *partition noise* because of emitter current division between the base and collector.
2. From zero Hz to a frequency $f_1$ (anywhere between 100 and 1,000 Hz, depending upon the transistor), a semiconductor noise due to leakage and surface phenomena, is added to the white noise. It rolls off at the rate of 3 dB per octave above zero Hz. This is referred to as the 1/f noise.

3. From a frequency $f_2$ in the high audio or low radio frequencies, on up to the limits of the circuit, there is an increase in noise with frequency at the rate of 6 dB per octave. This is not due to any particular source, but rather to roll-off of transistor beta at the high frequencies.

Curves are frequently available showing how the noise figure varies with the frequency, collector current and source resistance. One such set of curves is in Fig. 4-13. These indicate how the noise figure varies with two important factors at the same time—collector current and the resistance applied to the input at the base. Attempts should be made to design a circuit so that operation is within the lowest or 1 dB noise figure curve. Thus the resistance of the source, $R_S$ should preferably be between 3,000 and 10,000 ohms, when using the transistor represented by this curve. The preferred collector current is determined once the specific input resistance is known. Should $R_S$ be 4,000 ohms, noise is optimum when $I_C$ is between 20 $\mu$A and 70 $\mu$A. On the other hand, with a 10,000-ohm resistor, the preferred $I_C$ is between 12 $\mu$A and 32 $\mu$A. Circuit noise will generally be optimized when the resistors at the input are as small as practical while the collector current and voltage are minimized.

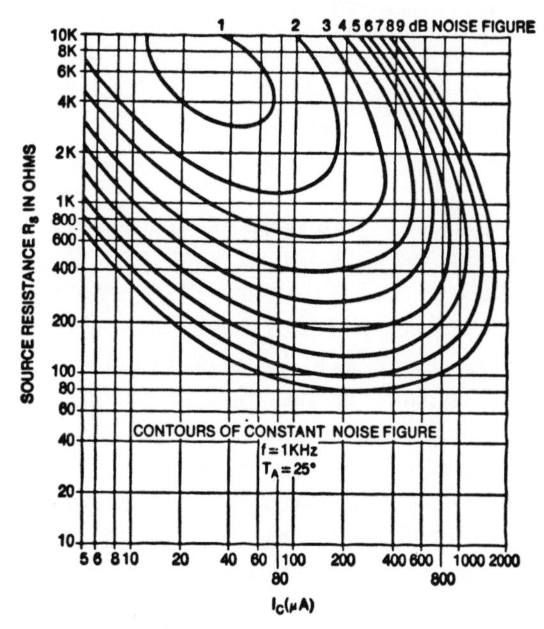


Fig. 4-13. This graph shows the noise figure curves for the GE 2N3391A transistor (courtesy of General Electric Semiconductor Dept.).

## Noise of FETs

At the current state of the art, the JFET is superior to the IGFET in its noise characteristic. Manufacturers frequently supply curves for both types of field-effect

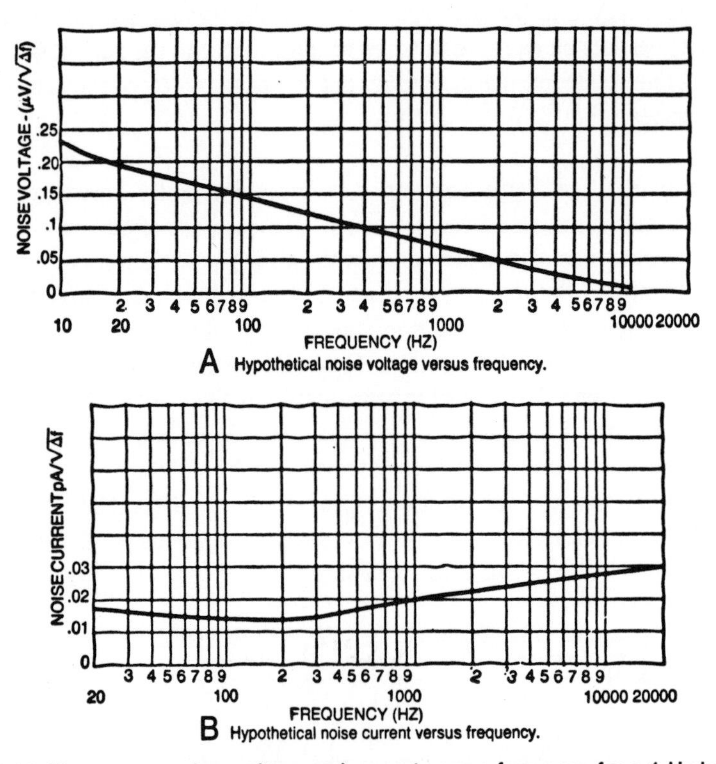


Fig. 4-14. These curves show voltage and current versus frequency for a 1 Hz bandwidth.

transistors, showing how the noise figure varies with the frequency being reproduced, the resistor at the input, $V_{DS}$, and $I_D$. A more informative set of curves shown in Fig. 4-14 relate $\overline{v_n}/\sqrt{\Delta f}$ and $i_n/\sqrt{\Delta f}$ to noise at the center frequency of the reproduced band. (Noise figures at the output of a circuit can be derived by assuming that there are two hypothetical noise generators at the input to that circuit. One of these generators supplies a voltage, $\overline{v_n}$. Noise at the output is due to $\overline{v_n}$ when the input to the circuit is shorted. When the input is an open circuit, the noise at the output is due to $i_n$, a hypothetical current generator at the input.)

The $\sqrt{\Delta f}$ in the denominator at the ordinate axis is the noise bandwidth of the amplifier. It is sometimes written $\sqrt{\sim}$ or $\sqrt{Hz}$. The curve in Fig. 4-14A, for example, shows the noise voltage with a 1-Hz bandwidth. If the bandwidth is increased to, let us say 3 Hz, $\sqrt{\Delta f} = \sqrt{3}$. The equivalent noise voltage for this bandwidth is the $v_n/\sqrt{\Delta f}$ read from the curve multiplied by $\sqrt{3}$.

Some curves are drawn for bandwidths from 1 Hz. In this case, the noise read from the curves should first be divided by the square root of the bandwidth used for the curve. The result is then multiplied by the square root of the actual bandwidth involved, as before.

Considering that all noise is reproduced over a specific bandwidth, it may not be evident which frequency to use to read $v_n/\sqrt{\Delta f}$ from the curve. As $f_H$ and $f_L$ are known factors, the significant frequency on the curve in Fig. 4-14 is $f_o = \sqrt{f_H f_L}$.

All the above also applies to the current noise generator, $i_n$, and to the curve in Fig. 4-14B.

As a rule of thumb, noise due to the input resistor is at a minimum when it is equal to $\overline{v_n}/\overline{i_n}$, although the resistor is seldom a major factor in producing noise in an FET circuit. Noise will usually be minimized when the transconductance of the device and the resistor at the input are large, and the FET is biased at about $V_{GS} = 0$ along the $I_{DSS}$ curve.

# 5

# Coupled Circuits

Practical amplifiers require more gain than can be derived from a single transistor stage. To achieve this, several transistors are connected together in one circuit. The first transistor in the circuit provides a specific amount of gain. The second transistor uses the amplified output from the first device, and increases the signal further. The overall gain from two devices is the product of the gains of the individual stages.

Three methods are frequently used to couple two transistor stages. They are RC coupling, direct coupling and transformer coupling. Minute differences exist between the coupling circuits used with bipolar devices when compared with those used with FETs. Gain relationships and impedances are about the only factors that differ significantly. Because it is more difficult to design a bipolar transistor circuit due to the low impedances presented by succeeding stages, emphasis is placed here on designs using these devices.

## RESISTANCE-CAPACITANCE COUPLING

A typical RC-coupled circuit is shown in Fig. 5-1. As a first approximation, assume C1, C2, and C3 are short circuits for the signal or ac voltages, and that they are open circuits for dc. Signal voltage, $V_{in}$, is fed to the base of transistor Q1 through C1. Dc bias conditions for this stage are established by $R_{B1}$, $R_{E1}$, and $R_{C1}$. Input voltage, $V_{in}$, is amplified by Q1. The voltage gain of this stage, when isolated from Q2, is approximately equal to $R_{C1}/R_{E1}$. The amplified signal is fed through C2 to the base of Q2. Ignoring $R_L$, the load resistor, the gain of Q2 is about $R_{C2}/R_{E2}$. The amplified signal is fed through $C_3$ to the load resistor, $R_L$.

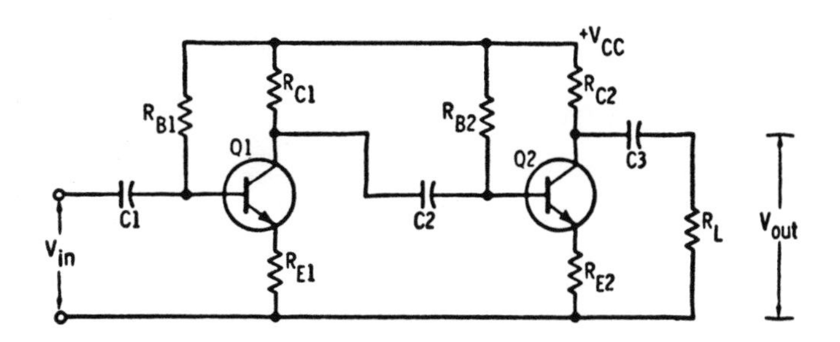


Fig. 5-1. A typical resistance-capacitance circuit.

When $R_L$ is not ignored but considered as the factor it is in the ac load resistance in the collector circuit of Q2, the resistance in the collector circuit becomes $R_L||R_{C2}$, or the resistance of $R_L$ in parallel with $R_{C2}$. Refer to this as $R_{P2}$. The gain of Q2 is then approximately equal to $(R_L||R_{C2})/R_{E2}$. Similarly, the load resistance of the Q1 stage of amplification is the input impedance of Q2 in parallel with $R_{C1}$, $R_{B2}$, $R_{E2}$, and $R_{C2}$ establish the quiescent dc conditions for Q2.

The design of an RC-coupled amplifier usually starts with the output stage. Circuit components around Q1 are determined later. If, for example, the output is to be 10 rms volts with an input of 0.1 rms volt, the overall ac voltage gain must be at least 10/0.1 = 100. The voltage gain of the first stage multiplied by the voltage gain of the second stage must equal 100 (or slightly more). We could make the voltage gain of each stage equal to 10. The gain of the second stage is usually made lower than that of the first stage due to the normally small load impedance, $R_L$, fed by the output of the final stage.

Assume that $R_L$ equals 5,000 ohms. Transistor Q2 must develop 10 volts rms or 28.2 volts peak-to-peak across $R_L$. Then 28.2 volts must also be developed across $R_{C2}$. If $R_{C2}$ is made large compared to the 5,000-ohm resistor, the voltage swing across both resistors is limited, as shown in Fig. 5-2. In this drawing, the actual transistor output curves have been omitted for clarity. Only the ac and dc load lines are shown.

When the 5,000-ohm resistor, $R_L$, is not in the circuit, the ac and dc load lines coincide and the collector voltage swing may range from $V_{CC}$ to 0. Connecting the 5,000-ohm resistor (through C3) across $R_{C2}$, reduces the ac load resistance to the parallel combination of $R_{C2}$ ohms and 5,000 ohms. $R_{C2}$ should be a maximum of 1/10 of $R_L$ in order to have a reasonably large ac voltage swing across the output load.

The impedance presented by the Q2 circuit to the output of Q1 is $R_{B2}$ in parallel with $\beta R_{E2}$. The voltage required across the impedance is the output voltage of Q2 divided by the gain of the stage.

Draw the dc and ac load line curves for Q1 as was done for Q2 in Fig. 5-2. Here too, the voltage swing across the output load resistor must not be restricted in any way and the transistor operation should be on the most linear portion of its characteristics.

It is common practice to design the first stage of an amplifier for low noise. In this

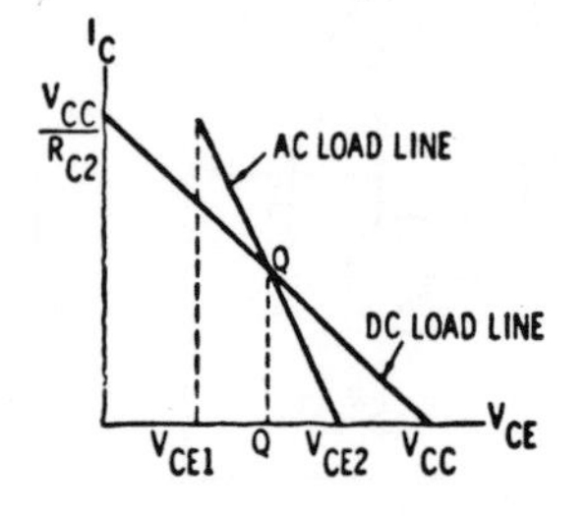


Fig. 5-2. Graph showing the theoretical ac and dc load lines for Q2 in Fig. 5-1.

case, the collector voltage, collector current, and the impedance seen by the input of the first stage, should be as small as practical.

## DIRECT-COUPLED AMPLIFIERS

Transistors may be connected to each other without coupling capacitors. A circuit of this type is shown in Fig. 5-3A. The design of this circuit is not very different from that of the capacitance-coupled type. $R_{C2}$ is determined as before. The voltage at the base of Q2, and hence the collector of Q1, is dependent upon the voltage across $R_{E2}$. Assuming that the base-emitter voltage of Q1 and Q2 is 0.6 volt, the voltage at the collector of Q1 is equal to 0.6 volt added to the voltage across $R_{E2}$. The quiescent collector voltage at Q1 must be determined as before, and the size of $R_{E2}$ is determined from this. The overall gain is still the product of the gains of the two individual stages.

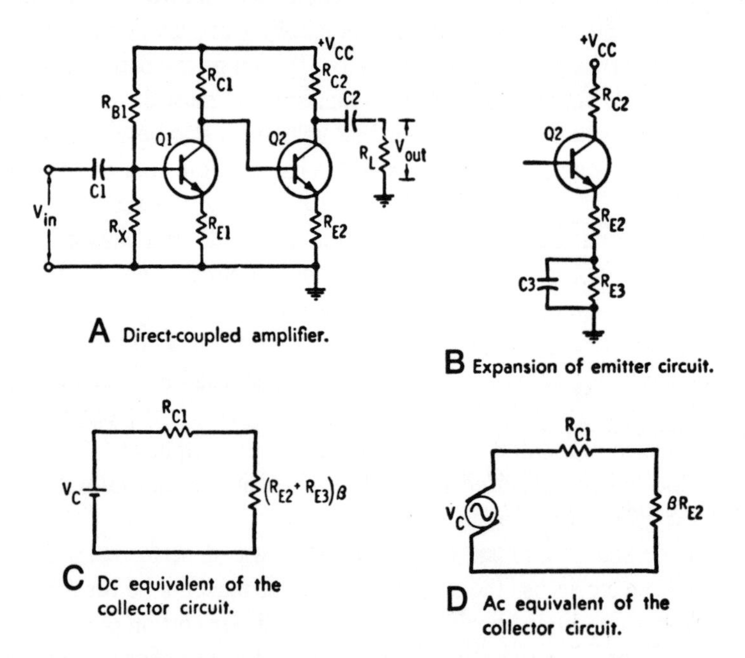


Fig. 5-3. Diagrams illustrate the operation of a direct-coupled amplifier circuit.

The voltage at the collector of Q1 should be at about the center of the load line or at about $V_{CC}/2$. $R_{E2}$ can be increased to accomplish this. This additional resistance in the emitter will produce a considerable loss of gain. The emitter circuit can be augmented, as shown in Fig. 5-3B, by adding a resistor, $R_{E3}$, in series with the original emitter resistor $R_{E2}$. Bypass $R_{E3}$ with a large capacitor. The impedance of the capacitor should be less than one-tenth of $R_{E3}$ at the lowest frequency to be amplified.

The circuit for Q1 is determined as before, except that this time, the voltage drop across $R_{C1}$ is due to the sum of the collector current flowing through Q1 and the base current of Q2. In analysis, the lead between the collector of Q1 and the base of Q2 should be broken. Calculate the collector circuit of Q1 according to the Thevenin theorem. Determine the voltage at the collector of Q1. It is $V_C$. Place this voltage in series with the parallel combination of $R_{C1}$ and $r_d$ (output resistance of Q1), which is usually equal to $R_{C1}$. This Thevenin equivalent circuit is placed across the input of Q2. The dc and ac equivalents of the collector circuit are shown in Figs. 5-3C and 5-3D, respectively. Find $I_C$ from Fig. 5-3C and the signal input voltage to Q2 from Fig. 5-3D. $V_C$ is the collector voltage at Q1 when Q2 is not in the circuit.

### Darlington Circuit

Commonly used direct-coupled circuits are shown in Fig. 5-4. Q2, an emitter follower, is the load on Q1, which is also an emitter follower. The voltage gain is approximately equal to 1, while the current gain is equal to the product of the betas of the two transistors. In all respect, this combination acts as a single transistor with the total beta equal to the product of the betas of the two devices.

$R_{E1}$ in Fig. 5-4A may be omitted or used as part of the resistance in the emitter of Q1. It is also across the base-emitter circuit of Q2. The voltage across $R_{E1}$ is equal to the base-emitter voltage across Q2 plus the voltage across $R_{E2}$. Voltage across $R_{E1}$ resulting from the emitter current of Q1, in conjunction with $R_{E2}$, determines the collector (or emitter) current flowing through Q2.

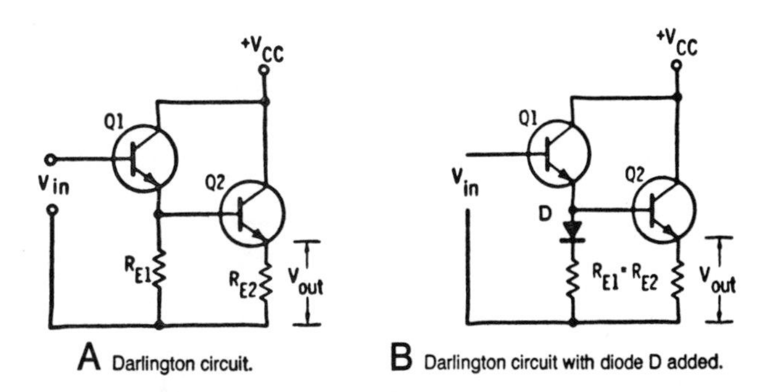


Fig. 5-4. Two versions of the Darlington circuit.

Should the circuit be changed to that shown in Fig. 5-4B, an interesting and useful relationship exists. Assuming that the diode has the same voltage characteristic as the base-emitter junction of Q2, and that $R_{E1} = R_{E2}$, then the currents flowing through both resistors are identical because of identical voltage drops. Assuming that there is a negligible base current in Q2, the emitter current of Q2 is identical with that of Q1.

## Complementary Transistor Circuit

Direct coupling using complementary transistors is commonplace. A circuit of this type is shown in Fig. 5-5. Signal is fed to Q1. From Q1, the amplified output appears across $R_{C1}$ in parallel with the input impedance of Q2. Q2 amplifies the signal further.

The size of the Q1 idling current in Fig. 5-5 is established by $V_{CC}$. The current is dependent primarily upon the magnitude of the supply voltage, $V_{CC}$, and the size of the base resistor, $R_{B1}$, in the circuit. A portion of the Q1 collector current flows through the base-emitter junction of Q2 as well as through $R_{C1}$. The collector current for Q2 flows through $R_{C2}$ to ground.

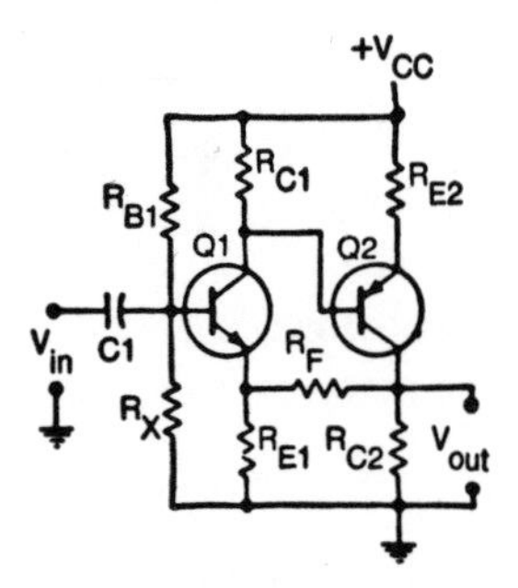


Fig. 5-5. A complementary transistor pair is often used in direct-coupled circuits.

The overall $\beta$ of the circuit is the product of the betas of the two transistors. Assume that $\beta_1$ and $\beta_2$ refer to the current gains of Q1 and Q2, respectively. The approximate equations for the circuit are:

$$R_{in} = \frac{R_{B1}R_X}{R_{B1} + R_X} = \frac{R_{B1}}{(R_{B1}/R_X) + 1} \tag{5-1}$$

$$R_{out} = \frac{R_F R_{C2}}{R_F + (\beta_2 R_{C2})} = \frac{R_{C2}}{1 + (\beta_2 R_{C2}/R_F)} \tag{5-2}$$

$$A_V = \frac{R_{E1} + R_F}{R_{E1}} = \frac{(R_{E1}/R_F) + 1}{R_{E1}/R_F} = 1 + \frac{R_F}{R_{E1}} \tag{5-3}$$

$$A_i = \beta_1\beta_2 \qquad (5\text{-}4)$$

$$G = \beta_1\beta_2 \left(\frac{R_{E1} + R_F}{R_{E1}}\right) = \beta_1\beta_2 \left[\frac{(R_{E1}/R_F}{R_{E1}/R_F}\right] = \beta_1\beta_2(1 + R_F R_{E1}) \qquad (5\text{-}5)$$

One variation of the circuit is when $R_X$ and $R_F$ are removed so that they are made equal to infinity in the equations while $R_{E1}$ and $R_{E2}$ are shorted or made equal to zero. Substituting these factors into the equations for the circuit, $R_{in} = R_{B1}$, $R_{out} = R_{C2}$, $A_v = 1$, $A_1 = \beta_1\beta_2$ and $G = \beta_1\beta_2$.

An alternate version is when $R_X$, $R_{C1}$ and $R_{E1}$ are infinite and $R_F$ is equal to zero. For this circuit, $R_{in} = R_{B1}$, $R_{out} = 0$, $A_v = 1$, $A_i = \beta_1\beta_2$ and $G = \beta_1\beta_2$.

Different versions of the circuit involve interchanging the pnp with the npn transistor while reversing the polarity of the supply voltage. Equations 5-1 through 5-5 still apply.

### Parallel Circuits

Another form of the dc-coupled circuit uses two transistors in parallel. In this type of arrangement, the collector currents in both transistors should be equal. The best method of accomplishing this is by using independent but equal resistors in each emitter before tying them together. It is also useful to have independent but equal resistors in the two collector leads and in the two base leads, but the effect is not as pronounced as when the resistors are in the emitter leads.

### Differential Amplifier

The differential amplifier in Fig. 5-6 is probably the most important dc-coupled configuration. Two voltages, $V_{in1}$ and $V_{in2}$, are fed to the circuit. The output voltage is

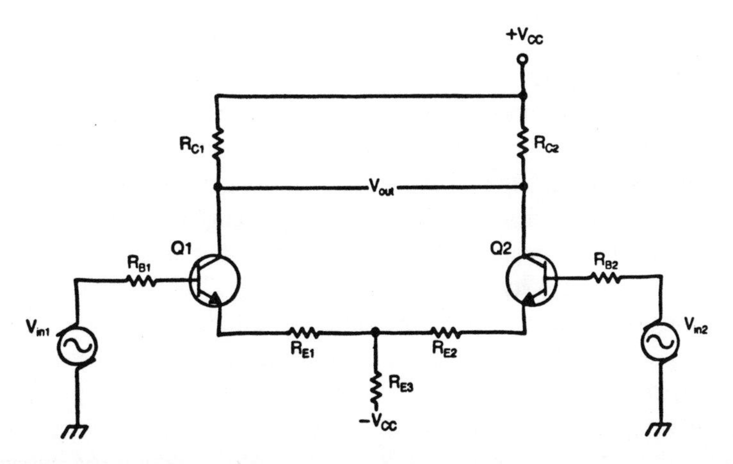


Fig. 5-6. A differential amplifier circuit.

proportional to the difference of $V_{in1}$ and $V_{in2}$. The output at the collector of Q1 is 180° out of phase with the output at the collector of Q2. $R_{E3}$ should be large or synthesized by a constant-current source.

The differential amplifier is used where good stability is required. To accomplish this, both transistors are usually placed in identical circuits where all corresponding components are equal. Transistors are also chosen so that their $V_{BE}$ and $I_{CEO}$ characteristics are identical at room temperature and track well at elevated temperatures. Under ideal circumstances, the two transistors are tied together thermally through the use of a common heatsink or are both formed in an integrated circuit. With this arrangement, $V_{BE}$ and $I_{CO}$ vary together. The effects of the two factors on collector current thereby cancel as far as the overall circuit is concerned.

Thermal runaway of each device is an important consideration. Stability is enhanced through the use of a large resistor, $R_{E3}$, in the common-emitter circuit. Should you determine from stability calculations that a specific size emitter resistor must be used when a single transistor is in a circuit, double the size of that resistor should be used for $R_{E3}$ when the transistor is a component in a differential amplifier.

When discussing the ac gain characteristics of a differential amplifier, it is assumed that identical components are in both halves of the circuit. Hence, $R_{C1} = R_{C2} = R_C$, $R_{E1} = R_{E2} = R_E$, and $R_{B1} = R_{B2} = R_B$. Output voltage, $V_{out}$, is usually taken between the collector of Q1 and Q2. Gain of the circuit is the same as the gain of either one of the transistors in the circuit, namely:

$$A_v = \frac{\beta R_C}{\beta(r_e + R_E) + r_b +} \approx \frac{R_C}{R_E} \qquad (5\text{-}6)$$

where $r_e$ and $r_b$ are the emitter and base resistances, respectively, of the transistor. Note that $R_{E3}$ is not in the gain equation because currents from both transistors pass through this resistor in opposite phases. Gain from either collector to ground is $A_v/2$.

The circuit is sometimes modified by adding a resistor directly from the emitter of one transistor to the emitter of the other. This arrangement is frequently applied to voltmeter or milliammeter circuits where a meter movement is used instead of this emitter-to-emitter resistor. In this case, current flowing through the meter is proportional to the voltage applied to one input of the differential amplifier while the other input is grounded through a resistor or connected to a bias voltage. Voltage gain of the circuit is determined using the formula stated in Equation 5-6 with the exception that a resistance equal to $R_E$ in parallel with one-half the resistance of the meter movement (or emitter-to-emitter resistor) is substituted for $R_E$ in the equation. Input impedance of the circuit in Fig. 5-6 is equal to the resistance in the base circuit of the transistor in question in parallel with $\beta$ times the sum of $R_{E1}$ and $R_{E2}$. If a resistor is connected between the two emitters, the input impedance just stated is modified by considering the product of $\beta$ and the emitter-to-emitter resistor to be in parallel with it.

A variation on the circuit in Fig. 5-6 is used where an output with respect to ground is required. In this case, $R_{C1}$ is set equal to zero ohms while the entire output is developed across $R_{C2}$. Gain is now one-half of that in Equation 5-6. Because of the lack of symmetry, the effects of variations of $I_{CEO}$ with temperature are not cancelled. Where input signals are small, the circuit in Fig. 5-6 must be used as is, or the difference voltage at the output may be masked by leakage current variations with temperature.

Gain in either case depends upon the signals (or lack of signals) applied to the two inputs. If the identical signal is applied to both inputs, the ideal output is zero. In practical situations, these identical signals may be in the form of noise. Gain of an amplifier, $A_{vc}$, due to this *common-mode* signal is:

$$A_{vc} = \frac{R_C}{2R_{E3}} \tag{5-7}$$

A *common-mode rejection ratio*, CMRR, has been established as a figure of merit to indicate how well the goal of eliminating this undesirable factor is accomplished.

$$\text{CMRR} = \frac{A_v}{A_{vc}} \tag{5-8}$$

It is best that the CMRR be as large as possible. CMRR can also be expressed in dB. The abbreviation for this ratio then becomes CMR(dB) or simply CMR and is equal to 20 $\log_{10}$ CMRR.

Rather than using the gain equations to compute CMRR, it should be determined experimentally in several simple steps. Use a single signal generator at $V_{in1}$. Remove any signal sources from the end of $R_{B2}$ shown connected to $V_{in2}$. Now find $A_v$ by feeding a known signal, $V_{in1}$, to Q1 while grounding the end of $R_{B2}$, which is shown connected to its generator. Measure $V_{out}$. $A_v$ is $V_{out}/V_{in}$. Next, disconnect $R_{B2}$ from ground and $V_{in2}$. Connect the end of $R_{B1}$ that is connected to $V_{in1}$ to the corresponding end of $R_{B2}$. Feed a large known signal, $V_{large}$, from generator $V_{in1}$ to the common inputs. The signal at $V_{out}$ between the two collectors is measured and referred to as $V_{out}$(Com Mode). $A_{vc}$ is the ratio of $V_{out}$(Com Mode)/$V_{large}$. Substitute $A_v$ and $A_{vc}$ into Equation 5-8 to determine the CMRR. It should be in the order of magnitude of several thousand.

CMRR is best when $R_{E3}$ is large. If the emitters of the transistors are to be at just about ground potential, $R_{E3}$ must be connected to a negative voltage supply, $-V_{EE}$. Should $R_{E3}$ be large, $-V_{EE}$ must be large. To minimize this voltage, a constant-current circuit is frequently substituted for $R_{E3}$. Here, the ac resistance is large while the dc voltage drop across the circuit is small. One such circuit is in Fig. 5-7.

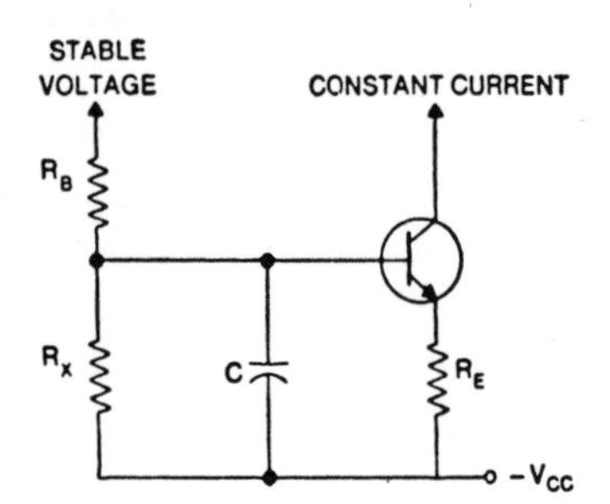


Fig. 5-7. A constant-current source.

A reasonably constant voltage is developed across $R_X$, as well as across $R_E$ and the base-emitter junction. There will be a constant current flowing through the transistor. The current will be equal to the voltage across $R_E$ divided by the ohmic size of that resistor. The voltage across $R_E$ is equal to the voltage across $R_X$ less the $V_{BE}$ of the transistor. The voltage across the junction and $R_E$ of the constant-current source can be held tighter if $R_X$ is replaced by a zener diode with the anode connected to $-V_{CC}$.

An FET can also be used as a constant-current source. Current flowing through the device is noted in Equation 2-8. Drain current is related to the gate-to-source bias voltage. If this voltage is made equal to zero and the gate is connected to the drain, the constant current becomes $I_{DSS}$. In the constant-current circuit in Fig. 5-8, bias voltage is either developed across $R_S$, or $R_S$ is shorted so that there is 0-volt bias for the transistor. $R_S$ can be determined by applying Equation 2-8 or through use of Equation 5-9 as follows:

$$R_S = \frac{V_P}{I_D}\left(1 - \sqrt{\frac{I_D}{I_{DSS}}}\right) \tag{5-9}$$

The output impedance is defined by Equation 2-15.

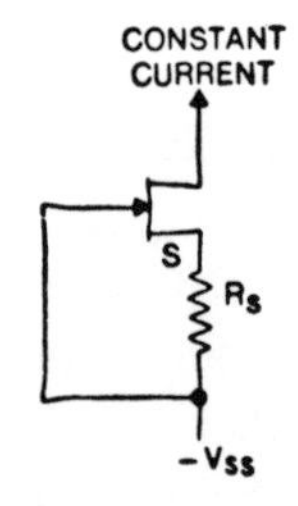


Fig. 5-8. A FET can also be used as a constant-current circuit.

Differential amplifiers can be coupled together, as shown in Fig. 5-9. The input signal is fed to and amplified by Q1, and it is further amplified by Q. The output is $V_{out1}$.

The signal at the collector of Q2 is 180° out of phase with, but usually made equal to, the signal at the collector of Q1. The output from Q2 is amplified by Q4. The output

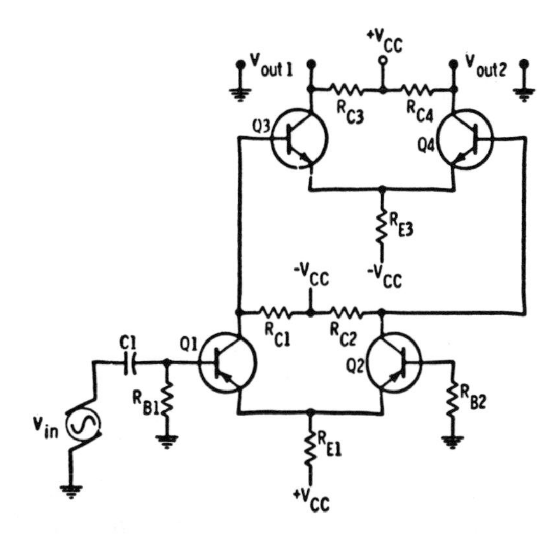

Fig. 5-9. An example of a direct-coupled differential amplifier.

here is $V_{out2}$. The output voltages, $V_{out1}$ and $V_{out2}$, are 180° out of phase and can be fed to the output stages of a push-pull power amplifier.

Field-effect transistors can also be used in differential amplifier circuits. One such arrangement is in Fig 5-10. Here, the constant current is equal to the $I_{DSS}$ of Q3. If both halves of the circuit are well matched, the differential voltage gain is that of a single common-source stage of amplification and equal approximately to $g_{m1}R_{D1}$ where $g_{m1}$ is the transconductance of Q1. Common-mode gain is approximately equal to $R_{D1}/2R_{Q3}$ where $R_{Q3}$ is the ac resistance presented by the constant-current transistor Q3. If a resistor is used instead of Q3, $R_{Q3}$ is the ohmic size of that resistor.

Drift with temperature has been a drawback of FET differential amplifiers. However, it does have advantages. For one, the input impedance is extremely high. Another advantage is that thermal runaway is almost impossible.

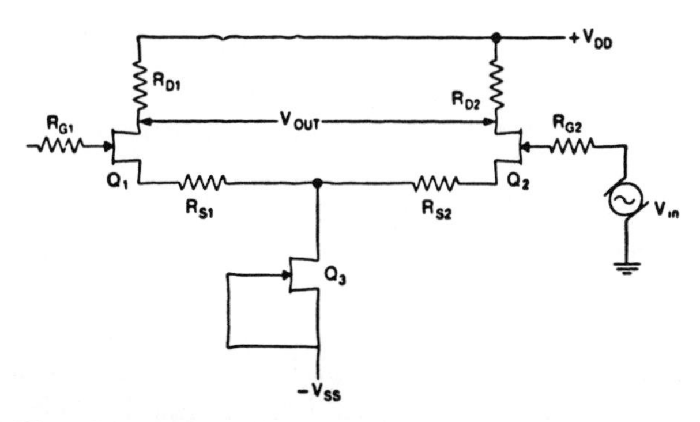

Fig. 5-10. A FET differential amplifier with a current source.

In order to minimize drift, the two transistors should be matched for transconductance, $g_m$, the drain currents $I_D$ at a specific gate-to-source voltage, $V_{GS}$, and for leakage currents, $I_{GSS}$. Drift can be minimized further by biasing the transistors at about:

$$I_D = \frac{g_{mo}^{\ 2}}{10I_{DSS}} \tag{5-10}$$

$$V_{GS} = \frac{g_{mo} + 3.2I_{DSS}}{1.6g_{mo}} \tag{5-11}$$

where $g_{mo}$ is the transconductance at the point where $V_{GS} = 0$ and $I_D = I_{DSS}$.

Experimentally, drift can be reduced to a minimum by shunting either $R_{S1}$ or $R_{S2}$ with a resistor. Just which one of the resistors is to be shunted and how large the resistor should be, can be determined by trial and error. Relative drain measurements are made as the temperature applied to the circuit is varied over the required operating range. Choose the resistor that keeps this drain current variation to a minimum. A similar experimental procedure can be used to compensate for offset voltage. Either $R_{D1}$ or $R_{D2}$ is to be shunted with a resistor until $V_{out} = 0$ volt when $V_{in1}$ and $V_{in2}$ are both equal to 0.

## TRANSFORMER COUPLING

An example of coupling two transistors using a transformer, is shown in the circuit in Fig. 5-11. Assume the betas of Q1 and Q2 are both equal to 100. The base resistance is 500 ohms for Q2 and 0 ohms for Q1. Each transformer is 75 percent efficient. The required voltage gain of the circuit is 30 with an output of 1 watt across a 9-ohm load resistor. A 12-volt supply is available.

First, consider Q2 and the output circuit. One watt is to be developed across the 9-ohm load resistor. If the transformer is 75 percent efficient, at least 1 watt/0.75 = 1.33 watts must be delivered to the primary of T2. After adding a safety factor of 25 percent because of losses in $R_{E2}$, saturation resistance or voltage, and $I_{CBO}$, the actual amount of power that is desirable at the primary is 1.33 + 0.25(1.33) = 1.66 watts.

A transistor that can deliver 1.66 watts must dissipate double this power, or at least 3.32 watts. Assuming zero dc resistance in the primary winding and a 12-volt supply, the quiescent current is 3.32 watts/12 volts = 0.277 amp. In Fig. 5-11B, the load impedance is 12 volts/0.277 amp = 43.4 ohms. If we wish to have a gain of 10 in this stage, $R_{E2} + r_{e2}$ must be less than 43.4 ohms/10 = 4.34 ohms ($r_{e2}$ is the emitter resistance of Q2). Expressing $I_{E2}$ in milliamps, $r_{e2}$ is $26/I_{E2}$. Hence, $r_{e2}$ is 26/277. This is very small. However, $r_{e2}$ is seldom less than 1 ohm. Letting $r_{e2}$ equal 1 ohm, $R_{E2}$ can be made equal to 4.34 ohms − 1 ohm = 3.34 ohms. Use a standard 3.3-ohm resistor for $R_{E2}$.

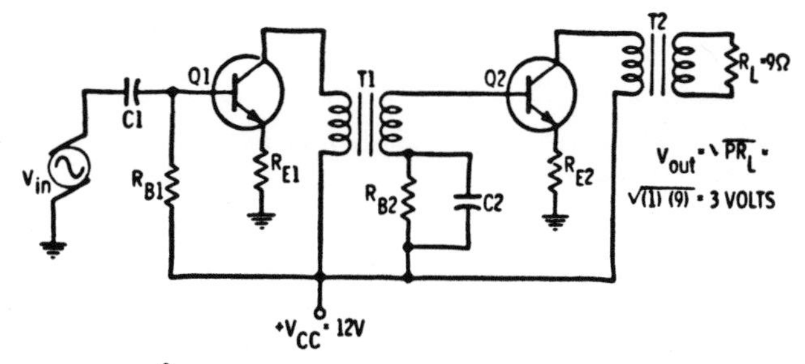


A Two-stage transformer-coupled amplifier.

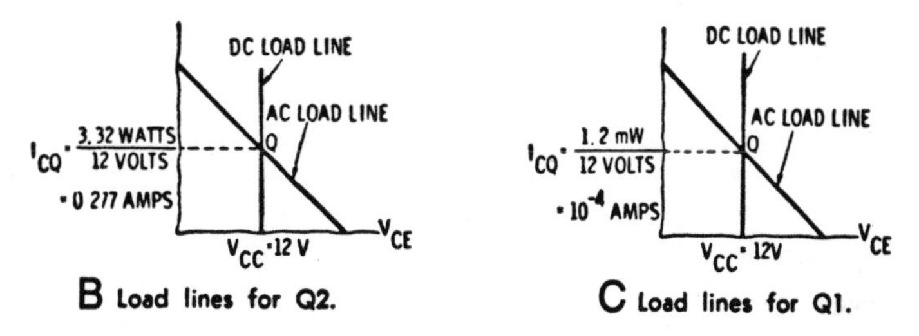


B Load lines for Q2.

C Load lines for Q1.

Fig. 5-11. A transformer-coupled circuit with its load lines.

In the base circuit, if the quiescent collector current of Q2 is 0.277 amp, the base current is 0.277 amp/$\beta$ = 0.277/100 = $2.77 \times 10^{-3}$ amps. As the supply for the base circuit is 12 volts, $R_{B2} + \beta R_{E2}$ = 12 volts/$2.77 \times 10^{-3}$ amps = 4330 ohms. Since $\beta R_{E2}$ = 100(3.3 ohms) = 330, $R_B \approx 4330 - 330$ = 4,000 ohms.

The impedance of C2 should be less than 1/10 that of $R_{B2}$, or less than 390 ohms at the lowest frequency to be amplified. Assume that this frequency is 20 Hz; hence $C = 1/2\pi(20)(390) = 20.4\mu F$.

Returning to the output circuit, if the ratio of the primary impedance of T2 to the secondary impedance is 43.4 to 9, the voltage ratio, and hence turns ratio, of the transformer is $\sqrt{43.4/9}$ = 2.2:1. Therefore, the gain of the output transformer is 1/2.2. The gain of Q2 is $43.4/(R_{E2} + r_{e2})$ = 43.4/4.3 = 10.1. The voltage gain from the base of Q2 to the 9-ohm load resistor is (1/2.2)(10.1) = 4.59. Since $V_{RL} = \sqrt{PR_L}$, it is apparent that $\sqrt{(9 \text{ ohms})(1 \text{ watt})}$ = 3 volts across the 9-ohm load resistor. It follows that there must be 3 volts/4.59 = 0.65 volt at the base of Q2.

The input impedance to Q2 is $(R_{E2} + r_{e2})\beta + r_b$ = (3.3 +1)100 + 500 = 930 ohms. There must be $(0.65 \text{ volt})^2/930$ ohms = 0.45 mW at the base of Q2 if there is to be 1 watt across the 9-ohm resistor at the output. Adding 25 percent for transformer losses, $0.45 + 0.25(0.45) \approx 0.60$ mW must be delivered to the primary of T1. Q1 must then dissipate $2 \times 0.60$ mW = 1.2 mW. Using a 12-volt supply, the quiescent collector current will be $(1.2 \times 10^{-3} \text{ watts})/12$ volts = $10^{-4}$ amps. Hence, the load impedance on Q1, according to Fig. 5-11C, is 12 volts/$10^{-4}$ amps = 120,000 ohms.

The ratio of the secondary to primary impedances of T1 is 930/120,000. The voltage ratio and voltage gain of the transformer is the square root of this ratio or 30.5/346 = 1/11.3. Multiplying this by the gain of Q2 and T2, the gain from T1 to $R_L$ is (1/11.3) × 4.59 = 1/2.46. Q1 therefore must have a gain of 30 × 2.46 = 74 if the overall voltage gain of the circuit is to be 30. This can be accomplished by making the ratio of the primary impedance of T1, or 120,000 ohms, to $R_{E1} + r_{el}$, to be equal to 74 ($r_{el}$ is the emitter resistance of Q1) $R_{E1} + r_{el}$ is then equal to 1600. If $r_{el} = 26/I_{E1} = 26/10^{-1}$ = 260 ohms ($I_{E1}$ is the emitter current of Q1 in mA), then $R_{E1}$ is 1600 − 260 = 1340 ohms.

The base current of Q1 is the collector current divided by $\beta$ or $10^{-4}/100 = 10^{-6}$ amps. Assuming that the 12-volt supply is used, $R^{B1} + \beta R_{E1}$ = 12 volts/$10^{-6}$ amps = 12 × $10_6$ ohms. As $\beta R_{E1}$ is negligible compared to $R_{B1}$, $R_{B1}$ can be made equal to 12 megohms. It is preferable to apply bias circuit 2, discussed in Chapter 3, so that a smaller resistor can be used at the input of Q1.

# 6

# High-Frequency Devices

Bipolar transistors and FETs can be used as rf amplifiers. External circuits limit their activity to specific frequency ranges. Equivalent circuits of both types of devices are useful when designing circuits in the upper frequency bands.

## BIPOLAR TRANSISTORS

In Chapter 4, two equivalent circuits, the equivalent T and the equivalent h, were discussed. A third equivalent circuit—the hybrid $\pi$—is useful in describing the high-frequency characteristics of the bipolar transistor. As you are aware, it is possible to change ordinary resistive or reactive circuits from delta (or pi) arrangements to the T form and back again. The same is true of circuits involving transistors. The various components in all equivalent circuits are interrelated.

### Hybrid $\pi$ Equivalent Circuit

A high-frequency equivalent circuit of the transistor is shown in Fig. 6-1. The external base lead of the transistor is b. Inside the base, there is a point that cannot be reached by a lead. This point is b′; $r_{bb'}$ is the *base spreading resistance* between these two points.

A resistance, $r_{b'e'}$, shunted by a capacitance, $C_{b'e'}$, connects the points inside the base to the emitter. The resistance, $r_{b'e'}$, decreases as the collector current and

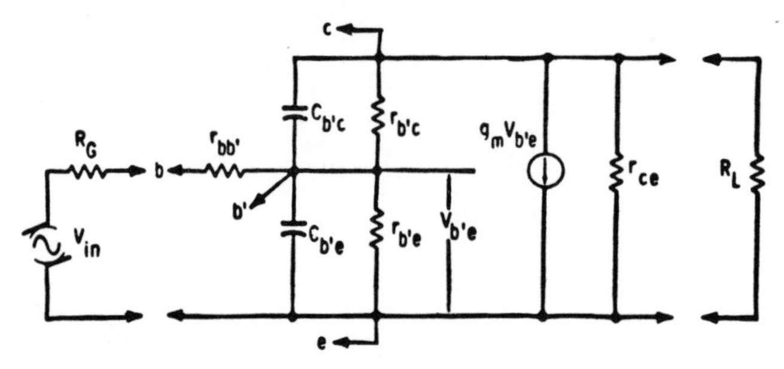


Fig. 6-1. A hybrid equivalent circuit for a transistor at high frequencies.

temperature increase, and it increases as the collector-to-emitter voltage rises. $C_{b'e}$, remains relatively constant under all conditions.

A resistance, $r_{b'c}$, shunted by a capacitance, $C_{b'c}$, connects the point inside the base to the collector. Resistance $r_{b'c}$ is usually large enough, approaching infinity, to be omitted from the equivalent circuit. $C_{b'c}$ decreases as the collector current and collector-emitter voltage increase.

Transconductance, $g_m$, applied to a bipolar transistor, is the change of collector current, $\Delta I_c$, to a change of base emitter voltage, $\Delta V_{be}$.

$$g_m = \frac{\Delta I_c}{\Delta V_{be}} \tag{6-1}$$

A current generator, $g_m V_{b'e}$, is at the output of the transistor. The resistance from the collector to the emitter, $r_{ce}$, shunts the current generator.

The $\alpha$ of the transistor varies with frequency. For this discussion, we are assigning the symbol $\alpha_0$ as the low-frequency common-base current gain normally found on specification sheets. At a frequency symbolized either by $f_{hbo}$,$f_\alpha$,$f_{\alpha b}$, or $f_T$, $\alpha$ has decreased 3 dB from its low-frequency value to $0.707_{\alpha_0}$. This is the $\alpha$ cutoff frequency. Once $\alpha_0$ and $f_0$ are known, $\alpha$ at any frequency can be found from the curve in Fig. 6-2. In order to use this curve, substitute f for $f_0$ on the horizontal axis. Note the ratio in dB at the frequency of interest. Find the voltage ratio represented by the dB ratio from Table 4-1 and multiply this fraction by $\alpha_0$ to determine $\alpha$ at the desired frequency.

At frequency $f_{ae}$ (or $f_\beta$), $\beta$ has decreased 3 dB from its low-frequency value at 1 kHz, $\beta_0$. It is approximately equal to $f_\alpha/\beta_0$. At $f_T$ (or $f_\alpha$), $\beta$ is 1. This frequency, $f_T$, is also known as the gain-bandwidth product of the transistor because it is equal to $f_\beta \beta$. As was the case with determining $\alpha$ at various frequencies, $\beta$ at a specific high-frequency can be determined from Fig. 6-2.

### Hybrid $\pi$ Parameters

Many of the hybrid $\pi$ parameters are difficult to measure. Instead approximate

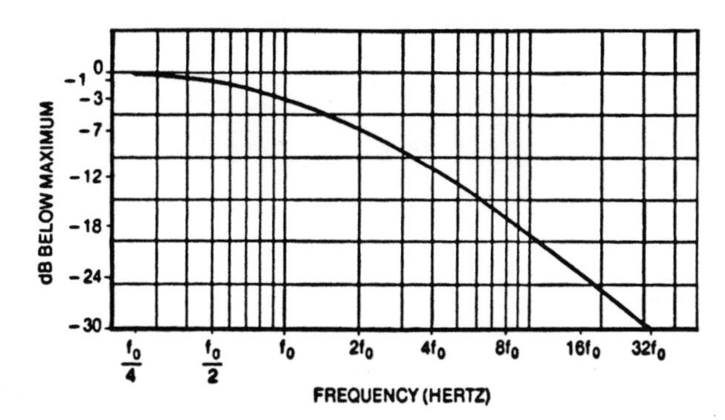

Fig. 6-2. Graph of a frequency-response curve for a low-pass filter.

relationships have been derived between the hybrid $\pi$ parameters and other factors with more readily determined parameters. Some of these relationships are:

$$g_m = \frac{I_C}{26} = \frac{1}{r_e} \qquad (6\text{-}2)$$

where,

$I_C$ is the quiescent collector current in milliamperes,
$r_e$ is the emitter resistance.

$$C_{b'e} = \frac{g_m}{2\pi f_\alpha} \qquad (6\text{-}3)$$

$$C_{b'c} \approx C_{ob} \qquad (6\text{-}4)$$

where,

$C_{ob}$ is the collector-to-base capacitance in the common-base equivalent circuit.
$f_\alpha$ is the $\alpha$ cutoff frequency.

$$r_{b'e} = \frac{\beta_o}{g_m} \qquad (6\text{-}5)$$

where,

$\beta_o$ is the current gain at low frequencies.

$$r_{bb'} = h_{ie} \qquad (6\text{-}6)$$

Equation 6-6 is true at high-frequencies. At low frequencies, $r_{b'e}$ should be subtracted from $h_{ie}$.

$$r_{b'c} = \frac{\beta_o}{g_m h_{re}} = \frac{r_{b'e}}{h_{re}} \quad (6\text{-}7)$$

$$r_{ce} = \frac{1}{(h_{oe} - g_m h_{re}} \quad (6\text{-}8)$$

See Chapter 4 for the definitions of the h-parameters.
The voltage gain, $A_V$, of a common-emitter amplifier is:

$$A_V = \frac{V_{ce}}{V_{be}} = \frac{I_c R_L}{I_b R_i} = \frac{\beta R_L}{R_i} \quad (6\text{-}9)$$

where,

$R_L$ is the load resistor,

$R_i$ is the input resistance,

$I_c$ and $I_b$ are the collector and base currents, respectively,

$\beta$ is the common-emitter current gain, or the ratio of the change of the collector current, $\Delta I_C$, to the change of base current, $\Delta I_B$, at the specific frequency in question.

A second method for determining gain makes use of the hybrid equivalent circuit in Fig. 6-1. As the output voltage is:

$$V_{ce} = g_m V_{b'e} R_L \quad (6\text{-}10)$$

Voltage gain from a point inside the base to the output is:

$$A_V = \frac{V_{ce}}{V_{b'e}} = g_m R_L \quad (6\text{-}11)$$

In a more general method the procedure is to find the gain at the midfrequencies using any of the equations studied in previous chapters. Then, find $f_o$, the frequency at which the midfrequency gain is down by 3 dB. The voltage gain at any frequency above $f_o$ can then be determined from the curve in Fig. 6-2. It must be recalled that the curve continues to roll off at the rate of 6 dB per octave or 20 dB per decade beyond the −30 dB point shown on the graph. The high frequency voltage gain is down 3 dB at the frequency $f_o$ when:

$$f_o = \frac{1}{2\pi[(R_G + r_{bb'})|\,|r_{b'e}]\,[C_{b'e} + C_{b'c}g_mR_L]} \qquad (6\text{-}12)$$

This is the same as the $f_o$ in Fig. 6-2.

Frequency response is normally extended if a resistor, $R_E$, is in the emitter circuit. Equation 6-12 then becomes:

$$f_o = \frac{g_m(R_G + r_{bb'+}\beta R_E) + \beta}{2\pi\beta[R_G + r_{bb'} + R_E]\,[C_{b'c}g_m(R_L + R_E)]} \qquad (6\text{-}13)$$

The ratio of the output current to the input current, the current gain $A_i$, is:

$$A_i = \frac{g_mV_{b'e}}{V_{b'e}/Z_{b'e}} = g_mZ_{b'e} \qquad (6\text{-}14)$$

where $Z_{b'e}$ is the impedance between b′ and e.

Response of transistors used in common-base and common-emitter configurations at high frequencies, can be approximated using the values of $\alpha$ and $\beta$ at the frequency involved. Once determined, these factors are substituted into the low frequency gain equations to estimate the behavior of the transistor at the high frequencies. Greater accuracy can be achieved if the various capacities are not ignored.

### Bipolar Power Transistors

Bipolar power transistors are usually limited to low-frequency applications because of the large capacitances normally associated with these devices. New designs have extended the usefulness of these transistors into the upper frequency ranges where $f_T$ is well above 0.5 GHz. However, saturation resistance and voltage at high frequencies are large. Collector voltage and current swings are consequently very limited.

It is difficult to achieve a completely stable rf circuit. There is always a tendency for an rf amplifier to become unstable because of feedback through the capacitors and resistors inside the transistor. The output signal is fed back to the base input circuit of the device through these capacitors. If the phase is proper, the fed back signal is amplified by the transistor. This process continues until the fed back signal at the input is sufficient to maintain an output signal without the application of a signal at the input from an external source. The transistor is then in a state of oscillation. The circuit is unstable.

In order to eliminate the feedback, some method of neutralization is used. Here, a feedback capacitor is connected externally to the transistor circuit. It is put into a circuit in such a configuration that the signal it feeds back is 180° out of phase with the signal fed back from the collector to the base inside the transistor. The capacitor of the external component is adjusted to just cancel the signal due to the transistor itself. It negates all effects of the internal positive feedback of the device. A circuit like that shown in Fig. 6-3 could be used for this purpose.

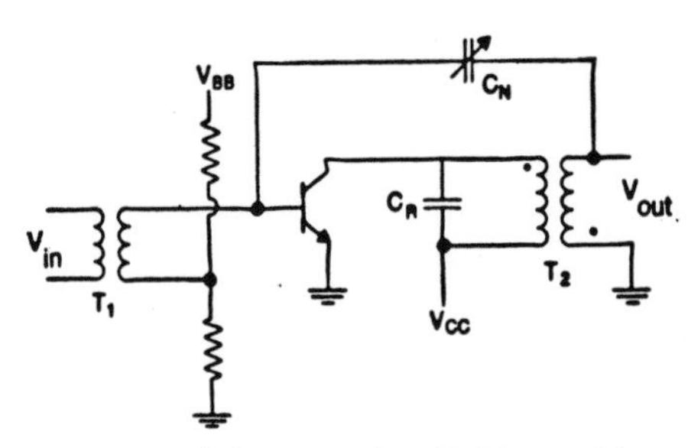


Fig. 6-3. Circuit showing how a neutralizing capacitor ($C_N$) is used in a common-emitter stage.

To adjust $C_N$, first disconnect the primary winding of T2 from the $V_{CC}$ supply. Now adjust the capacitor $C_R$ in the primary of T2 for a maximum or peak output at $V_{out}$, and adjust $C_N$ for a minimum output. Repeat these adjustments until more accurate adjustments of $C_R$ and $C_N$ can no longer be made.

While a common-emitter stage may at times be stable without being neutralized, this is not true of the common-base and common-collector circuits. Despite the high power gain of the common-base rf circuit (almost as high as the power gain of the common-emitter circuit), neutralization is critical. Neutralization affects only a narrow band of frequencies. Hence it is difficult to adjust the tuning controls when a common-base rf amplifier is being used.

Neutralization can also be accomplished by "tuning out" the inductance that exists in the emitter lead of the bipolar transistor. This is sometimes done by connecting a capacitor across a resistor in the emitter circuit. Neutralization arrangements of this type should be avoided for its ability to stabilize the circuit is very marginal. There is a tendency to oscillate, because when the capacitor across the emitter resistor is reflected back into the input circuit, it appears here as a negative impedance.

Rather than neutralizing a common-emitter circuit, it is best to stabilize it by design. The circuit will be stable if one resistor is placed in series with the base lead and a second resistor is wired from the collector to the emitter. The size of the resistors can be determined experimentally although the base circuit should see a resistance equal to or greater than $h_{ie}$ and the output circuit should see a resistance equal to or greater than $1/h_{oe}$. Similar stabilizing techniques can be used in common-collector and common-base circuits by just placing one resistor in series with an input lead and a second resistor across the output terminals. Make the impedances the transistor sees equal to or greater than $h_i$($h_{ib}$ for the common-base circuit and $h_{ic}$ for the common-collector circuit) at the input and $1/h_o$ ($h_{ob}$ and $h_{oc}$ for the common-base and common-collector circuits, respectively) at the output of the device.

## HIGH FREQUENCY FET AMPLIFIERS

The equivalent circuit used when working at the upper end of the rf band involves applying the short circuit admittances or y-parameters. These parameters are based on the definition of resistance, conductance, reactance, susceptance, impedance, and admittance as well as their relationships to each other and to the balance of the circuit. A review of these factors is in order if the meanings of the various y-parameters are to be clear and the equivalent circuit is to be useful.

### Review of Ac Circuit Relationships

As you know, conductance or G is equal to 1/R where R is the resistance of a circuit. If a circuit has a pure inductor or capacitor, the reactance, X, to the flow of ac or rf is $j\omega L$ for an inductor and $-j/\omega C$ for a capacitor. Here, L is the inductance expressed in henrys and C is the capacity expressed in farads. $\omega$ is $2\pi f$ or 6.28 multiplied by the frequency of the current flowing through the inductor or capacitor. Ac current flows through the capacitor and inductor. In the capacitor, this voltage goes through its cycle 90° after the current goes through its cycle. This is noted in the formula for the capacitive impedance by the $-j$ factor. As for inductors, the current lags the voltage by 90° and is noted in the formula by the $+j$ factor. The reciprocal of reactance, X, is susceptance, B, so that $B = 1/X$.

(Note that j is an imaginary number equal to $\sqrt{-1}$. Thus $j^2 = \sqrt{-1} \times \sqrt{-1} = -1$; $-(j)^2 = -(\sqrt{-1} \times \sqrt{-1}) = +1$. Manipulation of the j factor is necessary when analyzing the different ac circuit arrangements. The result will tell you if the voltage and current is the leading or lagging factor, or if they are in phase with each other.)

If a resistor is in series with an inductor or capacitor, the flow of ac current is impeded by both devices. The impedance of the circuit is assigned the symbol Z. The reciprocal of impedance is known as admittance, Y. $Y = 1/Z$. Z is related to both X and R by the formula $Z = R + jX$. Similarly, $Y = G + jB$. The phase relationship between V and I or the amount I lags or leads V depends upon the relative magnitudes of R and X or G and B. This is detailed in various books on ac circuit analysis. Only when the j factor in the impedance is negligible, are V and I in phase with each other.

The resistance of a circuit consisting of several resistors connected in series is equal to the sum of these resistors. Impedances and susceptances connected in series are also added together to determine the overall impedance of the circuit. Parallel combinations of resistors and parallel combinations of impedances are treated in a similar fashion where the reciprocal of the impedance of the circuit is equal to the sum of the reciprocals of the impedances of the various elements comprising the parallel-circuit arrangement.

The conductance or admittance of a combination of parallel-connected components is equal to the sum of the conductances or admittances of each individual component in the circuit. Of course, you cannot ignore all the j factors in the calculations.

All the basic ac circuit theory reviewed here is used when analyzing the y-parameter high-frequency equivalent circuit of the FET. When referring to this equivalent circuit, definitions of Y, G, and B discussed above apply. Here, however, the convention is to use lowercase letters.

## Y-Parameters

All circuit configurations of the FET are used in high-frequency applications. The tendency in the industry is to use the common-gate arrangement rather than the common-source circuit because the former is more stable. The neutralization procedures described above for bipolar devices can also be applied to the FET in the common-source arrangement. A common-source, y-parameter equivalent circuit is in Fig. 6-4. Each parameter has a g + jb term. They are defined as follows:

- $y_{is} = y_{11} = g_{is} + jb_{is} = I_{in}/V_{in}$. It is the input admittance when the output is short circuited, or $V_{out} = 0$.
- $y_{fs} = y_{21} = g_{fs} + jb_{fs} = I_{out}/V_{in}$. It is the forward transconductance when the output is short circuited, or $V_{out} = 0$. It is the ratio of the specific output current to an applied input voltage. Note that $g_{fs}$ is the $g_m$ transconductance we have been using in discussions of lower-frequency FET circuits.
- $y_{rs} = y_{12} = g_{rs} + jb_{rs} = I_{in}/V_{out}$. It is a reverse transadmittance when the input is short circuited, or $V_{in} = 0$. It defines the input current when there is a specific voltage at the output.
- $y_{os} = y_{22} = g_{os} + jb_{os} = I_{out}/V_{out}$. It is the admittance of the output when the input is short circuited, or $V_{in} = 0$.

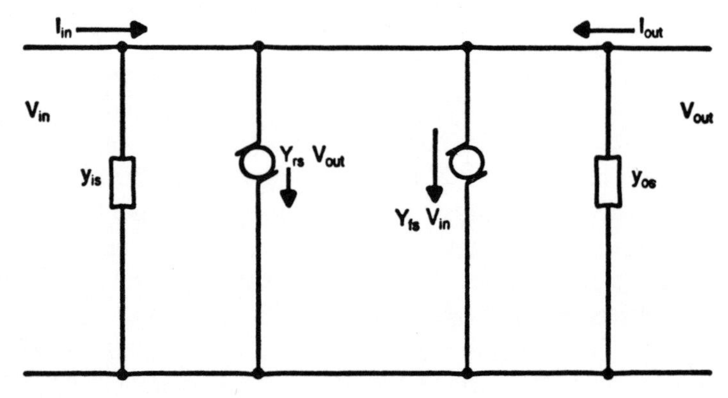


Fig. 6-4. The equivalent circuit for a common-source using y-parameters.

Two current generators are in the circuit. One is $y_{rs}$ $V_{out}$ equal to $I_{in}$. It is equal to $I_{out}$. Y-parameters vary with frequency. They increase as the frequency rises. Curves relating these parameters as well as their g and b components are usually provided by manufacturers.

Similar y-parameters are applied to common-gate and common-drain (source follower) equivalent circuits. The only differences in the symbols are in the second subscript. For example, the "s" in $y_{is}$ (or any of the other parameters defined above) indicates that the input admittance involved is in a common-source equivalent circuit. When the parameter refers to a common-gate circuit, the subscript "g" is substituted for the "s" and when the parameters refer to a common-drain equivalent circuit, a "d" is substituted for the "s" and the parameters are $y_{id}$, $y_{fd}$, $g_{id}$, $b_{id}$, etc. By convention, equations using y-parameters that are to be applied to all circuit arrangements omit the second subscript. The proper second subscript and value of the parameter are added when you apply the equation to a specific circuit.

## Rf Gain

Power must be delivered to a load resistor $R_L$ or impedance $Z_L$. The reciprocal of this impedance, $1/Z_L$ is a load admittance equal to $Y_L$ and the reciprocal of the resistor, $1/R_L$ is a load conductance, $G_L$. In rf circuits, the load usually consists of a tuned circuit. In Fig. 6-3, this consists of $C_R$ in parallel with the primary winding of the transformer. At one frequency, the reactance of the capacitor, $X_C$, is equal to the reactance of the inductor, $X_L$. When the two components are adjusted so that $X_C = X_L$ at the frequency applied to the circuit, the circuit is referred to as resonating at that frequency. When the resonant condition exists, the power gain of an FET is:

$$G(\text{resonance}) = \frac{|y_f|^2 G_L}{g_i(g_o + G_L)} \qquad (6\text{-}15)$$

where $|y_f|$ is the absolute value of $y_f$. This indicates that only the magnitude of $y_f$ is considered. Any negative sign applied to the number is ignored.

Maximum power is delivered to a load when the impedance of the load is equal to the output impedance of the source feeding the load. Considering a resonant circuit driven by a transistor, the maximum power gain is:

$$G(\max) = \frac{|Y_f|^2}{4 g_i g_o} \qquad (6\text{-}16)$$

When using the equation to solve for G, there are two terms emerging in the solution. One term is preceded by a j factor and one term is not preceded by a j. The magnitude of the quantity in the term preceded by the j factor is referred to as the gain of the *imaginary* power, and the magnitude of the quantity in the other term is referred to as the gain of the *real* power. Only real power gain is of concern to us. The j term is separated from the rest of the solution and discarded. Once derived, power gain relative to a 0 dB level of 1 mW can be stated in terms of dB. This can be determined from Table 4-1.

A stability problem exists in the common-source arrangement. It can be shown that the circuit is completely stable if:

$$2g_ig_o - (g_rg_f - b_rb_f) > |y_ry_f| \qquad (6\text{-}17)$$

where > means greater than.

This equation can be made more practical by remembering that a transistor does not operate in isolation, but must have input and output loads as well as a bias circuit. The conductance of the input generator or source, $G_s$, and the load, $G_L$, influence stability. For all practical purposes ($g_i + G_S$) should be substituted for the input conductance, $g_i$, of the FET in Equation 6-17 and ($g_o + G_L$) substituted for $g_o$, the output conductance of the FET. $G_L$ and $G_S$ can then be chosen to be certain that all stability criteria as determined from modified Equation 6-17 are met.

### Noise in High Frequency FETs

When the resistance of the source feeding a high frequency FET transistor circuit is high, the noise figure is extremely small. If the resistance of the source is $R_S$ and the transconductance of the FET is $g_{fs}$, the noise figure is equal to about:

$$\frac{(R_sg_{fs} + 1)}{(R_s)(g_{fs}}$$

for the common-source arrangement. As for the common-gate circuit, it is:

$$\frac{R_sg_{fs} + 2}{R_sg_{fs} + 1}$$

These relationships hold if the output impedance of the FET is equal to the impedance presented to it by a tuned circuit as its output load. Although the NF is extremely low when the transistor is used in the common-gate arrangement, it is even lower when the common-source circuit is used.

When comparing JFETs with IGFETs, the superiority of the JFET with respect to noise at high frequencies becomes obvious immediately. However, both devices are usually superior in this respect when compared to the bipolar transistor. Despite the higher noise factor, IGFETs are used at high frequencies where it is necessary that positive and negative input signal polarities are both to be applied to the input of the device.

## PASSIVE COMPONENTS IN HIGH FREQUENCY CIRCUITS

Rf circuits usually work into inductor, capacitor, and resistor loads, as well as into loads combining several of these components into resonant and other circuits. In order to use transistors effectively at these frequencies, a good understanding of circuits using reactive components is necessary.

### Resistor/Inductor Circuits

The equivalent circuit of an inductor can be drawn in two ways: as a resistor in series with the inductor, or as a resistor in parallel with the inductor. The equivalent circuits and their symbols are shown in Fig. 6-5.

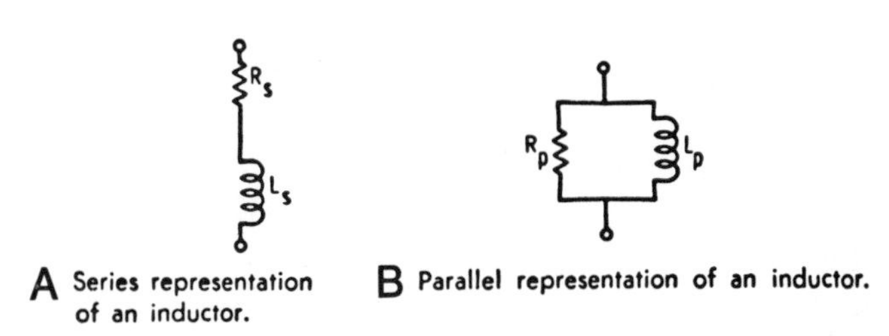


A Series representation of an inductor.

B Parallel representation of an inductor.

Fig. 6-5. The equivalent circuits for an inductor.

An inductor stores energy in the magnetic field around the wires, and a capacitor stores energy by putting a stress on (polarizing) the dielectric material between the plates. Most of the stored energy can be returned to the rest of the circuit. The portion of the energy returned to the circuit depends on losses in the component, such as the resistance of the wire of the inductor. A single number, Q, describes the characteristic, indicating a figure of merit relating the amount of energy returned to a circuit to the amount of energy dissipated in the inductor or capacitor. For the series equivalent circuit of the inductor:

$$Q_L = \frac{2\pi f L_s}{R_S} = \frac{6.28 f L_s}{R_S} \qquad (6\text{-}18)$$

where f is the frequency of the voltage impressed across the inductor and is normally referred to when specifying a coil. Similarly:

$$Q_L \quad \frac{R_p}{2\pi f L} = \frac{R_p}{6.28 f L_p} \qquad (6\text{-}19)$$

for the parallel equivalent circuit. For practical values, the Q of the parallel equivalent circuit is equal to the Q of the series equivalent circuit. Using these factors, the circuits in Fig. 6-5 can be converted from one form to the other. $L_p$ for all intents and purposes, is equal to $L_s$ if Q is of substantial size. In this discussion, they will be considered identical. $R_p$ is related to $R_s$ by the equation:

$$R_p = Q^2R_s \tag{6-20}$$

### Resonant Circuits

A resonant circuit consists primarily of a capacitor and an inductor. A circuit is said to be in resonance when the reactance of the capacitor is equal to the reactance of the inductor. This can occur at only one frequency for any resonant circuit. If the capacitor is connected across the inductor, a parallel resonant circuit has been established. When the component is in a series circuit, there is obviously a series-resonant circuit. Should the frequency at which the components are in resonance be applied to a parallel resonant circuit, the signal source sees a very high impedance. When applied to a series resonant circuit, the signal source sees a low impedance. The actual impedance the source sees is dependent on resistances in the circuit and the Q of the components.

In the circuit in Fig. 6-6A, assume that a coil is to resonate with a capacitor. The equivalent circuit is in Fig. 6-6B, where L is broken down into series components, $L_s$ and $R_s$. The parallel equivalent form is in Fig. 6-6C. Using this latter circuit, the resistance across $L_p$ is $R_{in}$ in parallel with $R_p$ or $R_{in}||R_p$. From Equation 6-19, $R_{in}||R_p = Q(2\pi)(f)(L_p)$. Assuming Q of the overall circuit is at about 10, $R_p$ can be determined from the relationship by using the equation for resistors in parallel if the resonant frequency and inductance of the coil, $L_p$, are known. Now the Q of the coil itself, $Q_L$, can be determined from Equation 6-19. Use Equation 6-20 to find $R_s$.

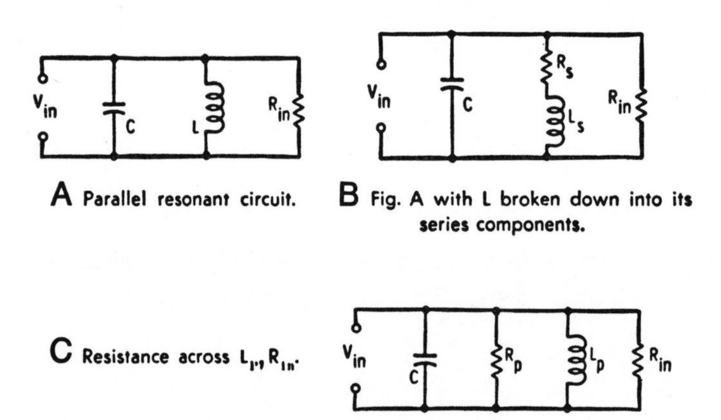


Fig. 6-6. Parallel resonant circuit and associated resistance.

The resonant frequency, $f_o$, of the circuit can be found by making $X_C$ equal to $X_L$ or:

$$f_o = \frac{1}{2\pi\sqrt{LC}} \tag{6-21}$$

where, L is the inductance expressed in henrys, C is the capacitance expressed in farads.

In a series resonant circuit, the voltage across the coil or capacitor is equal to $QV_{in}$ at the resonant frequency where $V_{in}$ is the input voltage to the series combination.

On either side of resonance, the voltage across the inductor and capacitor drop. At two specific frequencies, at $f_L$ below $f_o$ and at $f_H$ above $f_o$, the voltage will drop to 0.707 from the maximum output (drop 3 dB). The bandwidth $(f_H - f_L)$, resonant frequency, and $Q_L$ are related by the equation:

$$Q_L = \frac{f_o}{(f_H - f_L)} \tag{6-22}$$

## Resistor/Capacitor Circuits

The parallel $R_p$-$C_p$ capacitor circuit is mathematically related to its series equivalent, as shown in Fig. 6-7. The equations are similar to those used in the conversions of the L-R circuits. It is assumed that the Q's in Fig. 6-7A and 6-7B are relatively large and equal:

$$C_p = C_s \tag{6-23}$$

$$R_p = Q^2R_s \tag{6-24}$$

A Parallel $R_p$-$C_p$ circuit.

B Series $R_s$-$C_s$ circuit.

Fig. 6-7. The equivalent circuits for a capacitor.

## Matching Impedances Using Tuned Transformers

In order to get high Q, the resonant circuit must not be shunted excessively. This is seldom a problem where FET circuits are involved. Bipolar transistors do have relatively low input and output impedances and can present substantial shunts to the resonant circuit. Schemes like those in Fig. 6-8 are employed to alleviate the effect of excessive shunting

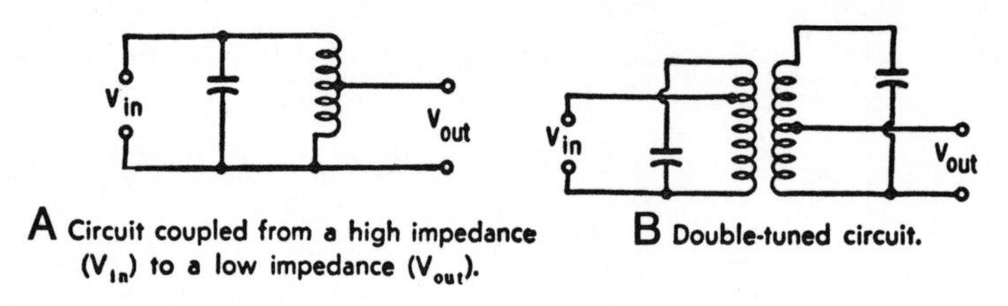


A Circuit coupled from a high impedance ($V_{in}$) to a low impedance ($V_{out}$).

B Double-tuned circuit.

Fig. 6-8. There are several possible schemes for attaining high Q.

by bipolar devices. These resonant circuits are normally connected between two transistor stages.

The double-tuned circuit in Fig. 6-8B is superior to the single-tuned one in Fig. 6-8A, because the double-tuned circuit provides a sharper bandpass. Tapped transformer windings are used to avoid excessive loading of the tuned circuits. Regardless of whether the coil is tapped or not, the proximity of the two windings has a bearing on the shape of the resonance curve.

A factor describing the degree of coupling of one winding to the other, is referred to as k, the coefficient of coupling. If coupling is perfect, $k = 1$ and all flux lines from the primary winding cut the secondary winding. If k is ⅓, ⅓ of the flux lines of the primary winding cut the secondary winding. When $k = 1/\sqrt{Q_pQ_s}$, where $Q_p$ is the Q of the primary winding and $Q_s$ is the Q of the secondary winding, k is assigned the symbol $k_c$, and is called the *critical coefficient of coupling*. When coupling is such that $k = k_c$, the signal induced at the resonant frequency from the primary into the secondary is at a maximum. Any closer coupling produces a dip in the response curve at the resonant frequency and a broadening of the overall resonance curve. Several broadband stages can be cascaded to properly reproduce a broader band than is possible when only one overcoupled stage is used.

## INSTABILITY

The inputs and outputs of transistor circuits are connected through the internal mechanism of the device. Oscillation may occur. One method of preventing oscillation is neutralization. Another method of reducing the probability of oscillation is to shunt the output terminals of the transistor with a resistor or make the load impedance small. A third method involves placing a resistor in series with one of the leads at the input of the transistor.

## MICROWAVE DEVICES

As we go up in frequency, the problems described so far in this chapter increase. Microwaves are increasingly used in a number of applications currently. A microwave signal is one in the Gigahertz region. One GHz equals 100 MHz. Most ordinary components are incapable of operation at such high frequencies. Silicon bipolar transistors can sometimes be used in the lower microwave region, but they become increasingly susceptible to thermal runaway and/or hot spots as the frequency rises. Ordinary FETs generally can't be used in microwave applications.

Since microwave signals are widely used today, it is obvious that new semiconductor devices have been developed for use in such extremely high frequency applications. In most cases their principles of operation are quite different from those of ordinary semiconductor devices.

Most of the microwave devices discussed in the next few pages are diodes, but they should not be confused with the standard PN junction diode discussed back in Chapter 1. The relationship between junction diodes and microwave diodes is very, very slim. The microwave devices are called diodes because they have two terminals. The word "diode" simply means two (*di-*) electr*ode*s.

### Negative Resistance

Many microwave devices are dependent on a property known as "negative resistance." A negative-resistance device does not behave according to Ohm's law.

Most substances normally exhibit regular (positive) resistance. This simply means that the relationships described by Ohm's law hold.

Let's return to very elementary electronics for just a moment. Ohm's law states that voltage equals current times resistance:

$$E = IR$$

This basic formula can be algebraically rearranged to solve for any of the variables:

$$I = E/R$$
$$R = E/I$$

Imagine a variable voltage source connected across a 1000 ohm resistor. Since the resistance is constant, the current flowing through the resistor will be determined by the voltage dropped across it;

$$I = E/R = E/1000$$

For example, if the voltage is 2 volts, the current works out to;

$$I = 2/1000 = 0.002 \text{ amp}$$

Now, if we increase the voltage to 5 volts, the current flow changes to;

$$I = 5/1000 = 0.005 \text{ amp}$$

According to Ohm's law, if the resistance is held constant, the current will rise with increases in the voltage.

A negative-resistance device behaves in just the opposite way. As you will learn shortly, this peculiar property is quite useful for generating microwave frequencies.

### Transit Time

Current cannot flow infinitely fast. It takes a finite amount of time for current to flow through the base region of a bipolar transistor. This transit time is extremely short, but it is not zero.

In low frequency applications, the transit time is negligible. It has no noticeable effect on circuit operation. But in the microwave range, the transit time can be a significant portion of a cycle, or perhaps even longer than a single cycle. Obviously such a condition will prevent the device or circuit from performing as desired.

There have been attempts to minimize this type of problem by decreasing the width of the base region. If the charge carriers (electrons or holes) don't have to travel as far, they can get through the region quicker. Unfortunately a thin-base transistor has additional problems. Interelectrode capacitance is increased. In addition, the thinner the base region is, the more fragile it will inevitably be. A thin-base exhibits decreased tolerance to reverse-bias voltages.

Reducing the thickness of the base region is just a partial solution at best. All semiconductor materials exhibit a property known as *electron saturation velocity* that seems to set a basic, inescapable limit on the high-frequency operation of a bipolar device.

### Tunnel Diodes

Taking advantage of negative resistance properties to permit microwave operation is not a new idea. As early as the late 1950s, it was suggested that the tunnel diode (discussed in Chapter 1) could have microwave applications, since it exhibited negative resistance under certain conditions.

### Gunn Diodes

In the early 1960s, John Gunn of IBM created another negative resistance diode device, that now bears his name. The Gunn diode is made from a rather unusual semiconductor material called gallium arsenide (GaAs). The special properties of this material will be discussed in somewhat more depth in Chapter 14. Some materials, including GaAs, allow electrons to exist in either of two conduction bands. Most materials permit only one conduction band at all times. In the lower conduction band, the properties described in Ohm's law hold. The electron mass and energy are low, but electron mobility is high.

By applying a higher voltage to the material, the electrons can be forced into a second conduction band, where electron mass increases, and electron mobility decreases. In this state the material exhibits negative resistance.

Many microwave oscillators are based on transferring electrons between the two conduction bands. This type of circuit is known as *Transferred Electron Oscillator*, or TEO.

The internal structure of a Gunn diode is illustrated in Fig. 6-9. Some technicians refer to almost any solid-state microwave oscillator as a Gunn device, but this is not really correct. Many microwave devices have very different structures from the device originally created by John Gunn.

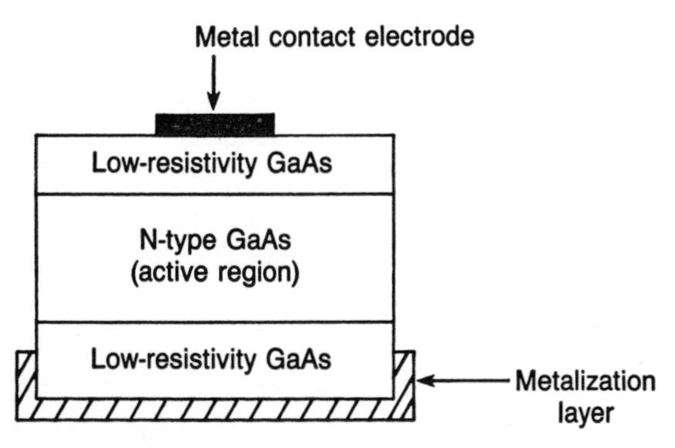


Fig. 6-9. The internal structure of a Gunn diode.

As the figure shows, a true Gunn diode is made up of a central slab of n-type gallium arsenide (GaAs). This slab is sandwiched between two layers of GaAs that have been doped to provide low resistivity. At the bottom is a thin metalization layer. A metal contact electrode is also placed on the upper side of the device.

Working our way from top to bottom, the first layer of semiconductor material primarily serves as an ohmic connection between the metal contact electrode and the central active region. The resistance of this contact layer is remarkably low. A resistivity of about 0.01 ohm per centimeter is typical. Besides providing a good contact with the end electrode, this section also blocks any migration of metallic ions from the metal electrode into the active region. This layer is grown epiaxially onto the main semiconductor slab (the central active section). It is usually about 1 to 2 microns thick.

The center section does all the work. It is somewhat thicker than the end regions. In most Gunn diodes the center section is between 6 and 18 microns thick. The exact thickness will determine the oscillating frequency of the device. The thicker the center region, the lower the frequency. For a center section that is 6 microns thick the oscillating frequency will be 18 GHz. Increasing the central slab to 18 microns reduces the oscillating frequency to 6 GHz.

The thickness of the central active section also controls the diode's threshold voltage for switching between the conduction bands. The threshold voltage typically ranges from just under 2 volts to a bit under 6 volts.

The final layer (at the bottom in the figure) is the substrate layer. It is basically similar to the topmost section, except it is metalized to permit bonding to the support structure of the diode. As in the top section, the GaAs material in the bottom layer is doped for very low resistivity.

A Gunn diode can function in either of two operating modes. The first of these is called the transit-time mode, sometimes the Gunn mode. The other mode is called the delayed transit-time, or limited space-charge mode.

### The Transit-Time Mode

The Gunn diode's basic operating mode is the transit-time mode. No external tank circuit is needed to construct a microwave oscillator from a Gunn diode in this mode. The operating frequency of the oscillator will be determined by the thickness of the central active region of the Gunn diode used.

If the Gunn diode is biased below its threshold voltage ($V_{th}$), the electric field will be uniform throughout the device. It will obey Ohm's law with positive resistance characteristics. If the voltage is increased (without exceeding $V_{th}$), the current will increase proportionately.

When the applied voltage reaches or exceeds $V_{th}$ electrons are injected into the diode's cathode faster than they can be fed out of the anode. As a result, the electrons start to pile up in a region or domain, as illustrated in Fig. 6-10.

The domain ("pile" of electrons) will now drift through the Gunn diode until it reaches the anode and is collected. Meanwhile, a new domain is formed at the cathode as the old domain leaves the anode. (This is shown in Fig. 6-11.)

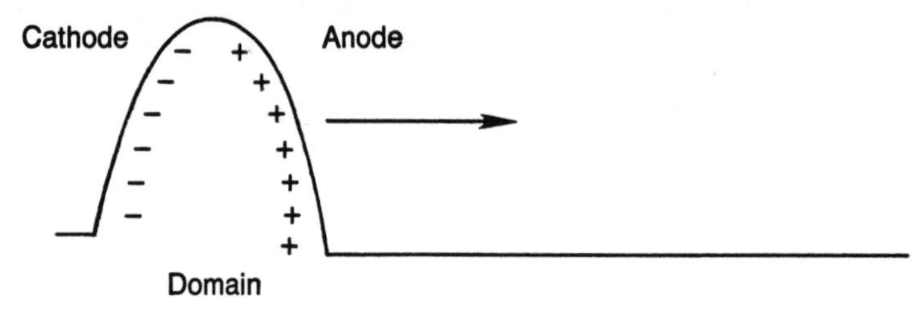


Fig. 6-10. When the threshold voltage is exceeded, electron domains build up.

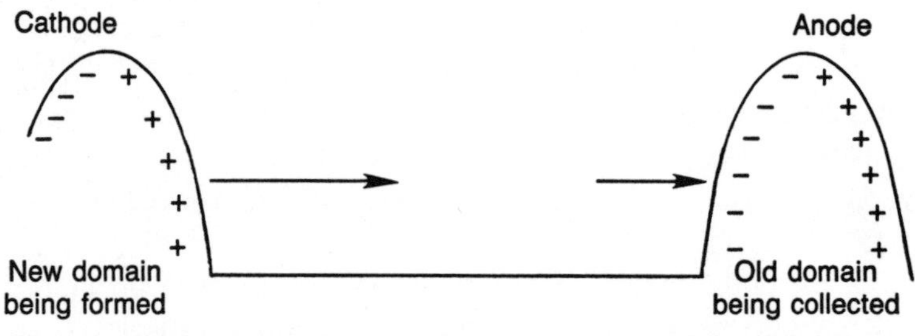


Fig. 6-11. When the old domain is collected at the anode, a new domain is formed at the cathode.

Naturally, these periodic "piles" of electrons affect the current at the output (anode) of the device. The output current will be at a relatively low continuous level, except when a domain is collected at the anode. When a domain is collected, a brief, but fairly large current pulse will be generated in the output circuit. The anode current of a Gunn diode being operated in the transit-time mode is illustrated in Fig. 6-12. As long as the proper voltage is applied, the Gunn diode functions as an oscillator all by itself.

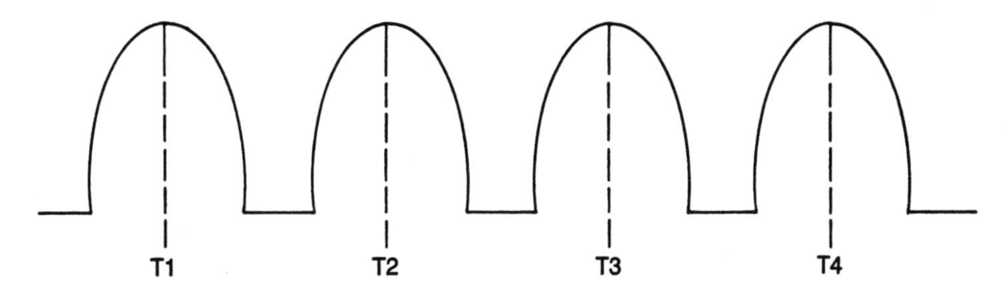


Fig. 6-12. The output of a Gunn diode is a series of current pulses.

There is a gap between current pulse peaks in the output signal. These gaps are referred to as the drift time and will be dependent on the width of the active region, and the particular characteristics of the semiconductor being used. The time between output pulses is determined by how fast the domains can move through the semiconductor material (drift velocity) and how far they have to go.

The transit-time mode is certainly simple and elegant, but it has its shortcomings. This mode is very frequency limited. The output frequency is determined by the specific characteristics of the Gunn diode used, and cannot be externally controlled.

A more serious limitation is that this mode is extremely inefficient. The efficiency is only between one and five per cent. Rather large amounts of dc power have to be used to generate even a small amount of microwave power.

Even with these inherent limitations, the transit-time mode can still be useful in certain applications, especially when a simple resistive load is to be driven.

### The Delayed Transit-Time Mode

The second operating mode for the Gunn diode is known as the delayed transit-time mode, or the limited space-charge accumulation (LSA) mode. This mode requires an external tank circuit (a tuned cavity). The addition of the tank circuit obviously increases the complexity and cost of the oscillator, but it will operate more efficiently than in the transit-time mode, and the output frequency can be varied according to the resonant frequency of the external tank circuit.

A simple LC resonant tank would be inadequate at such high frequencies, so a high-Q resonant cavity is used. To understand the functioning of the circuit, it is useful to imagine a LC tank in place of the tuned cavity. The equivalent circuit is shown in Fig.

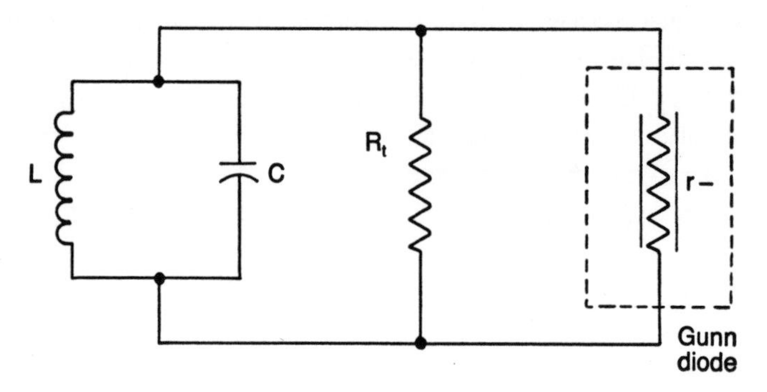


Fig. 6-13. An equivalent circuit for a Gunn diode oscillator in the delayed transit-time mode.

6-13. The Gunn diode is a negative resistance in parallel with the LC tank. Resistor R+ simply represents all the normal (positive resistance) losses in the tank circuit. In order for the circuit to oscillate, −r must be much less than +R.

Biasing the Gunn diode below the threshold voltage ($V_{th}$) will not have much effect. When the applied voltage exceeds $V_{th}$, domains are created, as in the transit-time mode. The output pulses cause the tank circuit to break into oscillations at its resonant frequency.

In order for this type of circuit to work properly, the dc bias should be precisely adjusted, so that the total bias applied to the Gunn diode drops below $V_{th}$ during negative peaks of the resonant waveform without dropping the applied voltage below the minimum sustaining level.

On negative peaks, the total bias (dc + rf) is less than the threshold potential, so the domains are stopped. If the current domain gets to the anode while the bias is below $V_{th}$, the next domain will be delayed until a sufficient threshold voltage is applied (on the positive portion of the tank cycle). In this way, the Gunn diode's current pulses are forced to follow the resonant period of the tank circuit.

The delayed transit-time mode can also be used in applications, such as Frequency Modulation (FM), and Automatic Frequency Control (AFC).

A delayed transit-time mode oscillator can put out about 100 times as much power as a comparable transit-time mode oscillator. Even in the delayed transit-time mode, Gunn diodes are not particularly efficient, but this mode is less inefficient than the transit-time mode.

## IMPATT Devices

Another microwave device is the IMPATT (*IMP*act *A*valanche *T*ransit-*T*ime) diode. The internal structure of this device is shown in Fig. 6-14. Compare this with the structure of the Gunn diode back in Fig. 6-9. The idea behind this device is that the phase delta in a PN junction with an applied rf voltage and an avalanching current can exhibit negative resistance at microwave frequencies.

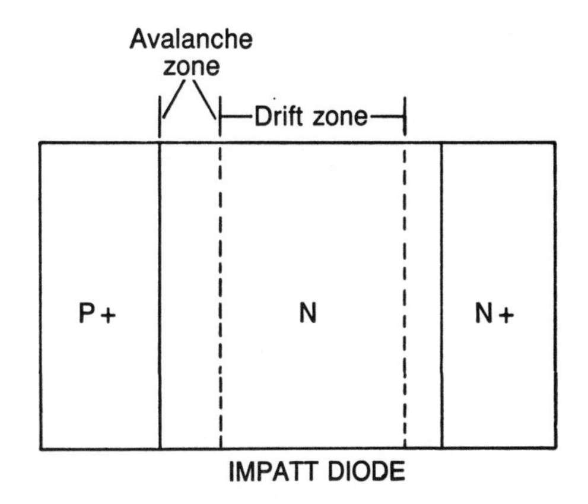


Fig. 6-14. Most of the activity in an IMPATT diode occurs in the central N-type region.

The main PN junction is on the left hand side of the structure illustrated in Fig. 6-14. The n+ region to the right serves as a low resistance contact with the electrode, and blocks metallic ion migration into the active region.

The active region is the n-slab in the middle. The semiconductor in this area is doped so that it is fully depleted (no free electrons or holes) at breakdown. As a result, a very small electrical field will cause velocity saturation of the electrons.

The electrons that are generated in the avalanche zone flow into the n-section drift zone. A very small increase in applied voltage will cause a substantial increase in current in this mode.

To understand how all this works, consider the current versus voltage curve for a PN junction diode shown in Fig. 6-15. We are only concerned with what happens when the diode is reverse-biased.

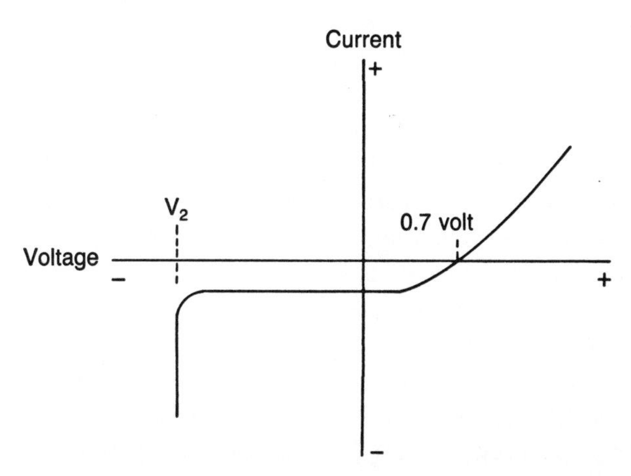


Fig. 6-15. If the applied voltage drops below $V_Z$, the diode suddenly starts to conduct very heavily.

If the diode is only slightly reverse-biased, virtually all current flow through the device will be blocked. There will only be a tiny leakage current. But, if the applied voltage is made increasingly negative, it will eventually exceed a critical breakdown voltage ($V_Z$). At this point, the current flowing through the diode will abruptly increase considerably. Under these conditions the PN junction is said to be operating in avalanche.

This avalanche current is the result of an effect called "secondary emission" or "avalanche multiplication." In this mode the electrons making up the ordinary leakage current have a high probability of colliding with other electrons, setting them into motion too, increasing the current.

The avalanche condition can cause an ordinary signal diode or rectifier diode to self-destruct. The breakdown voltage ($V_Z$) is analogous to the Peak Inverse Voltage (PIV) rating of the diode and should never be exceeded.

On the other hand, some types of special purpose diodes are designed not only to withstand, but to take advantage of avalanche effects. Zener diodes (discussed in Chapter 1) are an obvious example. IMPATT diodes are also doped to tolerate the high avalanche currents necessary for their microwave operation.

IMPATT diodes are available as both single-drift and double-drift devices. The differences will be explained shortly. For the time being, we will concentrate on the somewhat simpler single-drift version.

To use an IMPATT diode in a microwave oscillator, it is placed in parallel with a high-Q resonant tank circuit. As with any other microwave circuit, this means that the diode is installed in a resonant cavity.

Reverse-biasing the diode's PN junction creates a noise signal sufficient to shock-excite the resonant cavity into oscillation. The bias for the IMPATT diode is carefully chosen so that just a small additional applied voltage will force it into the avalanche mode. The oscillations from the tank circuit are fed back and combined with the original bias voltage. On positive peaks of the cycle, the IMPATT diode is driven into avalanche.

The amount of secondary emission that occurs in the avalanche mode is controlled by the applied voltage, and the number of charge carriers (electrons or holes) available within the semiconductor. These two factors interact so that the avalanche current pulse will keep increasing for a short time after the oscillation cycle has passed its peak value. There is an exponential growth of charge density at the avalanche point, while the avalanche charge current drifts to the far end of the drift zone.

The output of this type of circuit is in the form of a semi-square pulse waveform current that lags the applied (tank cycle) voltage by at least 90°.

An avalanching PN junction produces both types of charge carriers. Secondary emission results in an increase of both electrons and holes. In the single-drift IMPATT diode, only the electrons are used. The excess holes are simply returned to the cathode p-region. Obviously this results in a fairly low efficiency for the device. A typical IMPATT diode has an efficiency of less than 15 percent.

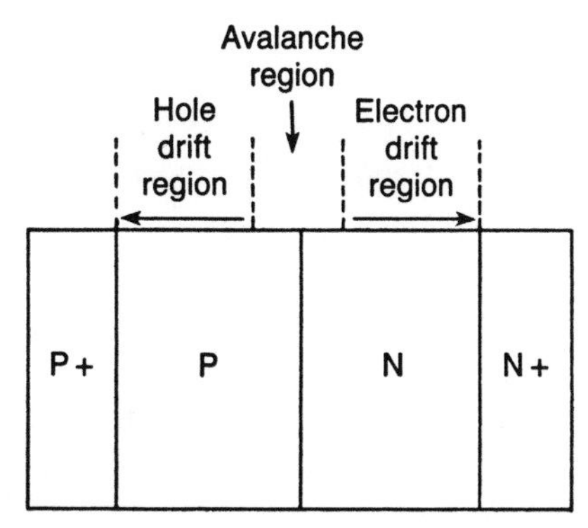


Fig. 6-16. The double-drift IMPATT diode has a structure similar to that of the single-drift IMPATT diode shown in Fig. 6-14.

When greater efficiency is required, a double-drift IMPATT diode can be used. The structure of this device is illustrated in Fig. 6-16. It is the same as the structure for a single-drift IMPATT device, except for the addition of a thick p-type slab to provide a drift region for the holes. The avalanche zone is the area surrounding the actual PN junction. The holes drift back through the p-type slab, and the electrons drift forward through the n-type slab. The total output efficiency is considerably higher than that of the single-drift device because the holes drifting through the p-type section are almost exactly in phase with the electrons drifting through the n-type section. The outside n+ and p+ sections simply serve as ohmic contacts for the appropriate charge carriers.

Neither type of IMPATT diode is really ideal for most microwave oscillator applications. For one thing, avalanching is inherently a noisy process, adversely affecting the purity of the generated signal.

It is difficult, if not impossible to generate signals in the low microwave region (below about 3 or 4 GHz) with IMPATT devices. On the other hand, this type of circuit has been used to generate signals up to 100 GHz.

Because of their relatively low efficiency, IMPATT devices demand large dc voltages for operation. A typical high-power unit might require as much as 150 volts dc. IMPATT devices are normally operated from a constant-current power source. In short, an IMPATT oscillator calls for a hefty, expensive power supply.

While their use in microwave oscillators is limited, IMPATT diodes are often used as amplifiers for microwave signals. They can also be used in a microwave frequency multiplier circuit.

### TRAPATT Diodes

The TRAPATT diode is essentially an IMPATT-like device that can be operated in a special mode permitting continuous tuning over a 0.9 GHz to 1.5 GHz range. The curious thing about this device is that no one really knows how it works.

The TRAPATT diode operates in the "anomalous mode." The inherent mysteriousness is reflected in the name. There are at least two major theories of how this mode might work, but currently no one can say which (if either) is ultimately correct.

According to the first theory, a trapped plasma appears within the device between sweeps of the IMPATT mode of operation. Supposedly the charge carriers are shielded by the trapped plasma from the external voltage field. As a result, the charge carriers drift out of the plasma at a relatively low velocity, explaining the high efficiency and lower operating frequency of this mode.

Currently this appears to be the more widely held theory. (Although that doesn't necessarily make it true.) The most common name for the device stems from this theory.

"TRAPATT" is an acronym for "*TRA*pped *P* lasma *A*valanche *T*ransit-*T*ime."

The second theory states that the special characteristics of this mode are due to a phenomenon called *A*valanche *R*esonance *P*umping. Because of this alternate theory, TRAPATT diodes are sometimes called ARP diodes.

Interestingly, an ordinary silicon PN junction diode can be forced to oscillate as a TRAPATT device under special conditions. Adjustments in such a circuit are exceedingly critical and touchy, so there aren't many practical applications.

As a rule, commercial TRAPATT diodes have the same basic structure as a single-drift IMPATT diode (refer back to Fig. 6-14). Most can be used as either an IMPATT or a TRAPATT device, depending on the applied bias voltage and other circuit conditions.

### BARITT Diodes

Yet another microwave device is the BARITT (*BAR*rier *I*njection *T*ransit-*T*ime) diode, shown in Fig. 6-17. This device is made up of a pair of back-to-back junctions. It is rather similar to a bipolar (pnp) transistor with no base connection. One of the junctions is slightly reverse-biased, while the other is slightly forward-biased.

If the bias voltage is less than a critical level (known as the punch-through voltage) the current flow through the diode will be no more than the ordinary leakage current of the reverse-biased junction. But once the bias reaches or exceeds the critical punch-through voltage, the device starts behaving quite differently. Under these conditions, the depletion region stretches across the entire n-type region, up to the forward-biased junction. All of the charge carriers (holes, in this case) at the forward-biased junction will flow across the n-type section. As a result, the current flow will rise very rapidly, as illustrated in Fig. 6-18. The bias voltage applied to a BARITT diode must be kept

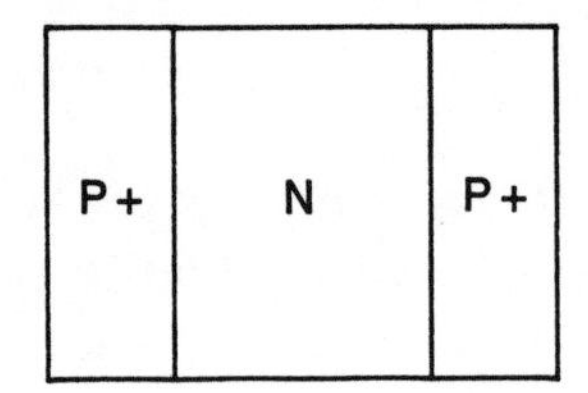


Fig. 6-17. The internal structure of a BARITT diode.

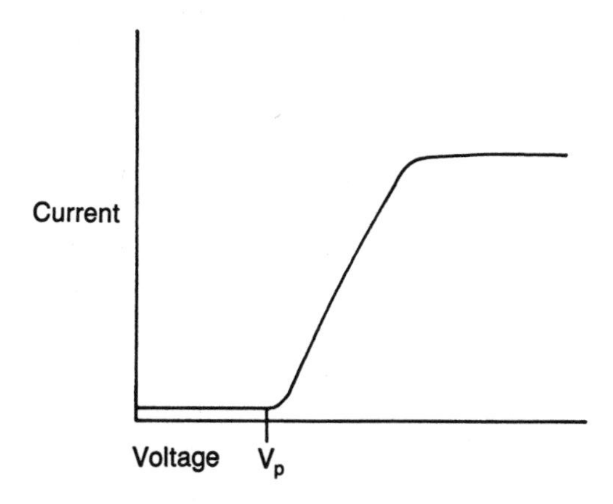


Fig. 6-18. The graph illustrates the operation of a BARITT diode.

between the punch-through voltage, and the threshold for avalanching for use in a microwave oscillator.

## PSIFETs

Ordinary FETs, especially IGFETs, are not well suited to microwave applications. They are primarily used only in switching and audio frequency applications. But electronics is a constantly developing field, and recent breakthroughs in MOSFET technology has given us FETs that can handle substantial power levels at microwave frequencies. Such microwave FETs offer numerous advantages over high frequency silicon bipolar transistors.

FETs offer higher input and output impedances. The dc bias is not affected by fluctuations in temperature, and thermal runaway is not a problem. FETs are voltage-driven, rather than current-driven, so they offer highly linear transfer characteristics.

A power FET for microwave applications is called a *Power SI* licon *FET*, or PSIFET. The basic internal structure of this device is illustrated in Fig. 6-19. Notice that it is not really dissimilar to an ordinary JFET. A short vertical channel is used, the gate and source connections are placed on the upper surface of the semiconductor slab, and the drain connection is at the bottom.

The gate voltage applied to a regular JFET controls the resistance (and therefore the current flow) of the lateral source-to-drain channel.

In the PSIFET, this channel is constantly kept fully depleted. There are no free electrons or holes in the channel. The electric potential in the depletion area is controlled by the applied gate voltage. This electric potential acts as a barrier that tries to block any electron flow between the source and the drain. By applying a large enough voltage to the drain, an electrostatic field will be created to counteract the depletion area's potential barrier. Under these circumstances, electrons can flow from the source to the drain.

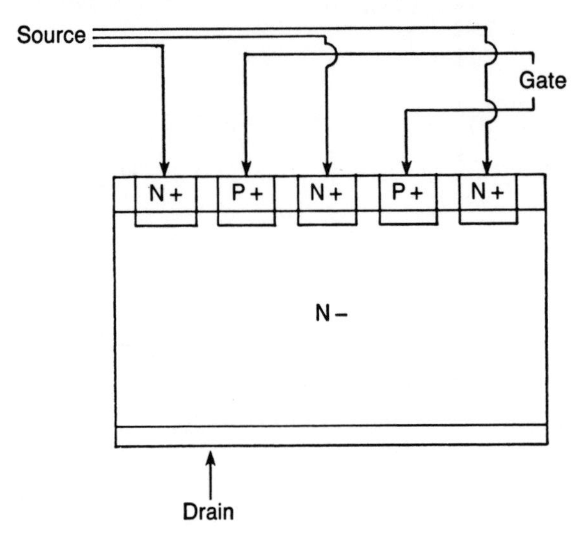


**Fig. 6-19. The PSIFET is especially designed for operation at microwave frequencies.**

In a PSIFET, the charge carriers travel by drift rather than diffusion, so the carrier velocity is significantly higher than that of a bipolar transistor. PSIFETs therefore feature a high breakdown voltage and a high gain-bandwidth product.

Because of their highly linear operation, PSIFETs are well suited for use in linear amplifiers for communications systems.

# 7

# Voltage and Current Feedback

Feedback judiciously placed around a transistor can provide an amplifier with characteristics relatively unaffected by many of the transistor's parameters. A feedback amplifier involves a sample of the output voltage or current fed back to the input. If a fed-back voltage is applied out of phase and bucking the input signal, the feedback is negative and the overall gain of the amplifier is reduced. If the voltage fed back adds to the input signal, the overall amplifier gain is increased and the feedback is positive.

The signal fed back to the input can be related to either the output voltage or output current. In the former case, the system is known as voltage feedback, while in the latter case, it is obviously referred to as current feedback. A combination of the two types of feedback around one amplifier is known as compound feedback.

## VOLTAGE FEEDBACK

The drawing in Fig. 7-1 represents a conventional transistor amplifier with a voltage gain of $A_V$. There is no feedback. Let $V_{in}$ be the input voltage to the amplifier proper. Here, it is equal to $V_s$, the voltage supplied by the signal source. The output voltage is $V_{out}$, and the voltage gain of the amplifier stage is:

$$A_V = \frac{V_{out}}{V_{in}} ; V_{out} = A_V V_{in} \tag{7-1}$$

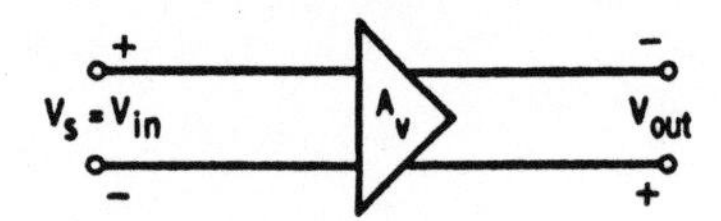

Fig. 7-1. A simplified representation of a conventional transistor amplifier.

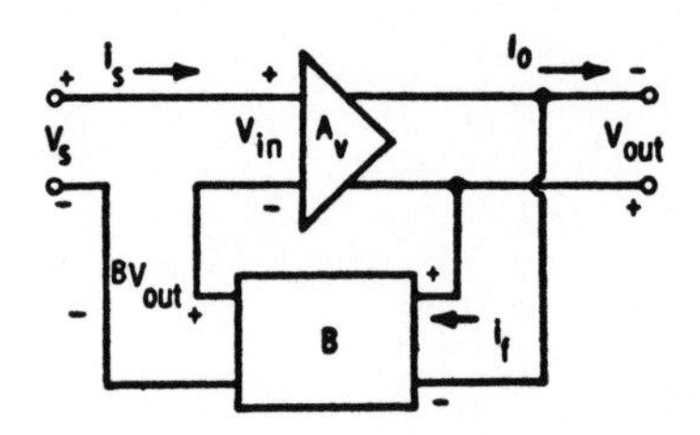

Fig. 7-2. Voltage feedback is used in many amplifier circuits.

Now, feed a portion of the output voltage back to the input (Fig. 7-2). This is known as series feedback since the signal fed back is applied in series with the input.

The portion of the output voltage fed back in series with the input is B. The portion of the input voltage, $V_{in}$, owed to the output voltage is $BV_{out}$. In the figure, this voltage bucks the input signal, $V_s$. The voltage at the input to the amplifier proper, ignoring polarity, is:

$$V_{in} = V_s + B(V_{out}) \text{ or } V_s = V_{in} - B(V_{out}) \tag{7-2}$$

The voltage gain of this circuit is:

$$A_f = \frac{V_{out}}{V_s} = \frac{A_V(V_{in})}{V_{in} - B(V_{out})} \tag{7-3}$$

by substituting Equation 7-1 for the numerator and 7-2 for the denominator. Now, division of the numerator and denominator by $V_{in}$ yields:

$$A_f = \frac{A_V}{1 - B(V_{out}/V_{in})} = \frac{A_V}{1 - B(A_V)} \tag{7-4}$$

which is the well-known feedback equation. The feedback factor is $\{1 - B(A_V)\}$. For negative or degenerative feedback, $B(A_V)$ is negative, while for positive feedback arrangements it is positive. With negative feedback, if $B(A_V)$ is much greater than 1, the gain of the overall circuit approaches 1/B. The noise and gain of an amplifier are reduced equally by a factor of $1/\{1 - B(A_V)\}$. Hence the signal-to-noise ratio remains unchanged with feedback.

There are many consequences of negative voltage feedback when it is applied to an amplifier as shown. The input impedance with voltage feedback, $Z_{if}$, is related to the input impedance without feedback, $Z_i$, by the relationship:

$$Z_{if} = Z_i\{1 - B(A_V)\} \tag{7-5}$$

The output impedance with voltage feedback, $Z_{of}$, is related to the output impedance without voltage feedback, $Z_o$, by the relationship:

$$Z_{of} = \frac{Z_o}{1 - B(A_V)} \tag{7-6}$$

The distortion of an amplifier circuit with feedback, $D_f$, is related to the distortion without feedback, D, by the relationship:

$$D_f = \frac{D}{1 - B(A_V)} \tag{7-7}$$

The bandwidth is likewise improved with inverse feedback. If the corner frequencies (high and low frequencies at which the midband gain is down 3 dB) of an amplifier are normally $f_H$ at the high-frequency end of the band and $f_L$ at the low-frequency end of the band, the respective corner frequencies with feedback around the circuit are extended to $f_{Hf}$ and $f_{Lf}$. The various factors are related to each other by the equation:

$$f_{Hf} = f_H\{1 - B(A_V)\} \tag{7-8}$$

for high frequencies. For low frequencies, Equation 7-9 applies:

$$f_{Lf} = f_L/\{1 - B(A_V)\} \tag{7-9}$$

Inverse feedback stabilizes the gain of an amplifier. Gain can vary radically (with parameters, temperature, components, etc.), in the amplifier shown in Fig. 7-1. This variation of gain is referred to as $\Delta A_V$. Should feedback be applied, the change in gain is reduced to $\Delta A_f$. Using the symbols previously assigned to gain with and without feedback, the variation in gain with feedback is related to the variation in gain without feedback by the equation:

$$\Delta A_f = \frac{A_f}{A_V}\left(\frac{\Delta A_V}{1 - B(A_V)}\right) \tag{7-10}$$

Amplifiers with feedback can become unstable or oscillate if the feedback around it turns positive. Thus in Equation 7-4, $B(A_V)$ can be equal to +1 at some high and/or low frequency. Should this happen, the denominator in the equation is equal to zero and the voltage gain with feedback is infinite. The amplifier becomes an oscillator. Probability of oscillation can be determined with some accuracy from the frequency response curve of the amplifier.

Using the two-stage amplifier in Fig. 7-3A as an example, assume that 20 dB of negative feedback is applied from the output at the collector of the second stage back to the emitter of the first. $R_1$-$C_1$ and $R_2$-$C_2$ are two networks producing high-frequency

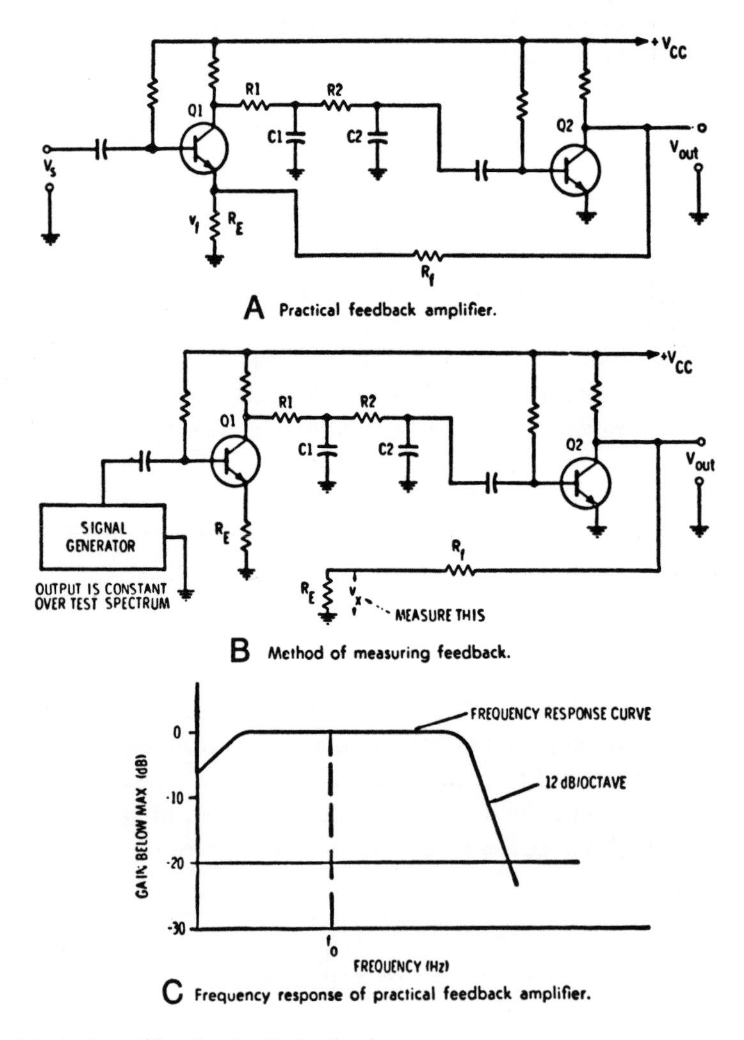


Fig. 7-3. A typical amplifier circuit with feedback.

rolloff at the eventual rate of 12 dB/octave. The feedback look is completed to the input through $R_f$, developing a voltage, $V_f$, across $R_E$ to buck $V_s$.

First, check the frequency response in the forward loop. To do this, disconnect $R_f$ from the junction of the emitter and $R_E$. Substitute another resistor equal to $R_E$ from $R_f$ to ground. Feed a signal at $V_s$ and measure the voltage, $V_x$, across the new $R_E$, as the frequency of the input signal generator is varied. The setup is shown in Fig. 7-3B. A plot of the frequency response of the forward loop using this setup is shown in Fig. 7-3C.

The feedback with the loop closed in Fig. 7-3A is, let us say, supposed to be 20 dB. It can be checked as follows. At a midfrequency, note $V_{out}$ in Fig. 7-3B. Now reconnect the feedback loop as in Fig. 7-3A. $V_{out}$ should have dropped 20 dB or to one-tenth of the value first measured.

On Fig. 7-3C, draw a line 20 dB below maximum gain or flat frequency response, to note this amount of feedback. The significant factor to be aware of now is the rate of rolloff as the frequency response curve crosses the line indicating −20 dB. If this rate is greater than 12 dB per octave, the amplifier will oscillate.

## CURRENT FEEDBACK

Here, voltage fed back to the input is related to the output current. A drawing of this is shown in Fig. 7-4. Current flowing through $R_T$ is sampled as a voltage across the resistor. A portion of the voltage, B, developed across the resistor, is fed back to the input. If $R_T$ is adjusted so that the required amount of voltage is fed back, the divider network determining B is not necessary. When there is current feedback, the voltage gain of the circuit can be derived as follows. The input signal voltage, $V_s$, is equal to:

$$V_s = V_{in} - BV_e \tag{7-11}$$

The voltage gain of the amplifier without feedback is stated in Equation 7-1. Voltage gain of the amplifier with feedback is:

$$A_f = \frac{V_{out}}{V_s} = \frac{A_V(V_{in})}{V_{in} - B(V_t)} \tag{7-12}$$

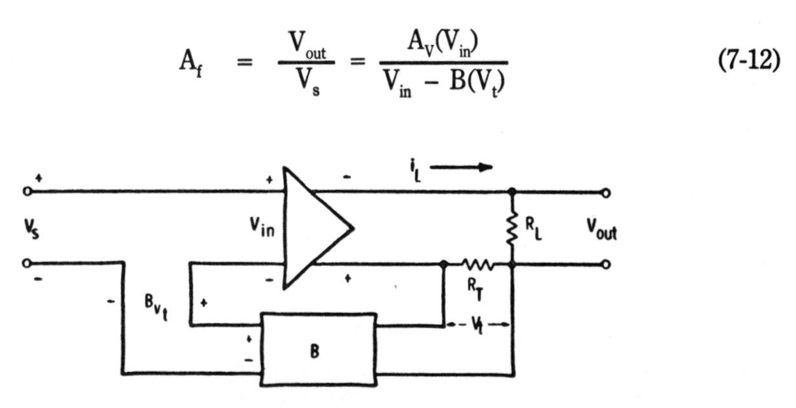


Fig. 7-4. Current feedback is employed in some amplifier circuits.

But:

$$V_t = i_L R_T = \left(\frac{V_{out}}{R_L}\right) R_T \tag{7-13}$$

so that Equation 7-12 becomes:

$$A_f = \frac{A_V V_{in}}{V_{in} - (R_T/R_L)B(V_{out})} \tag{7-14A}$$

Dividing the numerator and denominator by $V_{in}$ and letting $V_{out}/V_{in}$ equal $A_V$, the equation becomes

$$A_f = \frac{A_V}{1 - (R_T/R_L)B(A_V)} = \frac{A_V}{1 - \gamma A_V} \tag{7-14B}$$

If there is no divider network between $R_T$ and the input, B is made equal to 1. The quantity $(R_T/R_L)B$ in Equation 7-14B can be assigned the symbol $\gamma$ (gamma). If $\gamma(A_V)$ is much larger than 1, the voltage gain of the overall circuit is $1/\gamma$.

The effect of negative current feedback on an amplifier differs from the effect of negative voltage feedback only in that the former causes the output impedance to increase by the size of the feedback factor $\{1 - \gamma(A_V)\}$ while the latter causes it to decrease by the size of the feedback factor, $\{1 - B(A_V)\}$. Referring to Fig. 7-4, the load current, $i_L$, can be shown to be equal to the ratio $V_s/R_T$, when the amount of feedback is substantial.

## SHUNT FEEDBACK

Up to this point, series feedback was considered. The signal fed back to the input was applied in series with the input signal source. In the shunt feedback circuit, the signal is fed back and applied in parallel with the input signal. Voltage gain remains relatively unchanged when shunt feedback is added to the circuit while current gain varies considerably. A circuit using shunt feedback is shown in Fig. 7-5 where a feedback resistor is connected from the collector to the base.

Everything mentioned thus far with reference to negative or inverse voltage series feedback also applies to inverse shunt feedback. But there is one exception. In shunt feedback circuits input impedance is reduced by a factor $\{1 - B_i(A_i)\}$, where $A_i$ is the current gain of a circuit before feedback is applied and $B_i$ is the portion of the output current fed back to the input. A portion of the output signal is returned to the base. It may be considered in two ways. As a voltage-feedback circuit, the output from the

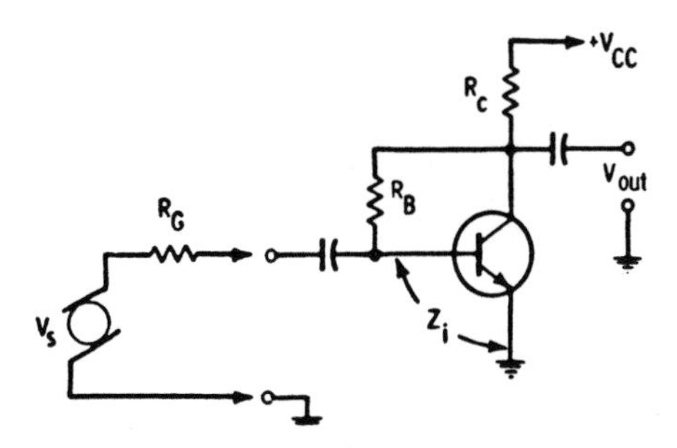


Fig. 7-5. This feedback circuit was previously used for bias stabilization.

collector is split between $R_B$ and the input impedance of the transistor, $Z_i$. In this case, B equals $Z_1/(R_B + Z_1)$. If $R_B$ is much larger than $Z_i$, as is almost always the case, B is $Z_1/R_B$. As a current-feedback circuit, the collector current is split between the load resistor, $R_C$, and the base resistor, $R_B$. The portion of the current fed back to the base, $B_i$, is $R_C/(R_B + Z_i)$. Assuming that $R_B$ is much larger than $Z_i$, $B_i$ becomes $R_C/R_B$. If $B_i(A_i)$ is large, $A_{if}$, the current gain with feedback, is $R_B/R_C$.

Shunt feedback of the type illustrated in Fig. 7-5 can be placed around three stages. The feedback resistor connects the collector of the third transistor to the base of the first. Equations are identical with those used in the one-stage example, except that here the current gain without feedback is equal to the product of the current gains of the three stages.

Another shunt-feedback arrangement is shown in Fig. 7-6. Here, $B_i$, the portion of the output current fed back, is equal to $R_E/R_B$. If $B_i(A_i)$ is much greater than 1, the current gain is equal to $R_B/R_E$. As in all shunt feedback arrangements, the input impedance with feedback is $Z_i/(1 - B_iA_i)$, where $Z_i$ is the input impedance without feedback.

Should it be necessary to measure the forward gain of the amplifier without feedback, disconnect $R_B$ from the emitter of Q2. Connect $R_B$ to a voltage equal to the emitter voltage of Q2, so that the bias on Q1 will not be upset. As in all shunt-feedback arrangements, the voltage gain here is only affected slightly be feedback. It is pronounced only when the feedback is very large.

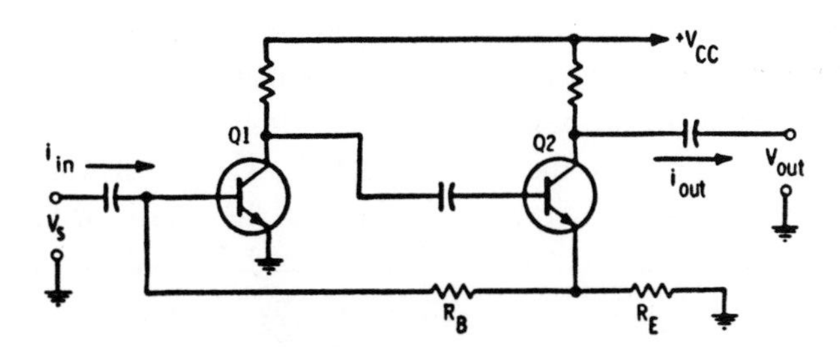


Fig. 7-6. This circuit demonstrates current feedback around two stages.

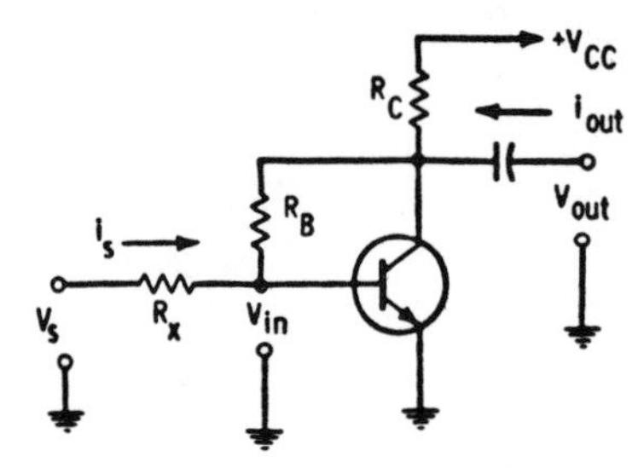


Fig. 7-7. A simple op-amp circuit.

The operational amplifier (op amp) in Fig. 7-7 is a significant variation on the shunt-feedback circuit. This circuit was originally designed for use in analog computers to perform several mathematical operations. It is well known in adder circuitry, but can also be used to subtract as well as to get averages of several numbers in a group. WIth the addition of proper feedback circuitry, it can perform basic calculus operations such as integration and differentiation. Its applications have now spread to uses in audio amplifiers, tone controls, buffers, oscillators, voltmeters, power supplies, and in countless other circuits. It is the basis of arrangements used in many linear integrated circuit chips.

### Single-Ended Operational Amplifier

Characteristics that describe the ideal op amp are infinite gain, infinite bandwidth from dc to infinite hertz, infinite input impedance, zero output impedance, and zero response time. Zero response time is essential if the signal at the output is to occur at the instant it is applied to the input. It is obvious that no amplifier—op amp or otherwise—fully meets all (or even any) of these requirements.

Any amplifier that can approach compliance with the specifications listed is known as an operational amplifier. An op amp is not necessarily an amplifier with a feedback loop. The open-loop circuit is an op amp. Its usefulness is increased tremendously when the amplifier is put into a feedback circuit. The amplifier in a feedback circuit is also referred to as an op amp or closed-loop op amp. We will refer to the circuit in Fig. 7-7 as an op amp. Although a bipolar transistor is used in the circuit, it could just as easily have used an FET. Some op amps have both an FET and a bipolar device. These, known as BiFET op amps, closely approach the ideal requirements for this type of circuit.

In the conventional op amp, the transistor is in its usual amplifying circuit. Input signal, $V_s$, is fed through $R_x$ to the base of the transistor in Fig. 7-7. Output voltage, $V_{out}$, at the collector, is fed back to the base through $R_B$. Current flowing through $R_X$ is $(V_s - V_{in})/R_X$, where $V_{in}$ is the voltage at the input of the transistor between base and ground. Theoretically, all this current also flows through $R_B$ as the input impedance of the op amp transistor is ideally infinite. Current flowing through $R_B$ is also equal to

$(V_{out} - V_{in})/R_B$. As both fractions stated in this paragraph represent the current flowing through $R_B$, they can be equated.

$$\frac{V_S - V_{in}}{R_X} = \frac{V_{out} - V_{in}}{R_B} \tag{7-15}$$

If the amplifier is to have "infinite" gain, $V_{in}$ must be extremely small or zero. At the same time, no current flows into the base circuit because of the "infinite" input impedance. As far as the voltage at $V_{in}$ is concerned, this point is at ground. It is referred to as a *virtual ground*. Setting $V_{in}$ equal to zero in Equation 7-15 yields the important Equation 7-16 where the actual gain has nothing to do with the gain of the active amplifier circuit.

$$A_V = \frac{V_{out}}{V_S} = \frac{R_B}{R_X} \tag{7-16}$$

The applications of the op amp are limited only by the imagination of the designer. Here, we will describe two circuits—one digital summing circuit and one linear tone control circuit. The actual amplifier is replaced in the drawing with its integrated circuit symbol, the triangle.

A *summing amplifier* is shown in Fig. 7-8. Here, different voltages, V1, V2, and V3, are applied to the input. The output, $V_{out}$, is proportional to the sum of these voltages. The sum of the current flowing through R1, R2, and R3 must also flow through $R_F$. Current $I_F$ must then be equal to (V1/R1) + (V2/R2) + (V3/R3). The output voltage, $V_{out}$, is equal to $I_F R_F$ and is proportional to the sum of the input voltages.

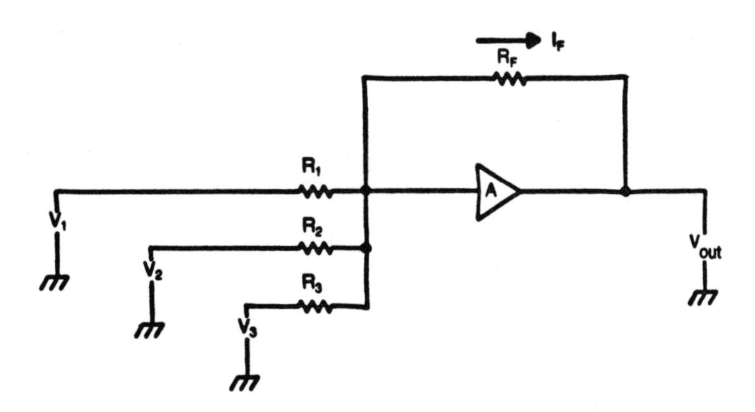


Fig. 7-8. An op-amp is often used as a summing amplifier.

The *tone control,* used in many audio amplifiers, has been designed by Mr. P. J. Baxendall. A representative circuit is in Fig. 7-9. The bass and treble control sections have been lifted out of the complete drawing for easy analysis. They are shown in Figs. 7-9B and 7-9C, respectively. In the practical circuit, C1 = C2 and R1 = R3. Equation 7-16 applies.

Starting with the bass control circuit, note that R4 has been omitted. It is in series with the infinite input impedance of the op amp and is thus negligible. R5 and C3 are high impedance at the low frequencies involved, and thus do not affect the operation of the bass control circuit. In the circuit as shown, when potentiometer R2 is at the cen-

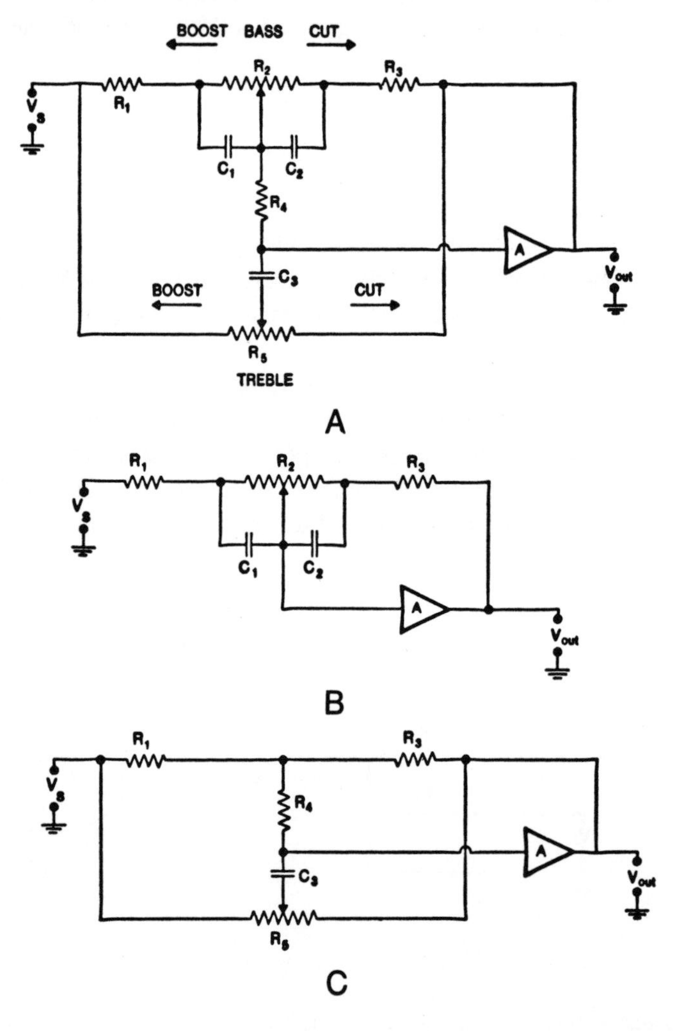


Fig. 7-9. This is a Baxendall feedback tone-control circuit. (A) Complete circuit. (B) Bass boost and cut circuit. (C) Treble boost and cut circuit.

ter of its rotation, $R_B$ in Equation 7-16 is equal to R3 + (R2/2‖C2). $R_X$ of the equation is equal to R1 + (R2/2‖C1). As R1 = R3 and C1 = C2, $R_B = R_X$ and the gain of the circuit at all frequencies is equal to 1 according to Equation 7-16. There is no bass boost or cut. In the maximum boost setting, C1 is shorted by the wiper arm. C2 shunts R2, shorting it all for practical considerations. The frequency response curve rises at the low frequencies without affecting the upper end of the audio band. $R_X$ of Equation 7-16 is then equal to R1, whereas $R_B$ is equal to the sum of R3 with the impedance of $X_{C2}$ or $R_B = (1 + j\omega C2R3)/j\omega C2$, where $\omega$ is equal to the product of $2\pi$ (or 6.28) and the frequency under consideration. Putting all this into an equation:

$$V_{out} = V_S \left[\frac{1 + (j2\pi fC2R3)}{j2\pi fC2R1}\right] \tag{7-17}$$

Gain at the low bass frequencies rises and is 3 dB above the gain at center or high frequencies when the numerator of the equation is equal to (1 + j) or is at $f = 1/2\pi C3R3$ Hz. Gain increases at the rate of about 6 dB/octave until a frequency determined by setting the denominator in the equation equal to zero is reached. Here, this extreme frequency is 0 Hz.

In the bass cut setting of the wiper arm, $R_B = R3$ and $R_X = R1 + X_c = (1 + j\omega C1R1)/j\omega C1, j\omega C1R1)$. From the op amp gain equation;

$$V_{out} = V_s \frac{j2\pi fC1R3}{1 + (j2\pi fC1R1)} \tag{7-18}$$

Gain is 3 dB below the gain at the center or high frequencies when the denominator of the equation is equal to (1 + j) or $f = 1/2\pi C1R1$ Hz. Gain rolls off at the rate of about 6 dB/octave until a frequency determined by setting the number equal to zero is reached. This frequency is 0 Hz.

The treble control circuit is in Fig. 7-9C. At the high frequencies, C1 and C2 are effectively shorts. The capacitors are shorts across R2. As shorting elements, C1 and C2 connect R1 to R3 and their junction to R4. This is as shown in the figure.

With the wiper on R5 set in the maximum boost position, the high frequencies pass easily through C3 to the input circuit of the amplifier while lower frequencies are attenuated. In the maximum cut position, C3 is in the feedback circuit. By feeding back the high frequencies more readily than the low frequencies, the upper end of the band is attenuated more than the middle- or low-frequency portion of the audio spectrum. In the treble boost setting of the control, the frequency at which the gain is 3 dB above the gain at the center frequency is $1/2\pi C3(R1 + 2R4)$. This relationship is also used to determine the frequency at which treble gain is 3 dB below the gain at the center frequency, in the treble cut setting of the control. In the boost setting of the control, output rises at the rate of 6 dB/octave, whereas in the cut setting it drops at the same rate. This rise or cut continues ideally to ∞ Hz. For intermediate settings of the control, the 3 dB attenuation and boost frequencies shift closer to the center frequency.

### Differential Operational Amplifier

The single-ended op amp is just a special case of the more general op amp circuit using differential amplifiers. An extremely high gain (infinite gain) differential amp with all the ideal characteristics noted above is an op amp. In addition to the ideal characteristics of the single-ended design, it should also have *zero offset*. Here, the output should be zero when zero signal is applied between the two inputs.

Op amps with two inputs can be represented as shown in Figs. 7-10A and B. The input marked with a minus sign is known as an *inverting input* for signal applied be-

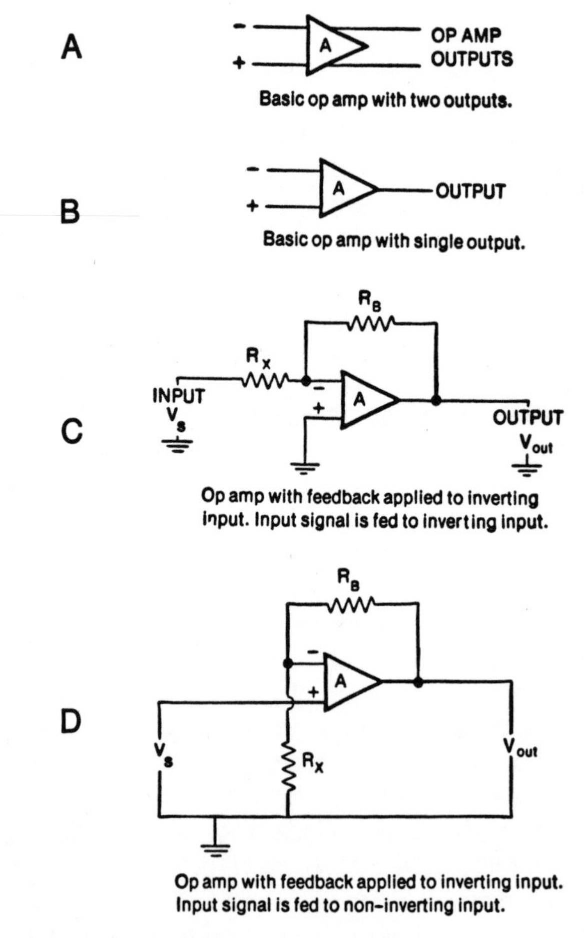


Fig. 7-10. These diagrams illustrate basic op-amp arrangement.

tween that input and ground goes through a 180° phase shift before it appears at the upper output terminal in Fig. 7-10A, or at the single output terminal in Fig. 7-10B. The terminal marked with a plus sign is a noninverting input because there is no phase shift between the input signal at this input and the signal at the upper output terminal in Fig. 7-10A or the single output terminal in Fig. 7-10B.

Most op amps consist of differential amps using the single output arrangement. Input signals are usually applied between one of the input terminals and ground and the output is taken between the single-ended terminal and ground. The output is proportional to the difference between the voltages at the two input terminals.

Feedback is usually applied from the output to the inverting input, as shown in Fig. 7-10C. In this type of arrangement, the + input is either connected to ground or is entirely absent. Where there is feedback, there is a tendency for the two input terminals to be at the same potential. As the + terminal is at ground, the − terminal is at its virtual ground. $R_X$ is the sum of all the resistances at the input, including that of the signal source. Consequently, gain Equation 13-15 applies to this circuit. A variation of the circuit involves a resistor wired between the + terminal and ground. This improves the stability of the circuit by reducing any of its tendency to drift with changes in temperature. It performs its function by making the impedances at the + and − terminals just about equal to each other. The resistor between the + terminal and ground should be made equal to the resistance of $R_B$ and $R_X$ when they are connected in parallel. If $R_X$ is less than several thousand ohms, the + terminal may be connected directly to ground.

Equation 7-16 describes the voltage gain of the inverting amplifier. There is also an output current. If the load is placed in the feedback loop (instead of the simple resistor such as $R_B$) rather than being placed between the output terminal and ground, the current flowing through the load is equal to $V_{in}/R_X$ regardless of the resistance of the load.

In Fig. 7-10D, the input signal is fed to the noninverting input while feedback is applied to the inverting input. While the input impedance is extremely high, the gain of the circuit is:

$$\frac{V_{out}}{V_s} = \frac{R_B + R_X}{R_X} \tag{7-19}$$

Should $R_X$ be made infinite and $R_B$ shorted or made equal to 0 ohms, the output voltage is equal to the voltage at the input or $V_{out} = V_s$. This circuit is known as a voltage follower as it isolates the output from the input without inverting or loading the input signal.

There are many variations on the basic circuit. Should the input voltages and resistors of Fig. 7-8 be used instead of $R_X$, you have a summing amplifier. If a capacitor C is used instead of $R_B$, an integrator circuit is formed where the output is the integral of the input and is equal to: $V_{out} = (1/R_XC) \int V_s\, dt$. Similarly, the output is the time derivative of the input if a capacitor is used instead of $R_X$. In this case, $V_{out} = R_BC(dV_s/dt)$. Should a diode be used in place of $R_B$ with the anode end connected to the – input terminal, the output voltage is logarithmic where $V_{out} = k\{1n(V_s/R_X) - \ln(I_o)\}$. In this equation, ln refers to the natural log to the base e of the function, k is a constant, and $I_o$ is the reverse saturation current of the diode ($\log_{10} X = 0.434\log_e X$).

In all applications, the op amp has a tendency to be somewhat unstable. To overcome this, a small experimentally chosen capacitor is usually placed in the negative feedback circuit, such as across $R_B$.

Several points should be considered when choosing and using an integrated circuit (IC) operational amplifier.

1. The open loop gain at the highest frequency of interest should be more than 10.
2. To avoid instability, frequency compensation must be provided so that the open circuit rolloff is slower than 12 dB/octave. Rolloff may be more than 12 dB/octave at frequencies where the output without feedback is less than the midfrequency output with feedback. Capacitors should be placed in the forward circuit or in the feedback loop to correct any unstable conditions that may exist.
3. Noise generated in the op amp should not interfere with the reproduced information.
4. The response of an op amp to an ideal square wave at its input, has been given the name *slew rate*. It is equal to the ratio of the change of the output voltage, $\Delta V_{out}$, to the period of time, $\Delta t$, it takes for the output voltage to make this change. High $\Delta V_{out}/\Delta t$ ratios are most desirable.
5. With zero voltage applied to the input of an op amp, the output voltage with current should be zero. If this is not the case, then an *offset* voltage or current is at the output. This is usually corrected by using a circuit similar to that in Fig. 7-11. The potentiometer is adjusted for zero volts at $V_{out}$. Some op amp ICs have special terminals for connecting the potentiometer.

Terminals are usually provided on IC op amps for connecting positive and negative voltage sources as power supplies. The two supplies are usually equal in voltage but opposite in polarity. In addition to these voltages, dc bias current must be supplied to

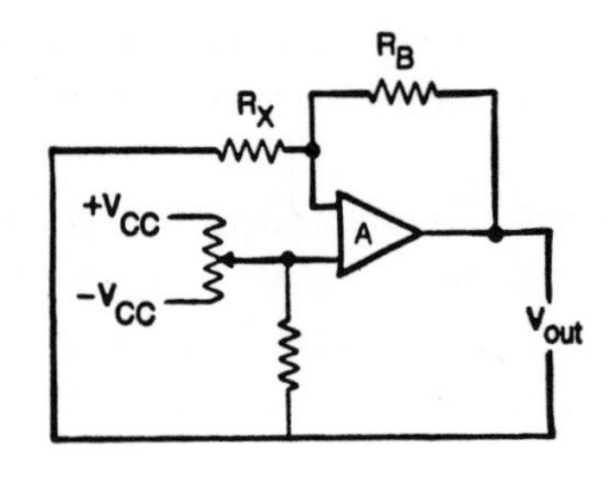


Fig. 7-11. Circuit arrangement permits adjustment for minimum offset voltage at the output.

both inputs. These currents are usually supplied through resistors connected to one of the voltage sources.

## SERIES FEEDBACK CIRCUITS

In Fig. 7-12, a portion of the collector current is sampled across the emitter resistor. A voltage caused by series feedback is developed across $R_E$. The portion of the signal fed back for use in Equation 7-14 is $B(R_T/R_L) = \gamma = R_E/R_C$. If $A_v$ is large, $A_f = R_C/R_E$.

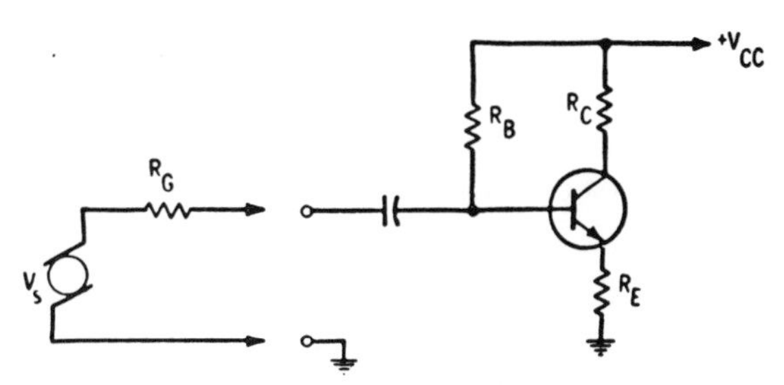


Fig. 7-12. Feedback circuit also used for bias stabilization.

The emitter resistor may be common to more than one transistor stage, as in Fig. 7-13. Here, $\gamma = R_E/R_C$. If $A_V$ is large, $A_f$ is $R_C/R_E$. Should there be a load, $R_L$, across the output, $\gamma$ becomes $R_E/R_C \| R_L)$ and the voltage gain is $(R_C \| R_L)/R_E$ where $R_C \| R_L$ is the equivalent resistance of $R_C$ in parallel with $R_L$. As before, the current gain is relatively unaffected by the presence of $R_E$.

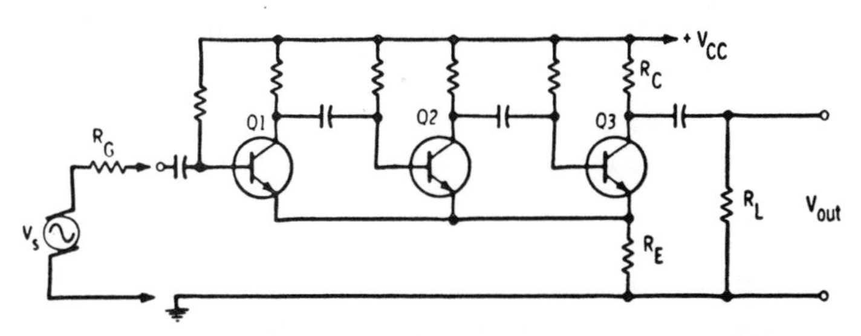


Fig. 7-13. Circuit features current feedback around three stages of amplification.

A popular circuit used in preamplifiers for magnetic phonograph cartridges is shown in Fig. 7-14. The voltage at the output of the second transistor is fed back to the emitter of the first. Equation 7-4 is used to determine the gain of the circuit with feedback, where $B = R_E/Z_F$. If the gain in the forward loop is large, the gain with feedback is $Z_F/R_E$.

$Z_F$ may be a resistor. In this case, gain is identical at all frequencies. Should it be desirable to have the gain vary with frequency, a reactive frequency-discriminating circuit can be used to replace $Z_F$. For example, if $Z_F$ is made into a capacitor, more high

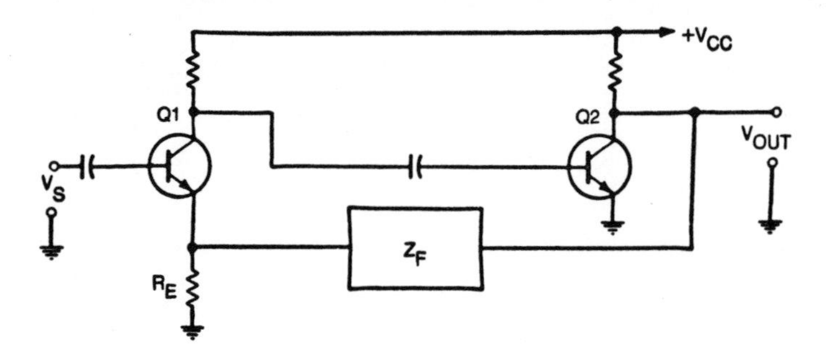


Fig. 7-14. This circuit is a feedback pair.

frequencies than low frequencies are fed back. Hence, the voltage gain of the overall circuit with feedback will be greater at the low frequencies than at the high frequencies.

## FEEDBACK AND THE RF CIRCUITS

Agc (automatic gain control) in a receiver is designed to maintain a constant average audio output or video contrast, regardless of the strength of the *rf* signal. Agc does not utilize feedback in the same sense as feedback has been discussed thus far. However, information is fed back from later stages in a receiver to the earlier *rf* and *i-f* amplifiers.

The signal strength is sensed and filtered after the second detector. It is fed back as a constant voltage to the bases of the mixer and amplifiers. The gain of a transistor varies with emitter current and collector voltage, when these are small. Hence, the rectified voltage applied to the base will control the *rf* gain.

Squelch circuits are used to silence a receiver until the received signal rises to a predetermined level. A squelch transistor is connected to the first amplifier stage in such fashion as to cut it off. When the signal on the squelch transistor reaches a predetermined magnitude, it turns on the first amplifier stage. The signal to do this job can be obtained from the agc line. The size of the signal required to turn on the circuit is adjusted with a threshold control.

The afc (automatic frequency control) provides for automatic readjustment of an oscillator frequency if it should shift from its original setting. In the conventional fm radio system, the received signal is mixed with that of the local oscillator signal to form a difference frequency. This should be at the i-f in a superheterodyne radio. A frequency detector at the end of the i-f chain produces an output proportional to the difference between the actual mixed signal and the required i-f signal. The polarity of the output indicates whether the difference frequency is greater or less than the i-f frequency. This voltage is then fed back to the oscillator where a transistor or diode is a capacitive element in the tank circuit. The voltage fed back controls the capacitor of the diode or transistor, affecting the resonant frequency of the tank circuit and hence the frequency of oscillation.

## BOOTSTRAPPING

The input impedance of a conventional common-emitter or common-collector circuit must always be less than the size of the resistor in the base circuit. The circuit in Fig. 7-15A provides a means of increasing the effective input impedance by using positive feedback, through C, from the emitter to the base. The gain of the overall loop is always less than one.

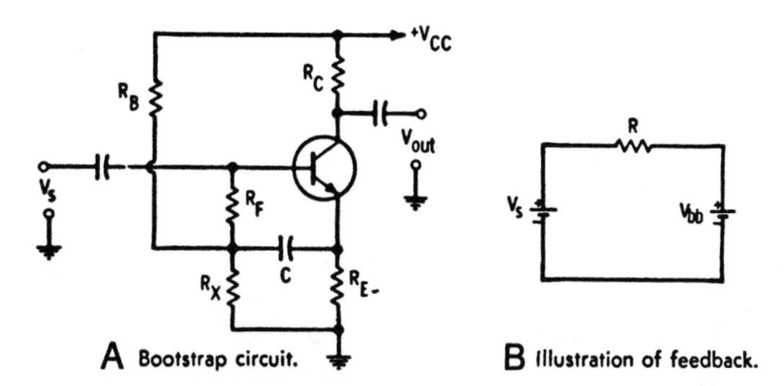


Fig. 7-15. A bootstrap circuit and an illustration of the feedback.

The operation of a bootstrap circuit can be understood with the help of Fig. 7-15B. If $V_{bb}$ is shorted, the signal voltage, $V_S$, sees an impedance equal to $R = V_s(1/i_s)$, where $i_s$ is the current flowing through the circuit. If a voltage, $V_{bb}$, is added to the circuit, the current in the circuit is $i = (V_s - V_{bb})/R$, so that the impedance seen by $V_s$ is $V_s(1/i) = V_s\{R/(V_s - V_{bb})\}$. If $V_{bb}$ is made approximately equal to $V_s$, the denominator approaches zero and the impedance presented to $V_s$ approaches infinity.

This is just what happens in Fig. 7-15A when the input signal voltage, $V_s$, and the voltage at the emitter (which is approximately equal to the voltage at the base) are applied at the two ends of $R_F$. It makes $R_F$ appear much larger than it actually is. The input impedance is equal to $\beta(R_B \| R_x \| R_E)$.

Bootstrapping may be applied to the FET to reduce the loading of the gate resistor. When applied to the source follower, the circuit will look something like that shown in Fig. 7-16. $R_S$ is made equal to:

$$R_S = \frac{V_P}{I_D}\left[1 - \left(\frac{I_D}{I_{DSS}}\right)^{1/2}\right] \tag{7-20}$$

where $V_P$ is the pinch-off voltage and $I_{DSS}$ is the drain current when the gate is shorted to the source. The input impedance is:

$$R_{in} = R_G\left(\frac{1 + g_m(R_S + R_T^m)}{1 + g_m R_S}\right) \tag{7-21}$$

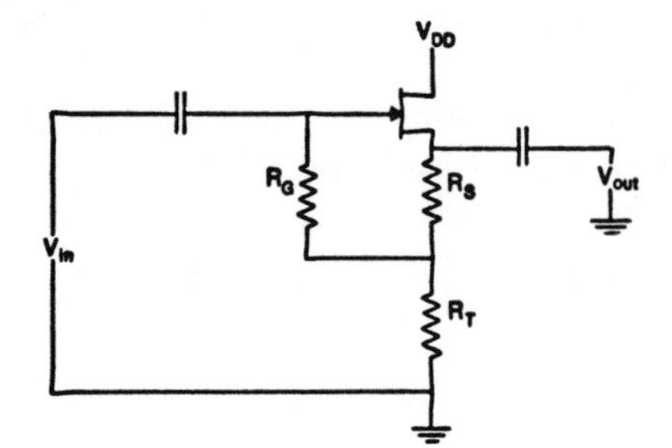


Fig. 7-16. A bootstrap circuit can also be constructed around a FET.

## FEEDBACK OSCILLATORS

If the feedback is positive and $B(A_V)$ is made equal to 1, then the gain with feedback is infinite. In this case, no ac input signal is required to produce an output. The amplifier is in a state of instability or sustained oscillation.

The magnitude of the sinusoidal oscillations is limited by the linear swing of the amplifier although *rf* oscillators are frequently biased class B or C for greater efficiency. In the latter instances, the transistor must be in a conducting mode to start the oscillations. It then transfers to a class-C mode of operation as the oscillations build up. The transfer from one mode to the other is accomplished through the use of an RC network in the emitter-base diode circuit. A voltage developed across the combination while the circuit is oscillating tends to reverse-bias the diode driving the transistor into the cutoff region for a large portion of the cycle.

### Bipolar Transistor Oscillators

In the feedback oscillator in Fig. 7-17A, a parallel resonant circuit formed by $C_R$ and $L_R$ is in the collector of the transistor. At resonance, $f_o = 1/2\pi\sqrt{L_R C_R}$, the impedance in the collector circuit is at a maximum and the phase shift used by the overall circuit is zero, or 360°. Any signal that may appear in the collector circuit (caused by random noise, voltage changes, etc.), is coupled inductively into the tickler coil, $L_F$, located near

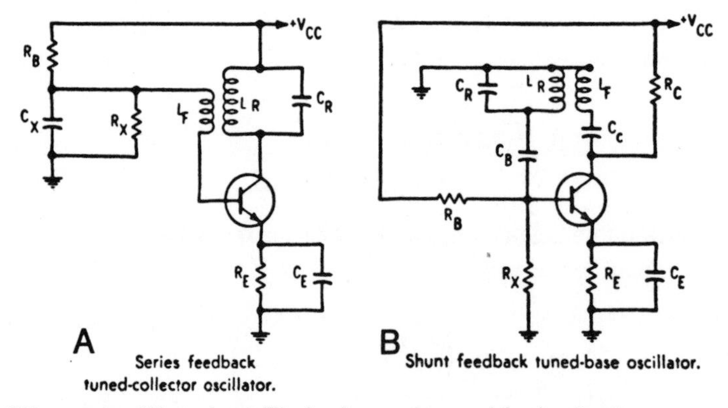


Fig. 7-17. Either series (A) or shunt (B) circuits can be used for feedback.

$L_R$. This signal is fed to the base of the transistor, where it is amplified and returned to the collector. The phases of the two inductors are arranged so that the feedback from the collector circuit to the base is positive. The circuit will oscillate at the resonant frequency. Resistors $R_B$ and $R_X$ establish the bias voltage, while $R_E$ is used for dc stabilization. $C_E$ and $C_X$ serve bypass functions.

The description of the circuit in Fig. 7-17B is identical to that just discussed, with the exception that the resonating components are in the base circuit. Other LC oscillator circuits are possible with capacitors or inductors to feed a portion of the output signal back to the input. A number of these arrangements are shown in Fig. 7-18. In each case, $R_B$ and $R_X$ are the base bias resistors, $R_C$ and $R_E$ are the collector and emitter resistors, respectively, and $C_R$, $C_{R1}$, $C_{R2}$, $L_R$, $L_{R1}$, $L_{R2}$, and M (mutual inductance) are elements of the resonant and feedback circuits. Other capacitors serve coupling and by-pass functions.

In the Colpitts circuits, the resonant frequency is

$$f_o = 1/2\pi \sqrt{L_R C_{R1} C_{R2}/(C_{R1} + C_{R2})}.$$

It will oscillate if the beta of the transistor is greater than $C_{R2}/C_{R1}$. The Clapp oscillator will oscillate if beta of the transistor is greater than $C_{R2}/C_{R1}$, but the frequency of oscillation is determined by the series combination of all capacitors in the resonant circuit. The primary advantage of the Clapp circuit over the Colpitts oscillator is better stability.

The Hartley oscillator is resonant at about $f_o = 1/2\pi \sqrt{C_R(L_{R1} + L_{R2} + 2M)}$ and it will oscillate if the beta of the transistor is greater than $(L_{R1} + M)/(L_{R2} + M)$.

Coils in Hartley and Colpitts circuits are usually tapped when the circuit is used at low frequencies. This is to allow them to be used with practical capacitors in the resonant circuits.

A piezoelectric crystal behaves as a high-Q inductance-capacitance parallel resonant circuit, and is highly stable. A crystal used in a Colpitts circuit appears in Fig. 7-19.

A single transistor in any circuit shifts the phase 180° from the base to the collector. In order for the circuit to oscillate, the phase shift between the base and collector must be 360°. A capacitive inductive circuit around the transistor can shift the phase by the additional required 180°. The RC phase shift networks in Fig. 7-20, can do this job. Here, the frequency of oscillation is $f_o = 1/2\pi C \sqrt{6R2 + 4RR_C}$. The circuit will oscillate if the beta of the transistor is greater than $(30R2 + 4RC^2 + 22RR_C/RR_C)$. Wien bridge or bridge T circuits may be used to replace the phase shift network shown here.

A variation of the RC circuit using two emitter-coupled transistors is shown in Fig. 7-21. The Wien bridge is used to select the frequency of oscillation. If the two resistors in the RC legs of the bridge are both equal to R1 and the two capacitors are C1, the circuit will oscillate at a frequency $f_o \approx 1/(6.28R1C1)$. Oscillation will occur only if the voltage gain of the amplifier without feedback exceeds 3.

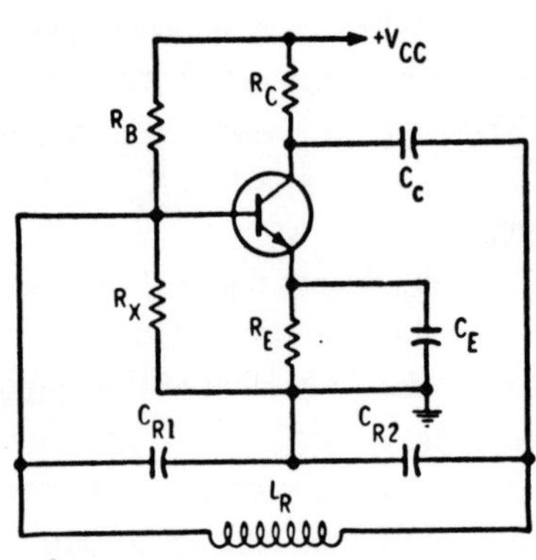


A Colpitts circuit with bias.

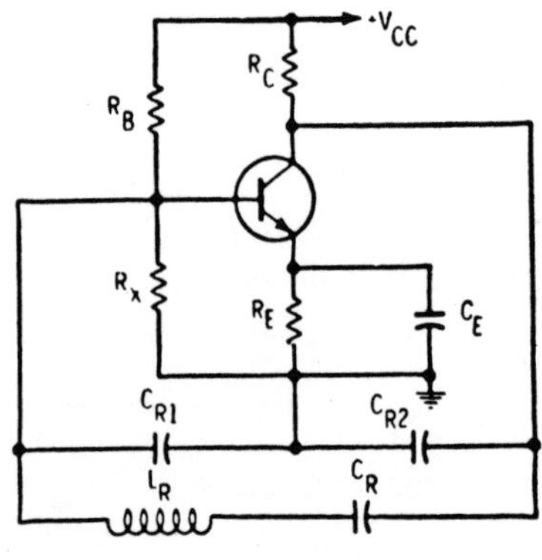


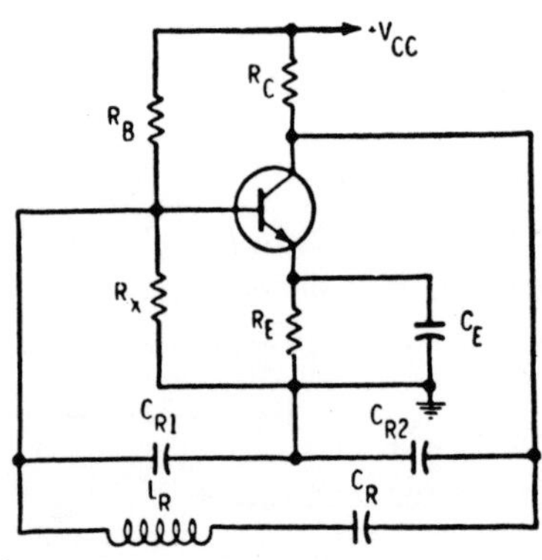

B Clapp circuit with bias.

Fig. 7-18. Here are some typical oscillator circuits.

C Hartley circuit with bias.

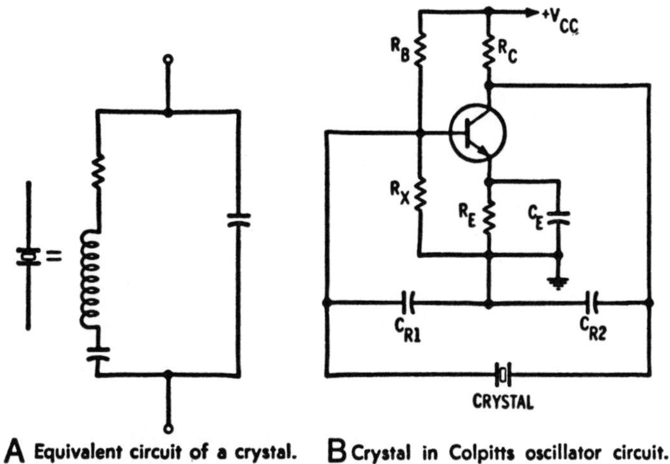


A Equivalent circuit of a crystal. B Crystal in Colpitts oscillator circuit.

Fig. 7-19. Crystals are used in many oscillator circuits.

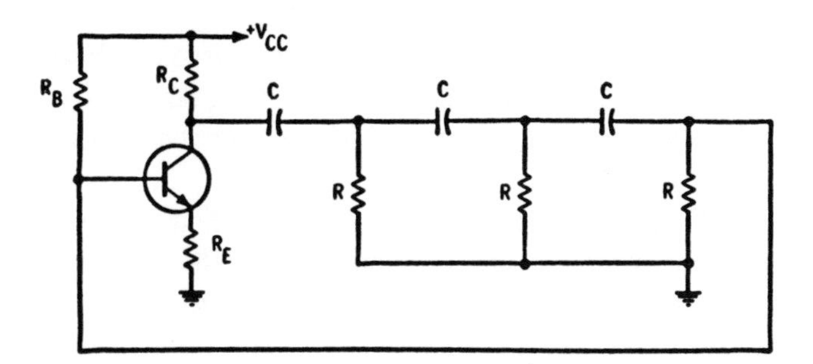


Fig. 7-20. This is an RC phase shift feedback oscillator.

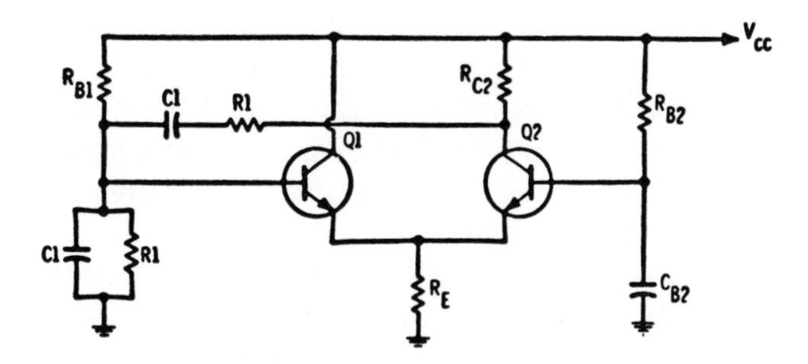


Fig. 7-21. This is a Wien-bridge oscillator in an emitter-coupled circuit.

## FET Oscillators

All circuits shown above using bipolar devices can also be used with FETs or with op amps. Because of the high input impedance of FET, these oscillators will be extremely frequency stable.

One of two methods is usually used to bias FET oscillators in *rf* applications. In one circuit, a parallel RC network is placed in series with the gate circuit. There is no voltage across the RC network at the instant voltage is applied to the oscillator circuit. The gate is biased at zero volts with respect to the source so that the gain of the JFET is at a maximum. A slight pulse is produced in the FET when the supply voltage is applied to the circuit. The amplitude of the initial pulse builds up (in the JFET circuit) to the point where the circuit is in a state of sustained oscillation. Rectified pulses of current at the oscillator frequency flow through the gate-source junction and through the resistor of the RC network in the gate circuit. Voltage is developed across that resistor, establishing the necessary bias conditions for the transistor to operate in its linear region. The resistor in the gate circuit is usually 1 MΩ and the capacitor impedance is about 100,000 ohms at the oscillating frequency.

A second type of bias arrangement involves placing the parallel RC network in the source circuit. Here, bias voltage is related to the source of drain current of the transistor. The resistor should be chosen using transconductance curves such as those shown in Figs. 3-15 and 3-16. Use the following procedure. Draw a straight line from the origin to the center of the relatively linear portion of the longer curve shown in a drawing similar to the one in Fig. 3-16. Note $V_{GS}$ and $I_D$ at the point where the straight line intersects the curve. Divide $V_{GS}$ by $I_D$. This is the resistance portion of the RC network that should be placed in the source circuit. The impedance of the shunting capacitor should be less than one-tenth the size of the resistor at the oscillating frequency.

## Phase-Locked Loop

Tank circuits in oscillators consist of LC components. If a varactor diode is used instead of C or in parallel with C, any change in the voltage applied to the varactor diode will change its capacity with a consequent change of the frequency of oscillation. An oscillator using a varactor diode as a voltage-dependent capacitor in its tank circuit is known as a voltage-control oscillator (VCO). This type of circuit is one of the basic elements of the phase-locked loop (PLL) arrangement. Another important element in the PLL is a solid fixed-frequency signal generator used as a reference frequency. This is usually a crystal-controlled oscillator.

Suppose you now take the two frequencies—the frequency from the VCO and the frequency from the solid fixed-frequency generator—and feed them to a phase detector. The dc output from the phase detector is related to the relative phases of the two signals. The only time they can be in phase is when their frequencies and phases are identical, because a frequency difference appears to the detector as if it were a phase difference.

Dc voltage at the output of the phase detector is consequently related both to the phase difference and the frequency difference of the two oscillators. Should the dc voltage from the output of the phase detector be fed back to the varactor of the VCO, it will shift the frequency of that oscillator until the frequency of the reference oscillator and VCO are identical. This is the basic arrangement of the PLL oscillator. Its prime function is to provide a stable frequency at its output.

PLL oscillators can be made more versatile by setting the fixed-reference oscillator at some basic frequency useful in a piece of equipment. Then add circuitry to divide that basic frequency down to the frequency needed as the reference for comparison with the VCO. Likewise, the VCO can be set at some high frequency that is divided down for comparison with the reference frequency or with some fraction of the reference frequency.

Other sections of the PLL include a filter to remove the ripple from the dc before it is fed back to the VCO and an amplifier to increase the magnitude of the dc. The various sections of the PLL are incorporated in ICs.

# 8

# Filters

Frequency discrimination exists in each circuit of every piece of electronic equipment. All circuits will pass or restrain one group of frequencies to a greater degree than another range. The most that can be hoped for is that a circuit will affect all frequencies within a specific band in identical fashion.

## BODE PLOTS

The reactance of a capacitor, C, in a circuit is:

$$X_C = \frac{1}{\omega C} = \frac{1}{2\pi f C} \text{ ohms} \tag{8-1}$$

where,

$X_C$ is the capacitance reactance,
$\omega$ is the angular frequency equal to $2\pi f$,
$\pi$ is a constant equal to 3.14,
f is the frequency in hertz,
C is the capacitance in farads.

Two basic filter circuits using a capacitor, are in Fig. 8-1. Should the input voltage, $V_{in}$, be at a very high frequency, the reactance of the capacitor is small, and $X_C$ for all practical purposes, is equal to zero. In Fig. 8-1A, the total input appears at the output.

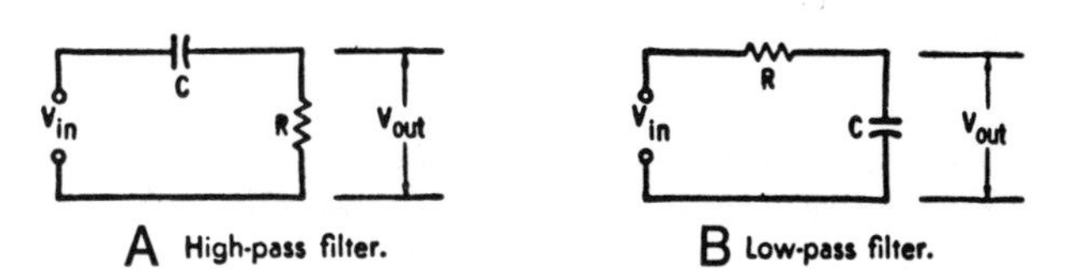


Fig. 8-1. RC networks can be used as simple filters.

$V_{in} = V_{out}$. When $V_{in}$ is a dc voltage, and the frequency is zero, the reactance of the capacitor is infinite. The total input voltage is across the capacitor. $V_{out} = 0$.

Between very high and very low frequencies, the capacitor behaves as an impedance that decreases as the frequency of the input signal rises. This frequency-discriminating phenomenon results in the gain curve shown in Fig. 8-2A. It is a plot of the ratio of output voltage to input voltage, expressed in dB.

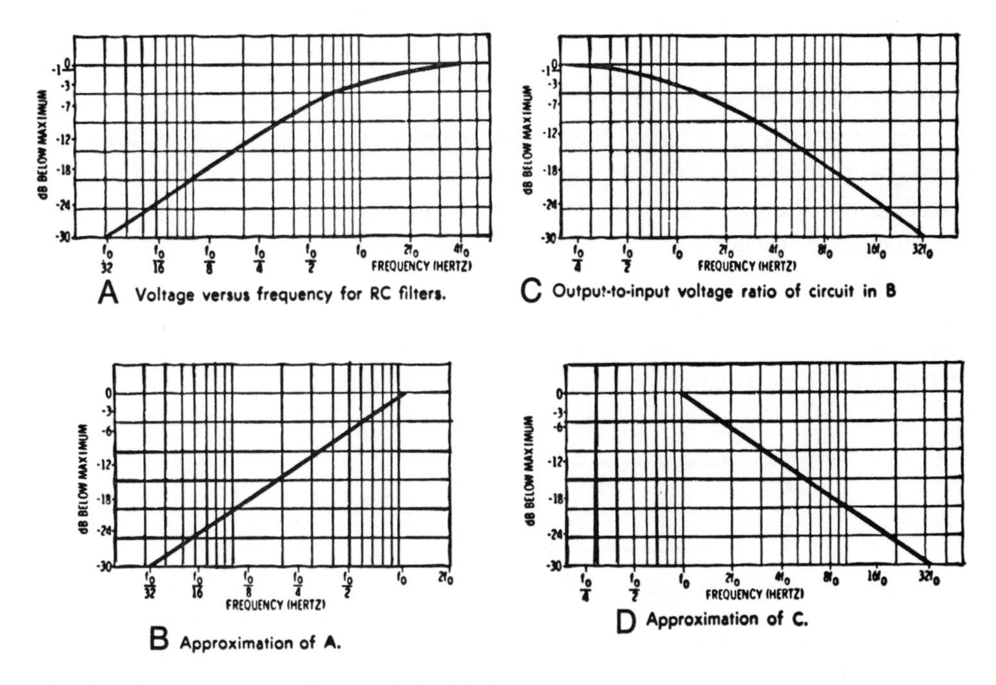


Fig. 8-2. These graphs are Bode plots for RC filters.

At the frequency, $f_o$, where $V_{out}$ has dropped 3 dB from its maximum, the reactance of the capacitor is numerically equal to the resistor in the circuit. This frequency is:

$$f_o = \frac{1}{2\pi RC} \tag{8-2}$$

The voltage has dropped 3 dB from its maximum at $f_0$. It dropped another 4 dB at $f_0/2$ and another 5 dB at $f_0/4$. From there, it drops 6 dB each time $f_0$ is reduced by a factor of 2. It can be considered in another way. After the output has dropped the first 12 dB, it continues to drop 20 dB each time the frequency is divided by 10.

The approximation to the curve in Fig. 8-2A is shown in Fig. 8-2B. Here, it is assumed that the output has dropped 0 dB at $f_0$, and drops 6 dB per octave, or 20 dB per decade, for all frequencies below $f_0$. This approximation is frequently used in design work.

Equation 8-2 can also be used to determine $f_0$ for the curve in Fig. 8-2C. This curve represents the output-to-input voltage ratio, expressed in dB, of the circuit in Fig. 8-1B. The eventual rolloff is at the rate of 6 dB per octave or 20 dB per decade. The approximate form is shown in Fig. 8-2D. Here, as well as for all other curves, the decibel is used to indicate the ratio between two voltages, Table 4-1 can be used to relate commonly used voltage ratios to their decibel equivalents.

## More Complex Circuits and Curves

The circuits in Figs. 8-1A and 8-1B are the simplest types of high-pass and low-pass filters, respectively, possible with one resistor and one capacitor. They both provide a maximum of 6 dB per octave of attenuation. Two or more such filters can be arranged in a circuit as shown in Fig. 8-3A. Here, $f_{01}$ for R1-C1 is $1/(2\pi R1C1)$, and $f_{02}$ for R2-C2 is $1/(2\pi R2C2)$. Assuming that $f_{01}$ is at 1,000 Hz and $f_{02}$ is at 2,000 Hz, the resulting approximate curves for the two values of $f_0$ are as shown in Fig. 8-3B. The attenuation curves are first drawn for each RC network and are then added together. Each network contributes 6 dB per octave to the rolloff. Two such networks add to a 12 dB per octave

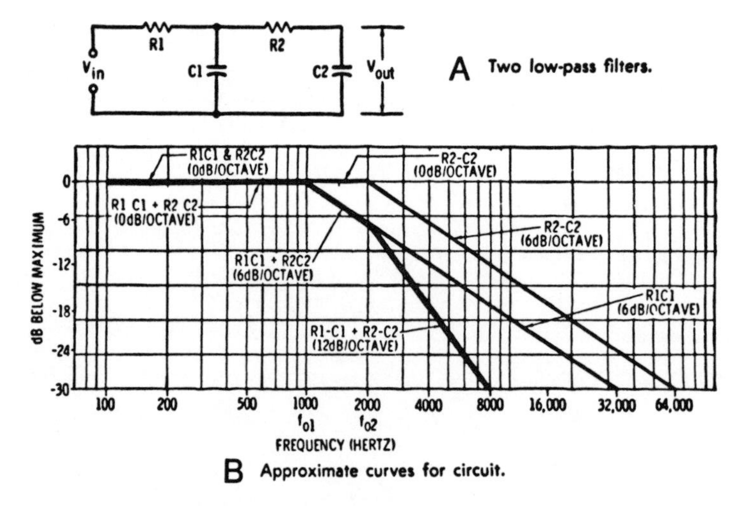


Fig. 8-3. Here we see a low-pass filter and its curves.

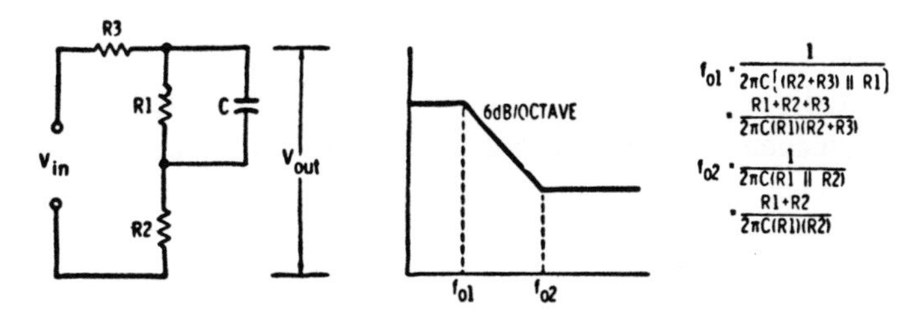


Fig. 8-4. A slightly more complex filter using only low- and high-pass circuitry.

eventual rolloff. A similar circuit arrangement can be drawn for the high-pass filter. Another RC filter network and its curve is shown in Fig. 8-4.

In the circuits discussed thus far, the output dropped to zero at either 0 Hz or ∞ Hz. The curve representing the characteristics of the circuit in Fig. 8-4 does not drop to zero at any time. Here, $V_{out} = V_{in}$ [(R1 + R2)/(R3 + R1 + R2)] until $f_{o1}$ is reached. Thereafter, it drops at the rate of 6 dB per octave until $f_{o2}$ and remains unchanged up to infinite hertz.

Equations in Fig. 8-4 specify the corner frequencies $f_{o1}$ and $f_{o2}$. The resistors and capacitors in the equations can be determined by following a few simple rules. Derived equivalent resistances and capacitances are then substituted for the R and C in Equation 8-2.

Rule 1. Short the input to the network at $V_{in}$ while letting $V_{out}$ remain open. You will usually get a parallel resistor-capacitor combination as a result of the short you created. Determine the net resistance of all resistors in the circuit that are in parallel with the capacitor. Substitute the values of R and C (R is the equivalent resistance of all resistors across the capacitor and C is the capacity of the capacitor expressed in farads) into Equation 8-2 to determine the corner frequency for the paralleling RC combination. If the capacitor in the original circuit is in series with the signal, the rolloff curve for each RC combination will be as shown in Figs. 8-2A and B. Should the capacitor be across the output or shunting part of the signal away from the output, the rolloff curve will be as shown in Figs. 8-2C and D.

Rule 2. Now break the short across the input circuit and short the output. Similar to the procedure in Rule 1, determine the equivalent resistance of all resistors in parallel with the capacitor. Substitute this information into Equation 8-2. The curves will rise at the rate of 6 dB/octave starting at the corner frequency determined from the equation. If the capacitor in the original circuit is in series with the signal, the curves will be the mirror image of those in Figs. 8-2A and B; if the capacitor is across all or a portion of the output, the curves will be the mirror image of those in Figs. 8-2C and D.

In Fig. 8-4, when the input is shorted, (R2 + R3) is in parallel with R1 and this combination is in parallel with C. Capacity of C and the equivalent resistance of R1 in parallel with (R2 + R3) are substituted into Equation 8-2 to determine the $f_{o1}$ shown in the figure. When the output is shorted, only R2 is in parallel with R1 and capacitor

C. $f_{o2}$ is determined by substituting C and the resistance equal to the R1 in parallel with R2 into Equation 8-2.

If an inductor, L, rather than a capacitor is in the circuit, Equation 8-6 would be used in place of Equation 8-2. Curves in Figs. 8-2A and B and their mirror images apply when L is across all or a portion of the output. Curves in Figs. 8-2C and D and their mirror images should be used when L is in series with the input signal.

## RC-Coupling

A typical RC-coupled stage is shown in Fig. 8-5A. The frequency response is affected by the various impedances in the circuit.

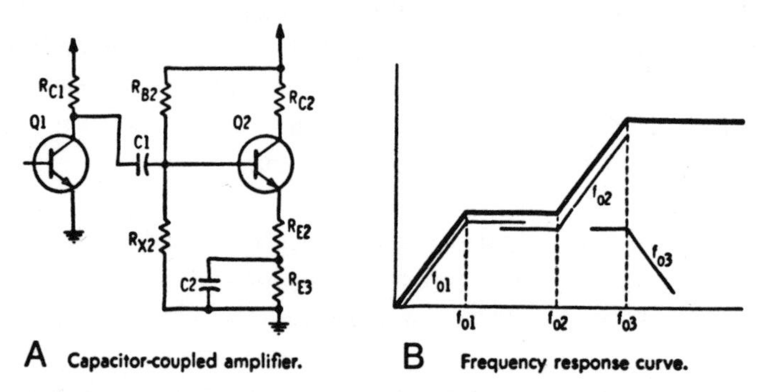


Fig. 8-5. A capacity-coupled amplifier circuit and its response curve.

One breakpoint is due to the coupling circuit involving C1, the output impedance ($R_{out}$) of Q1, and the input impedance ($R_{in}$) of Q2. It is low-frequency rolloff. The output voltage starts to drop at the rate of 6 dB per octave at $f_{o1}$ Hz. It continues dropping to 0 Hz.

$$f_{o1} = \frac{1}{2\pi C1(R_{in} + R_{out})} \tag{8-3}$$

The input impedance of Q2 is $R_{B2}$ in parallel with $R_{X2}$ in parallel with $\beta(R_{E2} + R_{E3} + r_3)$. The emitter resistance of Q2 is $r_e$. $R_{E3}$ is negligible if the impedance of C2 is much smaller than $R_{E2}$ at the frequency, $f_{o1}$. $R_{out}$ is the output resistance of Q1 and is usually equal to $R_{C1}$.

A second corner frequency, $f_{o2}$, occurs when:

$$f_{o2} = \frac{1}{2\pi C2R_{E3}} \tag{8-4}$$

At $f_{o2}$, the voltage starts to rise at the rate of 6 dB per octave. At $f_{o3}$, the voltage begins to drop at the rate of 6 dB per octave.

$$f_{o3} = \frac{1}{2\pi C2}\left(\frac{R_{E3} + R_{EM}}{R_{E3}R_{EM}}\right) \quad (8\text{-}5)$$

where,

$R_{EM}$ is the sum $R_{E2} + r_e +$ (resistance in base circuit)/$\beta$.

Typical curves of all these factors are shown in Fig. 8-5B. The thick line is the sum of the effects of the various curves. Response curves must be checked in the laboratory when precise results are required.

## Transformer-Coupling

The RC networks can replaced with their dual RL networks using a resistor and an inductor in each circuit. In Fig. 8-6, two RL networks are shown with their approximate curves. The corner frequency for an RL network is:

$$f_o = \frac{R}{2\pi L} \quad (8\text{-}6)$$

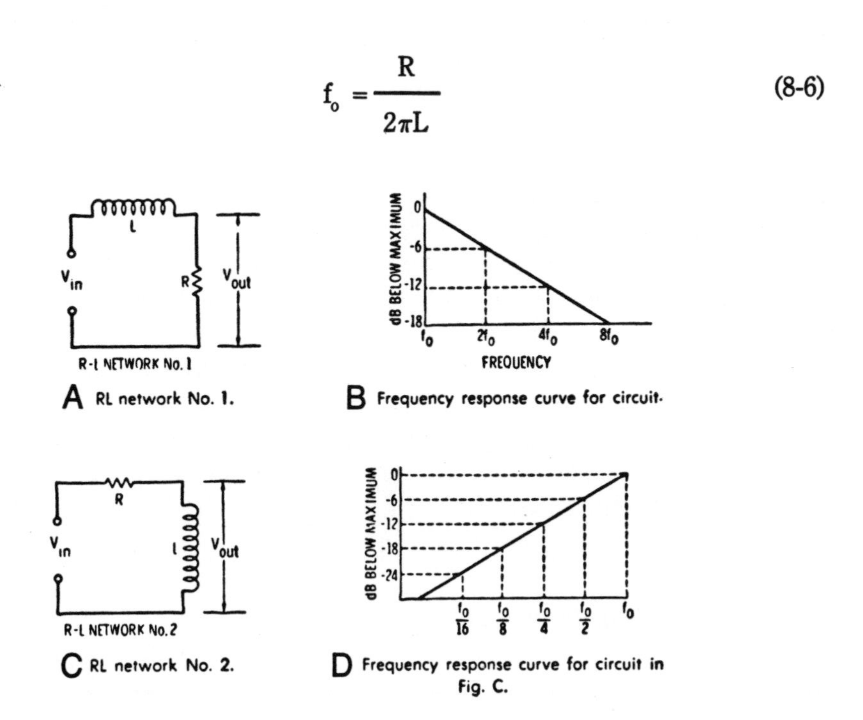


Fig. 8-6. Somewhat better filtering can be accomplished with RL networks, as shown by their characteristic curves.

A transformer is an inductance. The frequency rolls off at both ends of the band because of the transformer in a circuit. The corner frequencies can be found with the help of Fig. 8-7 and Equation 8-6.

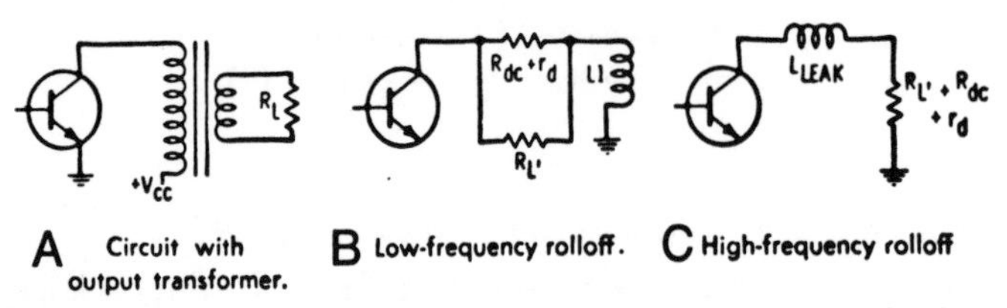

Fig. 8-7. Diagrams illustrating the characteristics of an output transformer circuit.

The major factors limiting the low frequencies are the inductance of the primary of the transformer and the load resistor, $R_L$, as reflected into the primary. The inductance of the primary winding of the transformer is L1. The inductance of a coil, when reduced by the presence of dc current in the winding, is known as the incremental inductance. Substitute the incremental inductance for L into Equation 8-6 when determining a corner frequency.

Let $R_L'$ be the resistance reflected into the primary due to $R_L$. $R_L' = (N1/N2)^2R_L$, where N1/N2 is the turns ratio of the primary to the secondary of the transformer. $R_p$ is all the resistance in the primary circuit. It is the sum of all the resistance presented to the transformer by the transistor, and the dc resistance, $R_{dc}$, in the primary winding. In this case, it is about equal to $r_d$. $R_L'$ in parallel with $r_d + R_{dc}$ in conjunction with the incremental inductance, are the components determining the low-frequency characteristics of the circuit. The parallel resistance combination should be substituted for R in Equation 8-6. L in the equation is the incremental inductance. The equivalent circuit is shown in Fig. 8-6C and the curve is that shown in Fig. 8-6D.

To determine the characteristics at the high-frequency end of the spectrum, the leakage inductance must be considered. It is a stray inductance in series with the primary winding of the transformer. The leakage inductance is measured by checking the primary inductance when the secondary winding is shorted.

Add $R_L'$ to $r_d$ and $R_{dc}$. Substitute this for R and the leakage inductance for L in Equation 8-6 to determine the corner frequency at the high end of the band. The equivalent circuit is in Fig. 8-6A and the curve is in Fig. 8-6B.

## CONSTANT-K AND M-DERIVED FILTERS

Filters are used in various circuits to either favor or exclude one frequency or a group of frequencies. They can readily be adapted to transistor circuits.

The constant-k and m-derived filters involve a considerable number of mathematical derivations. The components to be used in a circuit, the frequency characteristics, and how to use these filters to best advantage in a circuit, are all the data the designer requires.

Low-pass, high-pass, band-pass and band rejection filter circuits can be designed for use in the various applications. In each case, the m-derived filter has a sharper roll-off characteristic than does the constant-k filter. Design details are presented in the technical literature and will not be discussed here.

## CROSSOVER NETWORKS

In audio applications, it is frequently desirable to split the frequency band at some point. Filters are used to separate the low-frequency end of the audio spectrum from the higher frequencies. High- and low-pass filters are used to perform this function. The output from the low-pass filter is fed either to an amplifier designed to deliver power at these frequencies, or fed directly to a low-frequency loudspeaker known as a "woofer." Similarly, the high frequencies are fed either to an amplifier designed for that end of the audio band or fed directly to the speaker for reproducing the high end of the band, the "tweeter." The overall spectrum is reproduced by a loudspeaker system consisting of both a woofer and a tweeter. The filter system is further complicated when a three-way loudspeaker system is used by the addition of a midrange transducer to the woofer and tweeter combination. Should three loudspeakers be used in a system, the frequency ranges required from the woofer and tweeter are reduced so that each loudspeaker can perform its individual reproducing function better.

Crossover networks do not have to be sharp cutoff filters. Gradual frequency rolloff is frequently desired. Low Q inductors are thus used in filter circuits for this application. When the network is connected directly to the loudspeaker, large capacitors should be used. Examples of circuits using various filter arrangements are shown in Fig. 8-8. Sharpness of rolloff increases as we progress from circuit A through circuit C for arrangements where the output and input grounds are common, and from circuit D through circuit F where output and input grounds are not common. Formulas used to determine the value of the components are shown next to the components in the various circuit drawings. When the filter is between the power output stages of the audio amplifier and the loudspeaker, R in each circuit is the impedance of the loudspeaker at the chosen crossover frequency. Should amplifiers be connected to the outputs of the filter rather than a loudspeaker system, R is the input impedances of the amplifiers fed by the filter. The symbol for the crossover frequency is f.

## MODERN FILTERS

New approximate techniques have been developed to design filters using charts to simplify design procedures. Three different types of passive filters are currently in favor, each with a different group of characteristics.

### Passive Filters

The *Butterworth filter* has a smooth response in the passband and stopband. However, transition per number of components in the circuit, from the passband to the stopband, is not particularly sharp.

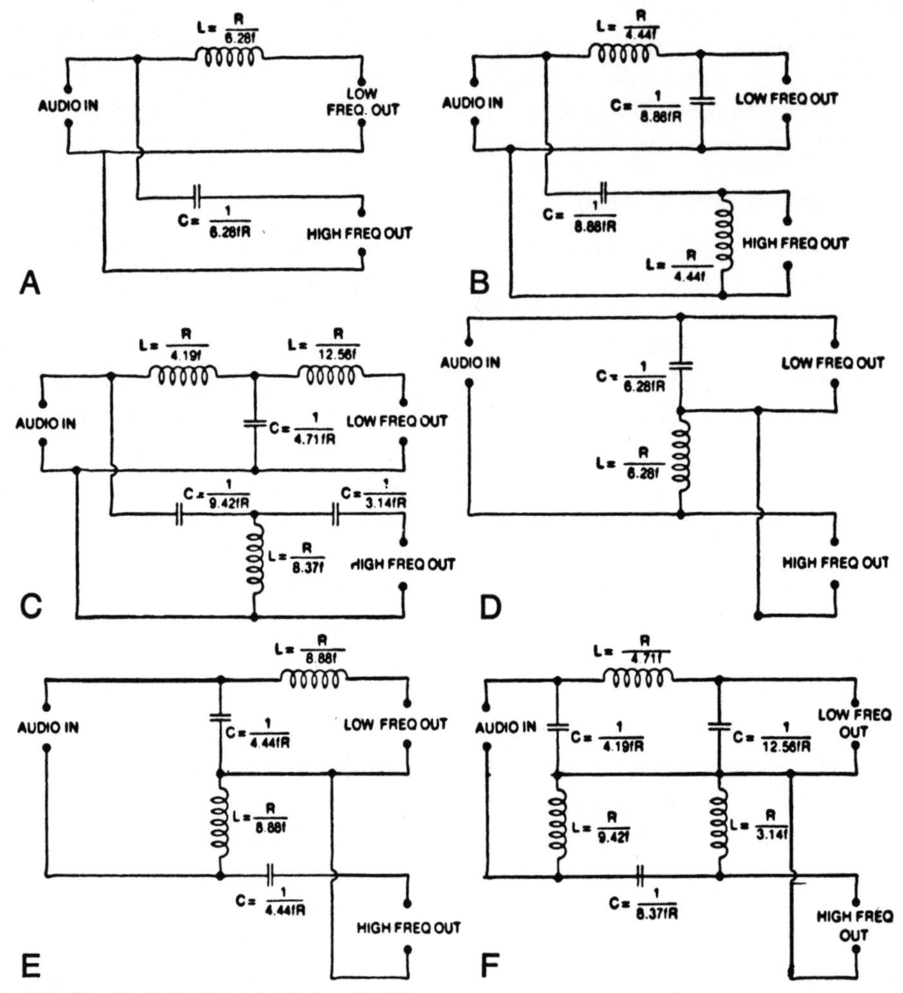

Fig. 8-8. Circuits featuring crossover networks.

The *Chebyshev filter* has a smooth response in the stopband, but there is ripple in the passband. Hence, not all frequencies in the passband pass through the filter without some attenuation. However, the frequency cutoff is sharper than in the Butterworth design.

The *Cauer (elliptic function) filter* has ripple in both the passband and the stopband. Here the transition between the two bands per number of components used is the sharpest.

The design of each of these filters is beyond the scope of this book. Detailed procedures and tables are available in the literature.

### Active Filters

Filters are frequently placed into several feedback loops around amplifier stages.

These are referred to as *active filters*. Some advantages over their passive counterparts are the possibility of extremely sharp transition between the passband and stopband sections of the response, the elimination of the need for inductors, and so on. Formulas and tables for design of active filters can be found in many of the publications available from filter manufacturers and in engineering manuals.

Drawings of low-pass and high-pass filters using the op amp are shown in Fig. 8-9A and Fig. 8-9B, respectively, along with the frequency, $f_c$, at which the passband gain has rolled off 3 dB. The bandpass filter circuit and the formula to determine the center frequency, are shown in Fig. 8-9C.

Sharpness of rolloff can be improved tremendously be placing several active filters into an arrangement where one filter circuit follows the other. If a properly chosen low-pass and high-pass filter are placed into one circuit, the net result can be a band-elimination arrangement.

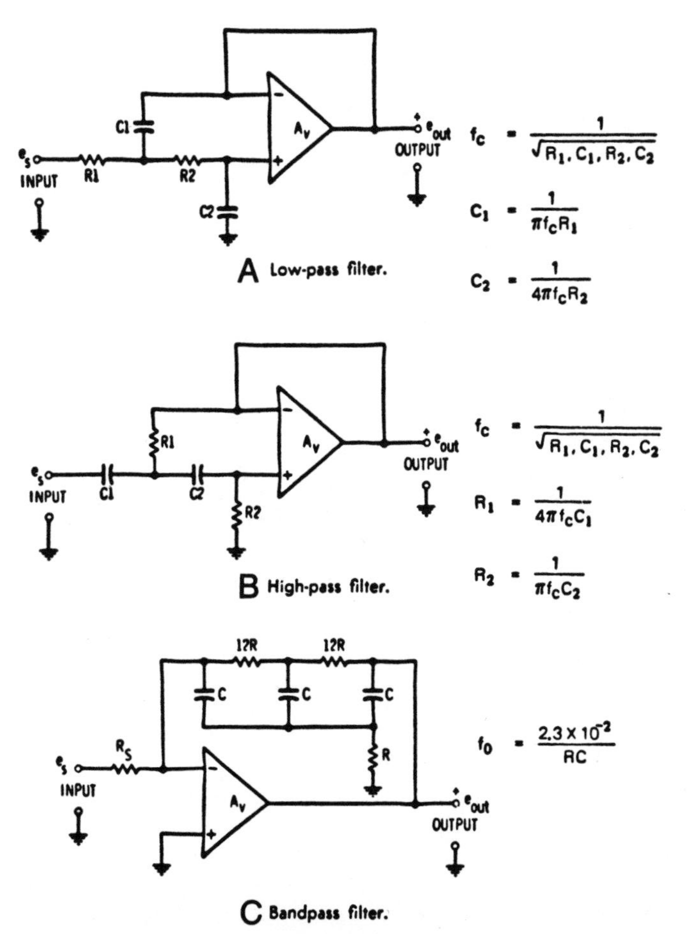


Fig. 8-9. Circuits using RC filter networks for feedback. (R is in ohms, and C is in farads.)

## FET Filters

Bipolar transistors load filter circuits, changing the $f_o$ frequency previously determined using the various formulas, and possibly also change the entire rolloff characteristic. The high input impedance of the FET is the ideal load on these filters for only then are the characteristics of the filters pretty similar to those derived from the calculations.

The twin-T and Wien-bridge notch or band-rejection filters have sharp response curves to attenuate a specific band of frequencies while passing all others. These filters are ideally suited to feed the high impedance at the input of an FET. Circuits are drawn in Fig. 8-10. The equation for $f_o$, the frequency with the greatest attenuation in both

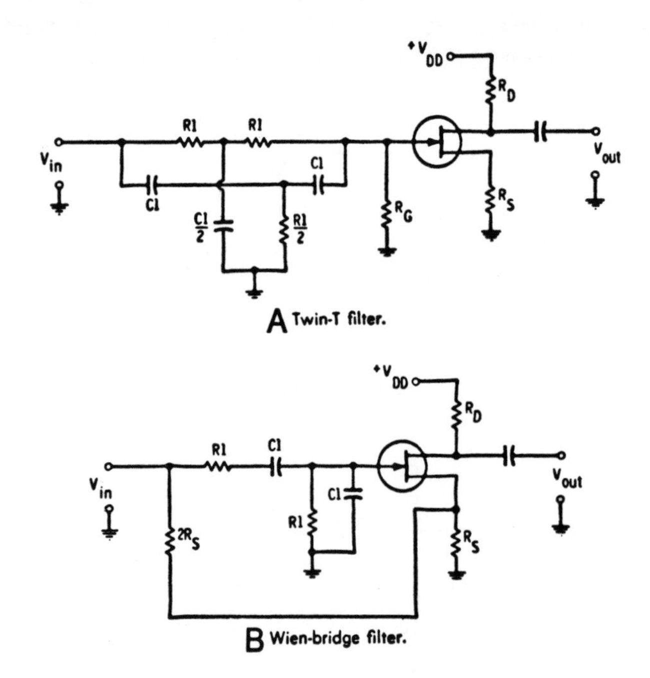


Fig. 8-10. Two typical notch, or band-rejection filters.

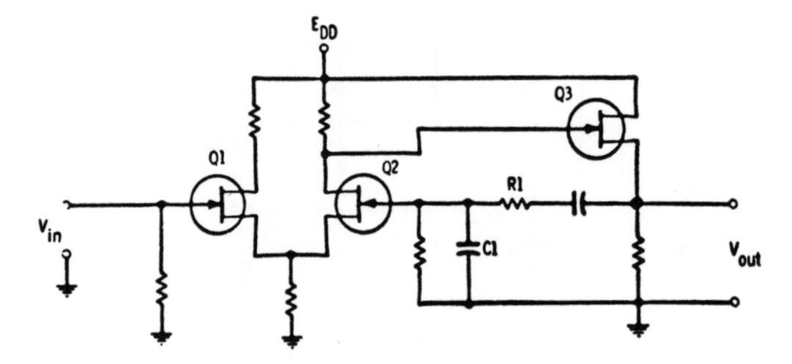


Fig. 8-11. An active high-pass filter circuit.

circuits, is $f_o = 1/6.28(R1)(C1)$. In Fig. 8-10B, $R_S$ should be small. This order of magnitude dictates that the input is best driven by a low impedance such as an emitter or source follower.

The FET is also ideal for use in active filter circuit arrangements. A possible high-pass filter circuit is shown in Fig. 8-11. In this circuit, a low-pass network consisting of R1-C1 is inserted in the feedback circuit between the output and the differential amplifier consisting of Q1 and Q2. Because more of the low frequencies are fed back than the high frequencies, the low frequencies are the primary ones to be attenuated. Similar arrangements can be drawn for the low-pass, bandpass, and band-rejection filters.

# 9

# Power Amplifiers

Small-signal equivalent circuits were presented in previous chapters. Many of the calculations performed can be accomplished without fully understanding the operation of the circuit. Here, we discuss the large-signal amplifier. Nebulous formulas cannot be applied. Information presented here is probably the most significant in design procedures.

Two groups of information are presented in this chapter.

- The application of transistors to power amplifiers.
- Limitations of the transistor, thermal considerations and the determination of heatsink requirements.

## CLASS-A AMPLIFIERS

Class-A amplifiers are devices in which there is always collector current regardless of the time in the cycle of the applied signal. This type of amplifier was discussed under small-signal devices. A typical class-A amplifier is shown in Fig. 1-A. The maximum power limitations and the load line are shown in Fig. 9-1B.

A transistor is capable of dissipating a specific maximum amount of power, $P_{CEM}$, under a particular set of conditions. Power, as always, is the product of the current flowing through a device and the voltage across it. Should a transistor be capable of dissipating a maximum of $P_{CEM}$ watts, the product of $V_{CE}$ and $I_C$ must never exceed $P_{CEM}$. A

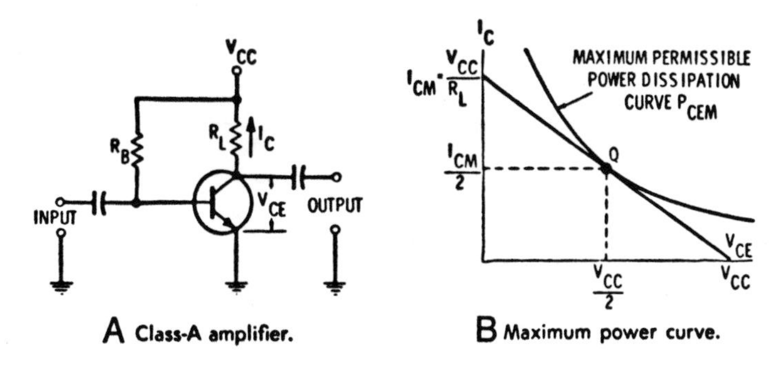


Fig. 9-1. A Class-A amplifier circuit and its response curve.

hyperbolic curve plotting all points where $V_{CE} \times I_C = P_{CEM}$ can be drawn on the set of collector characteristic curves, as shown in Fig. 9-1B. (The actual base current curves have been omitted so that the drawing will not be cluttered.) If the transistor is to operate within its power rating, the load line of the circuit must be below the $P_{CEM}$ curve and must never cross it. This statement will be modified in the discussion of class-B and class-AB power amplifiers.

In Fig. 9-1B, a load line that barely touches the hyperbola has been drawn. It connects a point equal to the supply voltage, $V_{CC}$, on the horizontal axis with a point equal to the collector current, $I_{CM} = V_{CC}/R_L$, on the vertical axis. Any load line that touches the hyperbola at one point, regardless of the size of $R_L$ or $V_{CC}$, must touch the $P_{CEM}$ hyperbola at $V_{CC}/2$ and $I_{CM}/2$. Since the load line does not touch the hyperbola at any other point, the maximum power, $P_{diss}$ (max), is dissipated by the transistor when the voltage at the collector is $V_{CC}/2$ and the collector current is $I_{CM}/2$.

$$P_{diss}(\text{max}) = \frac{V_{CC}}{2} \times \frac{I_{CM}}{2} = \frac{V_{CC}I_{CM}}{4} = \frac{V_{CC}^2}{4R_L} = \frac{I_{CM}^2R_L}{4} \qquad (9\text{-}1)$$

$P_{diss}$(max) is equal to $P_{CEM}$ when the load line is tangent to the maximum power dissipation hyperbola.

Now let us assume that a sine wave was fed to the amplifier. The construction in Fig. 9-2 shows the collector characteristics of a hypothetical transistor, the load line, and the effect of feeding a sine wave to the base of the transistor. The quiescent base current is set at $I_{B4}$. The sine wave swings the base current from $I_{B1}$ to $I_{B7}$ around $I_{B4}$. The collector current and voltage change during the cycle with the base current. These factors are tied to the load line.

As an example, at the instant the input signal swings the base current to $I_{B6}$, the collector current is at $I_{C6}$ because the load line and $I_{B6}$ intersect at this collector current. If a vertical line were extended from this point of intersection, we find it would intersect the $V_{CE}$ axis at $V_{CE6}$. This is the voltage between the emitter and collector, across the

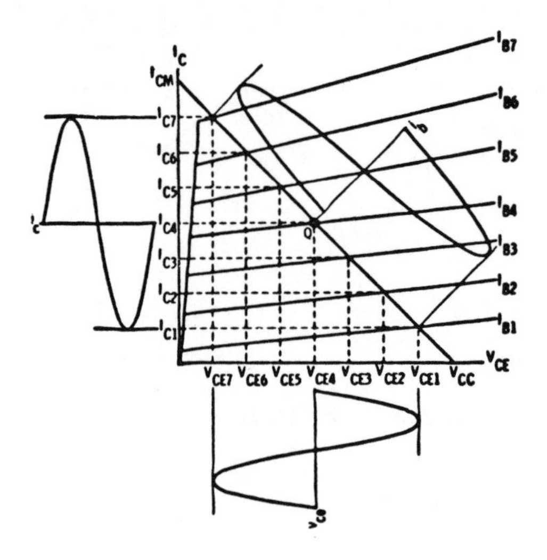


Fig. 9-2. The output characteristics of a Class-A amplifier.

transistor, when $I_B = I_{B6}$. The difference between this and the supply voltage is across the load.

Suppose that the input current is at the instant in the cycle when the base current is $I_{B2}$. Draw a horizontal and a vertical line from the point at which $I_{B2}$ and the load line intersect. The collector current, read on the vertical axis, is $I_{C2}$. The emitter-collector voltage across the transistor, read on the horizontal axis, is $V_{CE2}$. $V_{CC} - V_{CE2}$ is the voltage across the load at this moment.

As the sine wave swings the base current from $I_{B4}$ to $I_{B7}$, it swings the collector and load current from $I_{C4}$ to $I_{C7}$. ($I_{C7}$ is about equal to $I_{CM}$). At the same time, the voltage across the transistor swings from $V_{CE4}$ to $V_{CE7}$. ($V_{CE7}$ is essentially zero.) The voltage across the load swings from $V_{CC} - V_{CE4}$ to $V_{CC} - V_{CE7}$. ($V_{CC} - V_{CE7}$ is about equal to $V_{CC}$.)

Similarly, when the sine wave goes from $I_{B4}$ to $I_{B1}$, the collector and load current swing from $I_{C4}$ to $I_{C1}$. ($I_{C1}$ is about equal to zero) and the collector-emitter voltage swings from $V_{CE4}$ to $V_{CE1}$. ($V_{CE1}$ is about equal to $V_{CC}$.) The voltage across the load resistor swings from $V_{CC} - V_{CE4}$ to $V_{CC} - V_{CE1}$. Since $V_{CE1}$ is about equal to $V_{CC}$, $V_{CC} - V_{CE1}$ is very close to zero.

The overall base current swing is from $I_{B1}$ to $I_{B7}$. During this excursion, the approximate swing of the collector and load current is from zero to $I_{CM}$. When the collector current is zero, there is no voltage drop across the load resistor and the entire supply voltage is across the transistor. Voltage across the transistor swings from $V_{CC}$ to zero as the collector current swings from zero to $I_{CM}$. Voltage across the load swings from zero to $V_{CC}$ as the load or collector current swings from zero to $I_{CM}$.

The peak-to-peak current swing through the load cannot exceed $I_{CM}$. The rms current is $I_{Cm}/2\sqrt{2}$ and the rms voltage is $V_{CC}/2\sqrt{2}$. The power delivered to the load is the rms voltage multiplied by the rms current, or:

$$P_{R_L} = \frac{V_{CC}}{2\sqrt{2}} = \frac{I_{CM}}{2\sqrt{2}} = \frac{V_{CC}I_{CM}}{8} = \frac{V_{CC}^2}{8R_L} \times \frac{I_{CM}^2R_L}{8} \qquad (9\text{-}2)$$

This is the maximum sine wave power the transistor can deliver using the circuit in Fig. 9-1A. It is equal to half the maximum power the transistor may dissipate.

If there is no distortion, the quiescent collector current is $I_{C4}$, and the quiescent collector-emitter voltage is $V_{CE4}$. $I_{C4} = I_{CM}/2$ and $V_{CE4} = V_{CC}/2$, the average values of the dc current and voltage flowing through the circuit.

The power demanded from the power supply consists of the sum of the power dissipated by the transistor, the maximum instantaneous power dissipated by the load, and the power supplied to the base bias circuit. Compared to the first two, the last factor is negligible. Using only the first two items, the total input power delivered by the supply is:

$$P_{CC} = \left(\frac{I_{CM}}{2}\right)\left(\frac{V_{CC}}{2}\right) + \left(\frac{I_{CM}}{2}\right)^2 R_L = \frac{I_{CM}^2R_L}{2} = \frac{V_{CC}^2}{2R_L} \qquad (9\text{-}3)$$

The efficiency of the circuit is the ratio of the ac power that can be delivered by the circuit to the amount of power that must be supplied by the power source. When multiplied by 100, efficiency is expressed as a percentage:

$$\%\text{eff.} = 100\left(\frac{P_{R_L}}{P_{CC}}\right) = \left(\frac{V_{CC}^2/8R_L}{V_{CC}^2/2R_L}\right) 100 = 25\% \qquad (9\text{-}4)$$

The maximum efficiency of this circuit is 25 percent. It can be improved if the power dissipated by the load were reduced. In Equation 10-3, it would be desirable to eliminate the $(I_{CM}/2)^2R_L$ term. This power is dissipated in the load resistor and contributes nothing to the output. The load resistor in the collector circuit can be replaced by the primary winding of a transformer with the load connected across its secondary winding. A circuit of this type is shown in Fig. 9-3A.

A transformer that has N1 turns in the primary is magnetically coupled to the N2 turns in the secondary. If a voltage, V1, is placed across the primary, a voltage V2,

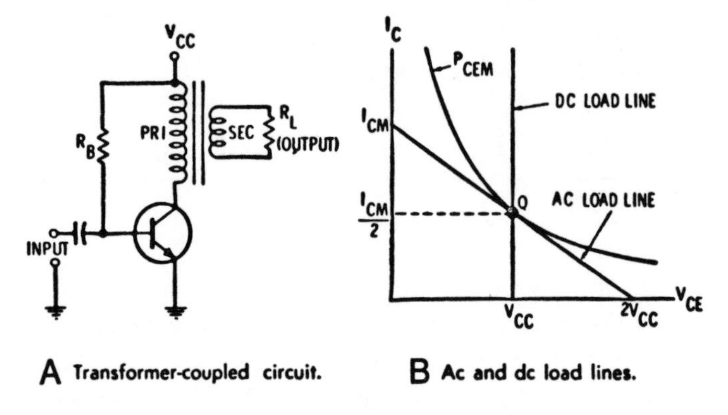


A Transformer-coupled circuit. B Ac and dc load lines.

Fig. 9-3. A transformer-coupled circuit and its response curve.

will appear across the secondary. Ignoring all transformer losses, the voltage and turns ratios are related by the equation:

$$\frac{V1}{V2} = \frac{N1}{N2} \tag{9-5}$$

A resistor load in the secondary, $R_L$, appears as a resistance, $R_L$, reflected into the primary. $R_L$ is equal to $R_L$ multiplied by the square of the turns ratio of the transformer, or:

$$R_L' = \left(\frac{N1}{N2}\right)^2 R_L \tag{9-6}$$

Let us now see just what happens when a perfect transformer with a zero dc resistance, is wired into the collector circuit of a transistor. A load line for this is drawn in Fig. 9-3B. One point on the dc load line is at $I_C = 0$ and $V_{CE} = V_{CC}$. The other point is at $V_{CE} = 0$ and $I_C = V_{CE}/R_P$. ($R_P$ is the dc resistance of the primary winding of the transformer.) Since $R_P$ here equals 0, $I_C$ is infinite. This determines a second point on the load line: $V_{CE} = 0$, $I_C$, $= \infty$. For all practical considerations, the line extends straight up and parallel to the collector current axis. Normally, there is some resistance in the transformer winding and the second point on the dc load line is finite.

Quiescent conditions are determined from the dc portion of the circuit, which has nothing to do with the ac load. Quiescent collector voltage is $V_{CC}$, as there is no voltage drop across the perfect transformer. The quiescent collector current is on the load line at Q. Maximum power can be delivered to the load when Q is at the intersection of the dc load line and the $P_{CEM}$ curve.

The ac load line is dependent on all the impedance in the primary of the transformer. If a perfect transformer is used, it is $R_L'$, determined from Equation 9-6.

The ac load line can be drawn on the curve, as follows. Refer to Fig. 9-4 in which the dc load line is not perfectly vertical. It is assumed that there is a dc resistance in the primary winding of the transformer.

1. Draw a dc load line using methods previously employed by connecting $V_{CC}$ on the $V_{CE}$ axis to $V_C/R_{dc}$ on the $I_C$ axis. $R_{dc}$ is the sum of all dc resistors in the collector (and emitter) circuit of the transistor.
2. Select a quiescent point on the dc load line and mark the point Q.
3. Draw a vertical line from Q to the horizontal axis. It intersects the voltage axis at $V_{CEX}$.
4. Determine the resistance reflected from the secondary of the transformer into the primary winding using Equation 9-6.
5. Draw an ac load line. Here, $V_{CEY} - V_{CEX} = I_{CX}R_L'$. Plot the point $V_{CEY} = I_{CX}R_L' + V_{CEX}$ on the horizontal axis. Connect Q to $V_{CEY}$ and extend the line to the vertical axis. This is the ac load line.

Several practical aspects should be noted in the design of this circuit. First, the dc load line in Fig. 9-4 is drawn in the usual way. One point must be $V_{CC}$. If the second point is too high on the vertical axis, falling off the graph, it may be calculated for another value of $V_{CE}$. For example, if $V_{CE}$ is $V_{CEX}$, the dc collector current at this point is $I_{CX} = (V_{CC} - V_{CEX})/R_{dc}$.

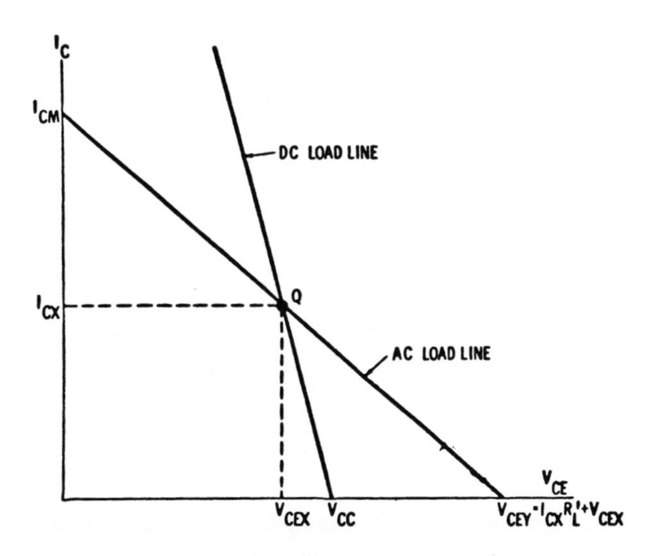


Fig. 9-4. One method for plotting an ac load line.

Next, the ac load line must pass through the Q point, which is on the dc load line. The Q point should be chosen so that the ac load line does not cross the peak power dissipation curve at any time.

As a third factor, the ac load resistance is not merely the reflected resistance, $R_L'$; it is equal to the sum of the reflected resistance, the dc resistance of the primary winding of the transformer, and any other resistance in the collector and emitter leads of the transistor. Furthermore, the secondary winding of the transformer has been assumed to lack resistance. This is not absolutely true. The ac reflected resistance from the secondary winding, $R_L'$, is not due merely to $R_L$, but to the sum of $R_L$ and the resistance of the secondary winding of the transformer.

Finally, the ac output travels up and down the ac load line just as it traveled along the load line in Fig. 9-2. If the output signal from the transistor is to be symmetrical, the voltage and current must swing equal amounts on either side of the quiescent value. The ac load line is usually chosen so that the one point is on the horizontal axis where $V_{CEY} = 2V_{CEX}$ in Fig. 9-4.

Referring to Fig. 9-3, the maximum possible voltage swing is $2V_{CC}$ and the maximum current swing is $I_{CM}$. Because the rms ac voltage is $2V_{CC}/2\sqrt{2}$, and the rms ac current is $I_{CM}/2\sqrt{2}$, the power delivered to the load is:

$$P_{R_L} = \left(\frac{2V_{CC}}{2\sqrt{2}}\right)\left(\frac{I_{CM}}{2\sqrt{2}}\right) = \frac{V_{CC}I_{CM}}{4} = \frac{V_{CC}^2}{4R_L'} = \frac{I_{CM}^2R_L'}{4} \tag{9-7}$$

while the average dc power dissipate by the transistor, or the power dissipated at the Q-point is:

$$P_{diss} = V_{CC}\left(\frac{I_{CM}}{2}\right) = \frac{V_{CC}I_{CM}}{2} \tag{9-8}$$

When comparing Equation 9-7 with 9-8, it becomes obvious that the transistor can deliver, at best, a power equal to half the power it dissipates. This is true of all class-A arrangements.

The power demanded from the supply is:

$$P_{CC} = \left(\frac{I_{CM}}{2}\right)V_{CC} \tag{9-9}$$

since no additional power is lost in the load resistor and the power dissipated in the base circuit is negligible. The percent efficiency is now 50%.

Class-A amplifiers, as a rule, should be distortion-free. The impedance or resistance presented to the base-emitter junction of the output transistor (source impedance) has a decided effect on the distortion. The most practical method for determining the optimum source impedance, is by trial and error. Zero ohms in series with the generator (a perfect voltage source) is usually optimum for minimum distortion. In some cases, however, the preferable source impedance is quite large.

The input signal may force $I_C$ and $V_{CE}$ to swing over the full length of the load line. There will be distortion at the extreme ends. The useful swing cannot extend to below $I_{CEO}$ at one end of the load line. Reasonable swing is limited to the saturation voltage at the other extreme. This is the lowest collector voltage that can be used at a specific collector current, to reproduce the linear signal present at the input.

## CLASS-B AMPLIFIERS

In this type of amplifier arrangement, each transistor conducts for one half-cycle only. Two transistors in a push-pull circuit are required to reproduce an entire sine wave.

A circuit is shown in Fig. 9-5A. A sine wave, fed to the driver transformer, appears in the secondary winding. An equal voltage is across each half of the secondary. Voltage at the base of Q1 is 180° out of phase with the voltage at the base of Q2. Each transistor conducts only during the portion of the cycle when its base is positive with respect to the emitter. Currents flowing through the collectors are shown as $I_{C1}$ and $I_{C2}$. They combine within the output transformer to recreate the composite signal which is delivered to the load, $R_L$.

The voltage between the collector and ground of the transistor that is turned off is equal to $V_{CC}$ plus the signal voltage across the half of the output transformer to which it is connected. The maximum voltage across the transistor during the time it is turned off is $2V_{CC}$. This voltage can exist across the transistor even when it is turned on because of the phase shift of a reactive load. In class-AB operation, $2V_{CC}$ can be across the transistor despite the existence of a quiescent idling current.

Voltage between the base and collector is greater than $2V_{CC}$. It is equal to $2V_{CC}$ plus the peak voltage across the base-emitter junction, when the transistor is not conducting. This latter voltage is determined by the maximum signal voltage at the base.

The load reflected from the secondary to the entire primary of the output transformer is $R_L'$, as above. For half the primary, the reflected load is $R_L''$. Because the ratio of the reflected load to the actual load is proportional to the turns ratio squared of the transformer, $4R_L'' = R_L'$. Resistance seen by one transistor is one-fourth the resistance seen by the two transistors combined.

In analysis and design of a push-pull circuit, it is convenient to consider one-half of the circuit at a time. This is drawn in Fig. 9-5B. The load that each transistor sees is $R_L''$. The load seen by both transistors is $4R_L''$. The load lines in Fig. 9-5C are drawn as before. The dc load line is drawn vertically starting at $V_{CC}$ on the horizontal axis. It is assumed that there is zero dc resistance in the transformer windings.

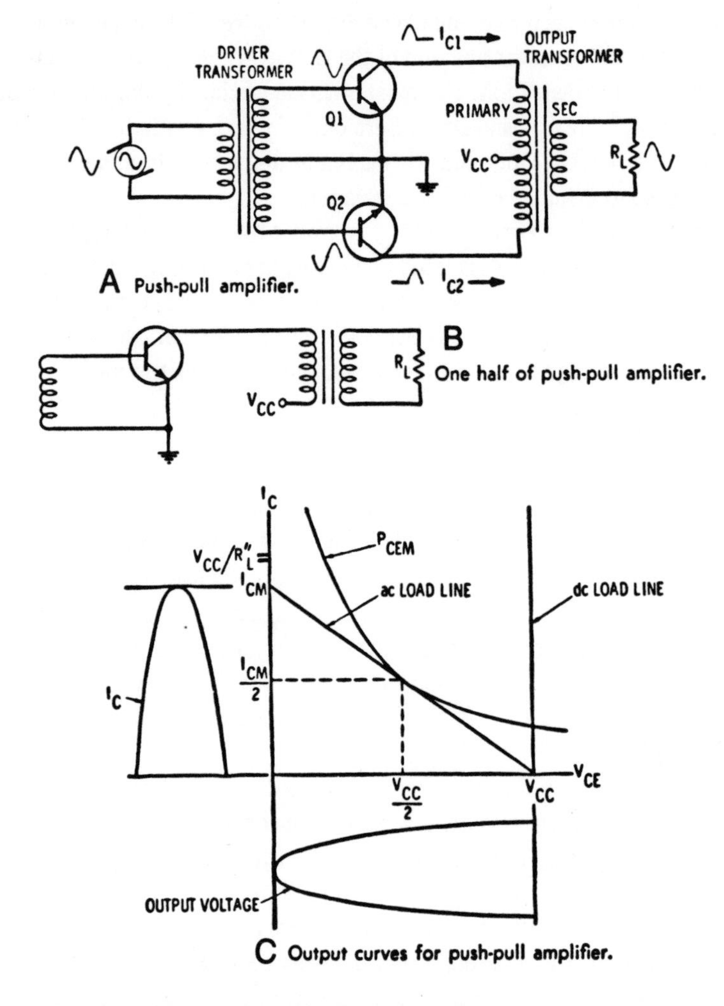


Fig. 9-5. A push-pull amplifier circuit and its response curves.

The ac load line for half of the push-pull circuit starts at $V_{CC}$ on the $V_{CE}$ axis. A second point is $I_{CM} = V_{CC}/R_L''$ on the vertical axis. The line connecting these two points is the ac load line. If the load line is tangent to the hyperbola, it touches it at $V_{CC}/2$ and $I_{CM}/2$.

Assuming a sinusoidal input, each transistor conducts during the half-cycle when its base is positive with respect to the emitter. The collector current swing in each transistor is as shown. The collector-emitter voltage is drawn next to the $V_{CE}$ axis.

The output voltage and collector current (for each transistor) are each one-half of a sinusoidal cycle. The rms value of half a sinusoidal cycle of voltage or current is the peak voltage or current divided by two. The average value is the peak voltage or current divided by $\pi$.

The maximum power in the half sine wave that can be delivered to the load by one transistor of a push-pull pair is:

$$P_{R_L}'' = I_{rms}V_{rms} = \left(\frac{I_{CM}}{2}\right)\left(\frac{V_{CC}}{2}\right) = \left(\frac{I_{CM}}{2}\right)^2 R_L'' = \frac{(V_{CC}/2^2}{R''_L} = \frac{V_{CC}^2}{4R_L''} \quad (9\text{-}10)$$

This assumes that the voltage swings from $V_{CC}$ to 0 volts and the current swings from zero to $1_{CM}$.

The maximum power dissipated by the transistor is at the center of the current or voltage swing, where the load line either touches or comes closest to the $P_{CEM}$ hyperbola. This point is where $I_C = I_{CM}/2$ and $V_{CE} = V_{CC}/2$. The power is:

$$P_{diss}(max) = \left(\frac{I_{CM}}{2}\right)\left(\frac{V_{CC}}{2}\right) = (I_{CM})^2 \frac{R_L''}{4} = \frac{V_{CC}^2}{4R_L''} \quad (9\text{-}11)$$

$P_{R_L}''$ is equal to $P_{diss}(max)$. This means that the maximum sinusoidal rms power a transistor delivers to a load is equal to the maximum power the transistor dissipates at the one point in the cycle when the voltage and current are at $V_{CC}/2$ and $I_{CM}/2$, respectively.

Current delivered by the power supply to the transistor and its load is the average dc current in one half-cycle. This is $I_{CM}/\pi$. The supply voltage is $V_{CC}$. The power delivered to the circuit by the supply is:

$$P_{CC} = \frac{I_{CM}}{\pi} V_{CC} = \frac{V_{CC}^2}{\pi R_L''} = \frac{I_{Cm}^2}{\pi} R_L'' \quad (9\text{-}12)$$

so that the efficiency expressed in percent is:

$$\%eff = \left(\frac{P_{R_L}''}{P_{CC}}\right) 100 = \frac{(I_{CM}V_{CC}/4)}{I_{CM}V_{CC}/\pi)} 100 = \frac{\pi}{4} 100 = 78.5\% \quad (9\text{-}13)$$

In class-B operation, the load line may be permitted to cross the $P_{CEM}$ curve. The average power dissipated by the transistor, $P_{diss}$, should be averaged over the half-cycle. However, the $P_{CEM}$ rating of a transistor must not be exceeded by $P_{diss}$ for any size or type of signal.

If the input signal causes the output voltage across the transistor to swing over a portion of $V_{CC}$, the output power is less than maximum. Letting k be a fractional part of the possible full range voltage swing, power dissipated by the transistor is:

$$P_{diss} = \frac{kV_{CC}^2}{\pi R_L''} - \frac{k^2 VCC2}{4R_L''} \qquad (9\text{-}14)$$

From Equation 9-10, the power delivered to the load when a transistor swings its full range is $V_{CC}^2/4R_L''$. The average power dissipated by the transistor over the half-cycle in this case can be shown from Equation 9-14 to be $(V_{CC}^2/R_L'')(1/\pi - 1/4)$, because k equals 1. This is also the average power dissipated by one transistor of the push-pull pair over a complete cycle for a full swing, because in a push-pull arrangement power is dissipated by one transistor of the pair during one half-cycle only and no power is dissipated during the second half-cycle. The transistor may dissipate a higher average power over the cycle if the swing is less than from zero to $V_{CC}$.

The maximum average power is dissipated when the voltage and current swing of the transistor is such that the output power it delivers is about 40 percent of the transistor's maximum capability. The maximum average power dissipated, $P_{diss}$, when the delivered output power is about 40 percent of the maximum, is $V_{CC}^2/\pi^2 R_L''$.

In other words, if the signal is sinusoidal and the output swing along the load line is, at its maximum, from 0 to $V_{CC}$ and from $I_{CM}$ to 0, the transistor delivers 2.5 times the maximum average power it can dissipate in a sinusoidal cycle. In any cycle, the maximum average power a transistor can dissipate safely, is less than $P_{CEM}$ watts. The load line may be above the $P_{CEM}$ hyperbola if the average power dissipated throughout the cycle is equal to or less than $P_{diss} = P_{CEM}$. This applies only to pure sinusoidal power and not to any other waveforms. Do not confuse the maximum *average* power a transistor dissipates *over the entire cycle* with $P_{diss}$ (max) in Equation 9-11. Here it is the maximum power that a transistor can dissipate at a *particular* instant in the cycle.

The discussion on class-B amplifiers thus far concerns itself with only one transistor of the push-pull pair. For the complete push-pull circuit, double the numbers for the power dissipation, power output, and power delivered to the circuit. The power ratios remain unchanged.

## CLASS-AB AMPLIFIERS

Class-AB is a cross between class-A and class-B operation. The bias is so designed that each transistor conducts for less than a full cycle and more than half the cycle. The primary advantage of class-AB operation is to minimize distortion—specifically cross-over distortion.

A plot of the base-emitter voltage against the collector current for the 2N3055 is shown in Fig. 9-6. This describes the transconductance, $g_{fe}$, of the transistor. The dc

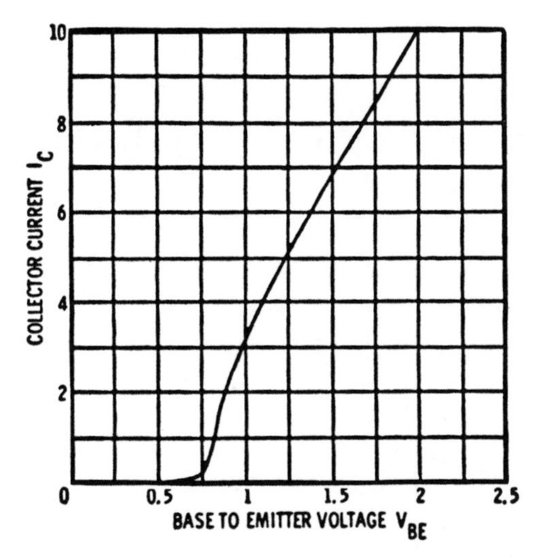


Fig. 9-6. Graph showing the transconductance curve for 2N3055 (courtesy of RCA).

transconductance at any one point, $g_{FE}$, is equal to:

$$g_{FE} = \frac{I_C}{V_{BE}} \tag{9-15}$$

so that, for example, at $V_{BE} = 1.5$ volts and $I_C = 6.7$ amps on the curves, $g_{FE} = 6.7$ amps/1.5 volts = 4.47 mhos. Thus, the dc collector current is 1.5 volts × 4.47 mhos when the base-emitter voltage is 1.5.

The ac transconductance is the variation of collector current and base-emitter voltage about one particular base voltage:

$$g_{fe} = \frac{\Delta I_C}{\Delta V_{BE}} \tag{9-16}$$

so that at $V_{BE} = 1.5$, assuming a base-emitter swing of ± 0.5 volts, $g_{fe}$ will be (10 amps − 2.9 amps)/(2 volts − 1volt) = 7.1 mhos.

The input signal to a power transistor is the base-emitter voltage, as discussed above. The collector current and output power are functions of the base-emitter voltage and transconductance. When two of the 2N3055 transconductance curves are placed back-to-back for the composite representation of the overall push-pull arrangement using two transistors, the output signal in class-B operation appears as shown in Fig. 9-7. There is no conduction when $V_{BE}$ is somewhat below 0.65 volt. Hence there is crossover distortion. If the transistor were biased to conduct a slight amount of quiescent or idling

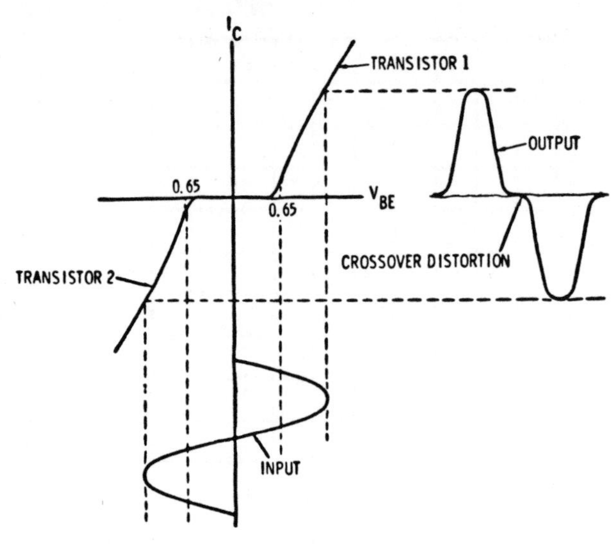


Fig. 9-7. Graph illustrating the effects of crossover distortion at the output.

collector current, the crossover distortion would be reduced or entirely eliminated. A composite curve for two transistors illustrating this is shown in Fig. 9-8.

The desirable minimum base-emitter voltage for the 2N3055 can be determined from the curve in Fig. 9-6. Extend the upper and straight portion of the curve to the axis. The point where it crosses the axis is the minimum desirable $V_{BE}$ or base-emitter volt-

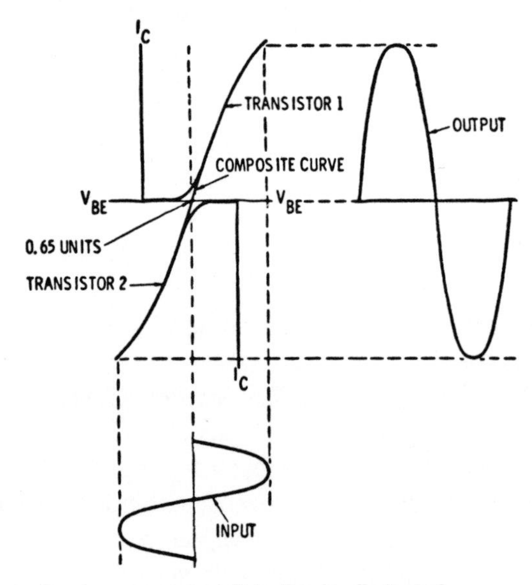


Fig. 9-8. Graph illustrating how crossover distortion is eliminated.

age can be read from the curve. Performance of the circuit can be improved if the collector idling current is raised to possibly 200 or 300 mA for large power transistors, although many manufacturers have found 10 to 25 mA satisfactory for the level of distortion they accept.

Distortion can be reduced further if the crossover point is not sharp, but curved. Diodes in series with the collector or emitter leads of the push-pull transistors will provide this curved characteristic.

Distortion in class-B or class-AB circuits is a function of the source impedance, as was the case with the class-A arrangement. The optimum impedance is close to zero and should be determined experimentally.

Output circuits used for class-AB and class-B amplifiers can involve input and output transformers. Transformerless amplifiers are most commonly used in high fidelity applications.

At this writing, the audio industry has more or less settled on three basic circuits used in an output-transformerless arrangement. The three circuits are analyzed in the following discussion.

The transistor is a power amplifier. Power must be fed to the input if power is to be delivered to a load at the output. Power is delivered to the output transistors from the preceding stage, the driver. The driver is not required to deliver as much power to the output transistors as they must deliver to the output load. Hence, the driver may be a smaller power device than the output stages. Small power transistors, often called voltage amplifiers, feed the drivers.

In all push-pull arrangements, the transistor circuit is designed so that each output device conducts primarily for one half-cycle. Phase inverters or complementary transistors are used to deliver signals to the output devices, so that they will conduct in this manner.

### Circuit Using a Driver Transformer

One arrangement uses a driver transformer for phase inversion. The basic circuit, shown in Fig. 9-9, uses two identical output transistors. Both are biased through the secondary windings of the driver transformer. Resistors $R_B$ and $R_X$ establish a quiescent collector current through the transistor so as to minimize crossover distortion.

Emitter resistors $R_{E1}$ and $R_{E2}$ serve several functions. While establishing local ac feedback in the emitter of each transistor, they help in dc bias stablization. A third function is to limit the transistor current should the output load, $R_L$, be shorted. The resistors are usually a small fraction of an ohm so that the power meant for the output load will not be wasted here. In single-ended class-A amplifiers, about 0.5 to 1 volt is usually developed across the emitter resistor at quiescent conditions. The push-pull amplifier is frequently designed so that a voltage within this range is developed across the resistor $R_{E1}$ or $R_{E2}$, when the output transistor delivers its peak current.

The two secondaries on the driver transformer are phased so that when signal is

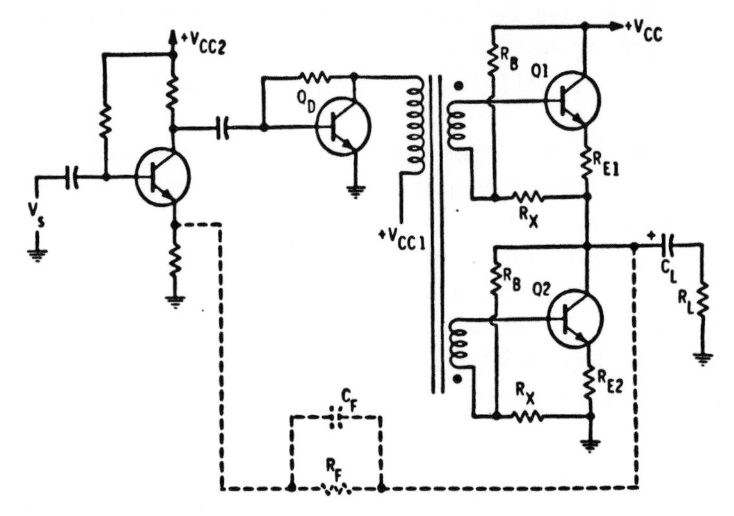


Fig. 9-9. Output circuit using a driver transformer for phase inversion.

applied, one output transistor will conduct while the other is idling. When a sine wave is applied, each transistor will conduct while the other is idling. When a sine wave is applied, each transistor conducts for one half-cycle, since the base-emitter junction is forward biased by the signal only for this period of time. The composite signal is reconstructed across $R_L$, the load.

A dot has been placed at one end of each secondary winding to indicate phasing. During the half of the cycle when the ends of the secondary windings with the dot are positive with respect to the other ends, only the upper transistor, Q1, conducts. The peak of the input signal may be large enough to drive the transistor into saturation. During this instant in the cycle, the voltage across Q1 is close to zero. The transistor may be considered as shorted and the top of $R_L$ is effectively connected to $+V_{CC}$. The peak voltage across Q2 is $V_{CC}$ minus the small voltage drop across Q1 and $R_{E1}$.

During the alternate half of the cycle, Q2 conducts. When Q2 is driven into saturation, the voltage across the load resistor is equal to the voltage drop across Q2 and $R_{E2}$. The peak voltage across Q1 when it does not conduct is $V_{CC}$ minus the small voltages across Q2 and $R_{E2}$.

The turns ratio of the transformer is designed so that a desirable ac load impedance is presented to the driver transistor, $Q_D$. Frequently, an impedance ratio of 9:1 is used. It should be modified if the distortion in the circuit is excessive.

The transformer should be capable of delivering enough power over the entire required frequency range, so that it will not limit the output. In some designs, the power the driver can deliver through the transformer to the output transistors, is limited to a specific value by the $+V_{CC1}$ supply voltage. This limits the amount of power the output stage can deliver to the load resistor or to an accidentally shorted output load. However, the protection this circuit affords is limited. If the supply voltage is made too small, the amplifier will not be capable of delivering its rated power into a normal load without distortion. If the supply voltage is too large, the protection is nonexistent. A true protective circuit usually has some device sensing the output current.

The amplifier is completed through the use of a feedback resistor, $R_F$, from the output back to the emitter of the first voltage amplifier. $C_F$ is used for phase correction and stability. The size of the capacitor can be determined experimentally. Feed a 10- to 20-kHz square wave to the input and note the output on a wideband oscilloscope. The capacitor is trimmed to the point at which there will be not ringing on the output signal while there is no rounding of the leading edge.

An alternate arrangement around the power output stages in Fig. 9-9 uses diodes to replace both $R_X$ resistors in the base circuits of Q1 and Q2. Diodes are also used to shunt emitter resistors $R_{E1}$ and $R_{E2}$. Cathodes of both diodes associated with Q2 are connected to ground and cathodes of diodes associated with Q1 are connected to the collector of Q2—the audio output. The diodes replacing $R_X$ are used to stabilize the quiescent transistor collector currents so that they will remain relatively constant despite variation of $V_{BE}$ with temperature. All resistors and diodes in the circuit are instrumental in establishing the idling current.

Diodes shunting $R_{E1}$ and $R_{E2}$ are virtually shorts across the respective emitter resistors when large current swings are present, so that the resistors will not limit the output power delivered to the load. Because of the presence of the shunting diodes, $R_{E1}$ and $R_{E2}$ can be made large without affecting the maximum output power.

### Quasi-Complementary Circuit

The conventional quasi-complementary amplifier is shown in Fig. 9-10. Q2 and Q4 form a Darlington amplifier. Q3 and Q5 are a complementary amplifier pair. $R_{B3} + R_{B2} + R_{X2}$ add up to the output load resistor of transistor Q1. $R_{X2}$ is small and may be considered as a short circuit in the ac portion of the analysis that follows. Note also that the voltage at X is $(1/2)V_{CC}$ when there is no signal applied to the amplifier.

Under quiescent conditions, the collector of Q1 is adjusted so that it is at the same potential as point X. When the collector at Q1 is positive with respect to X (due to an input signal), Q2 and Q4 conduct, and output power is developed across $R_L$. When an input signal swings the collector of Q1 negative with respect to X, Q3 and Q5 conduct. The two outputs combine across $R_L$ to recreate the input signal.

The dc collector currents flowing through Q4 and Q5 are determined by their respective base currents, which are, in turn, functions of the collector and base currents of Q2 and Q3, respectively.

The quiescent base-emitter currents of Q2 and Q3 are functions of the voltage developed across $R_{X2}$. This current will cause an operational shift toward class B when signals of different amplitudes drive the amplifier. The consequential crossover distortion

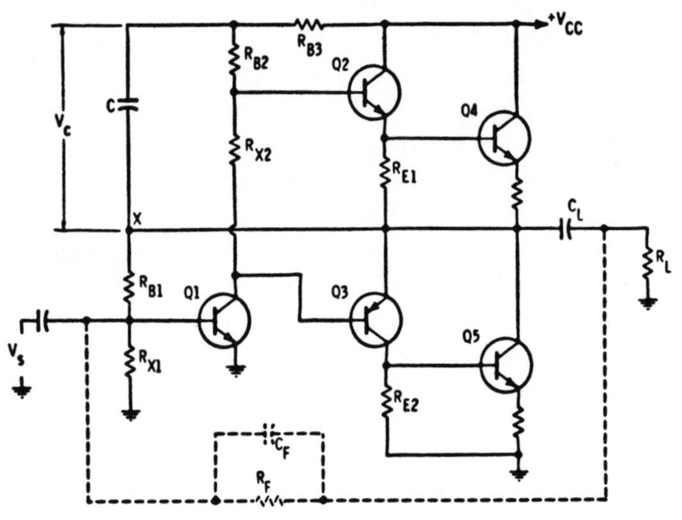

Fig. 9-10. A quasi-complementary amplifier.

can only be overcome by placing a considerable amount of feedback around the amplifier. In order to be capable of accommodating a large amount of feedback, the voltage gain of the circuit must be high. This can be accomplished by making the load resistor in the collector of Q1 large so that the ratio of the load resistor to the emitter resistance (ratio is voltage gain) is large. Capacitor C, in a bootstrap circuit, provides positive feedback so that $R_{B2}$ will appear to the signal much larger than it actually is.

Capacitor C also serves a second function. As the output voltage across $R_L$ reaches its positive peak, point X is essentially at $V_{CC}$. Without capacitor C in the circuit, the emitter and base of Q2 would both be at $V_{CC}$ and there would be no current flowing through the transistor.

Assume now that C is in the circuit. The voltage across the charged capacitor C remains constant at the voltage to which it was charged, $V_C$. As the lower end of C is at point X and the upper end is at the junction of $R_{B2}$ and $R_{B3}$, it maintains the junction at $V_C$ volts above the voltage at X. This voltage is higher than $+V_{CC}$ during positive signal peaks. As $V_C$ is across the sum of $R_{B2}$ and the base-emitter junction of Q2, it keeps this transistor conducting at all times—even when the point X or the emitter is at $+V_{CC}$.

The time constants of $R_{B2}C$ must be such as to maintain the voltage across $R_{B2}$ constant, even at the lowest frequency to be amplified. $R_{B1}$ and $R_{X1}$ bias Q1. The voltage across $R_{X2}$ is used to bias the base-emitter circuits of Q2 and Q3. $R_{X2}$ may be replaced by diodes to establish the bias while stabilizing the circuit against temperature fluctuations. $R_{E1}$ and $R_{E2}$ are small resistors shunting the base-emitter junctions of Q4 and Q5, respectively. This serves to raise the collector-to-emitter breakdown voltage of the output transistors, since the breakdown voltage is lower when this junction is open-circuited.

### Fully Complementary Circuit

The big disadvantage of the quasi-complementary amplifier is that the two halves of the push-pull circuit are different. While the two upper transistors form a Darlington amplifier, the two lower transistors are a compound combination of a complementary pair. The gains of both halves are almost identical. However, not all characteristics remain the same over the entire operating range. Amplifiers with precise characteristics have been designed using two Darlington amplifiers or two fully complementary circuits rather than one of each. Examples of these are shown in Fig. 9-11.

The capacitor coupling the amplifier to the load, $R_L$, is undesirable. Although it is used here to keep dc out of the load, its prime disadvantage is to cause low-frequency rolloff. In addition, if the capacitor is included in the feedback loop, there will be a corner frequency that may make the amplifier unstable. Furthermore, electrolytic capacitors are nonlinear and contribute to distortion. Finally, the capacitor must be charged through the output transistor. If, in the process, the transistor will handle more energy than it can dissipate safely, the transistor will be destroyed.

If two power supplies were arranged as in Fig. 9-12 for the circuit in Fig. 9-11B, the capacitor is not required. If both halves of the amplifier are well balanced, the voltage at both ends of the load resistor under quiescent conditions, will be at the same zero dc potential with respect to ground.

One of the requirements of this circuit is that the voltage drop across both output transistors be equal when no signal is applied. This necessitates that the quiescent currents flowing through both Q4 and Q5 be maintained at their relative values regardless of line and temperature changes. The single transistor driving this output circuit as in Fig. 9-11 will let the relative output currents flowing through Q4 and Q5 change under these variable conditions.

The differential amplifier provides a circuit in which dc conditions remain relatively constant despite environmental fluctuations. Transistors, carefully matched for tracking of $\beta$ and $V_{BE}$ variations with collector current and temperature, should be used. A typical circuit using differential amplifiers, based on the circuit in Fig. 5-9, is shown in Fig. 9-13. This circuit can be used to drive the phase inverter and output transistors in Fig. 9-12, or in any other amplifier similarly arranged.

The input signal is fed to the base of Q6. The output appears at the collector, shifted 180°. Since Q6 and Q7 are a differential pair, the output also appears at the collector of Q7. The two outputs, 180° out of phase with each other, are direct-coupled to a second differential amplifier pair composed of Q8 and Q9. The signal swings the collector of Q8 from almost $+V_{CC}$ to ground. Thus, the drive for transistors Q2 and Q4 of Fig. 9-12 is taken from the collector of Q8.

The output of Q9 is 180° out of phase with the output of Q8. In order to supply the drive to transistors Q3 and Q5 in Fig. 9-12, the drive signal must be in phase with that supplied to Q2 and Q4. Hence, the output of Q9 is fed to Q10 for an additional 180°

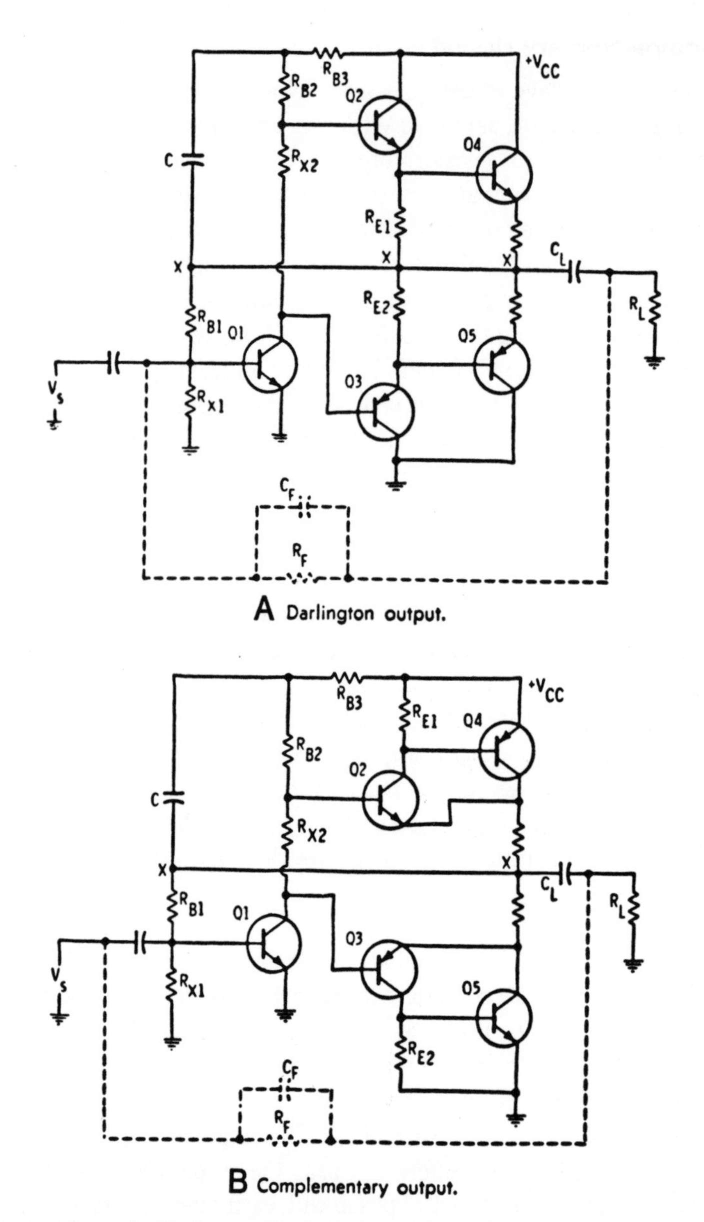


Fig. 9-11. A comparison of a Darlington circuit (A) and a complementary output circuit (B).

phase shift. $R_{C4}$ is of the proper magnitude to reduce the signal fed to Q10 so that the output signal from Q10 will be equal to that of Q8. The phases of the signals at the collectors of Q8 and Q10 are identical because of the phase inversion in Q10. However, the signal swings the collector of Q10 from somewhat more than $-V_{CC}$ to ground. This is the proper signal required to drive Q3 and Q5 in Fig. 9-12.

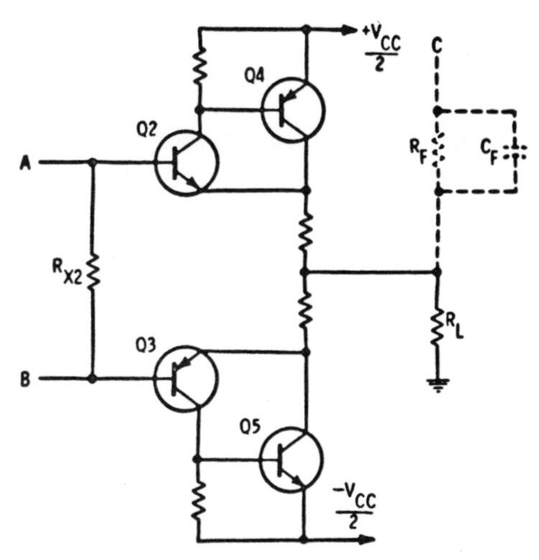


Fig. 9-12. This circuit is used to eliminate the capacitor from Fig. 9-11B.

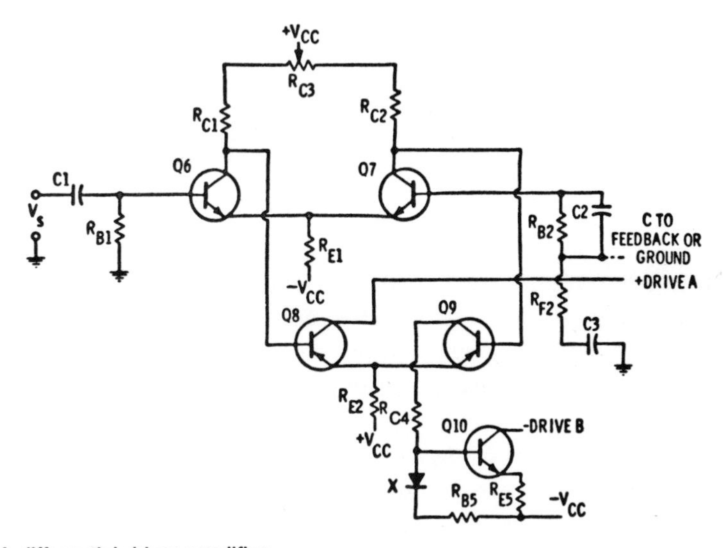


Fig. 9-13. A differential driver amplifier.

In a balanced differential amplifier, $R_{C1}$ equals $R_{C2}$, $R_{B1}$ equals $R_{B2}$ and a common resistor is used for the two emitters. Precise balance is assured through use of potentiometer $R_{C3}$. It should be noted that $R_{B2}$ must be connected to a dc ground return of some type. It may be connected directly to ground or to a point in Fig. 9-12 at feedback resistor $R_F$ and capacitor $C_F$.

The emitters of Q6 and Q7 are at about 0 volts dc or at ground potential. The combined current that flows through the two transistors also flows through $R_{E1}$. $R_{E1}$ is returned to the $-V_{CC}$ supply so that it can be made large. Since the emitters should be at about ground potential, $R_{E1} = V_{CC}/2mA$. A constant-current source exhibiting an extremely high impedance, can be used in place of $R_{E1}$. Consequently $-V_{CC}$ can be reduced in magnitude.

In Fig. 9-14A, $R_B$ and $R_X$ are connected across a large voltage, $2V_{CC}$. The current remains relatively constant through these resistors because of the large size of $R_B$. Hence, the voltage across $R_X$ is constant. As this voltage appears across $R_E$ and the base-emitter junction, the voltage across and current flowing through $R_E$ will not vary. The collector current will remain constant because it is closely related to the emitter current.

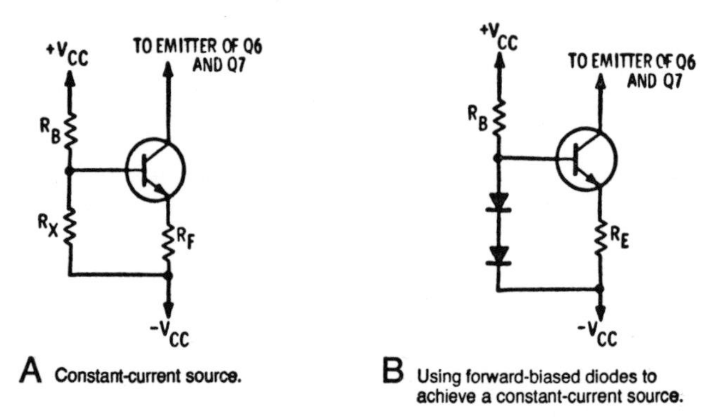


A Constant-current source.

B Using forward-biased diodes to achieve a constant-current source.

Fig. 9-14. A constant current source does not require a complex circuit.

The constant-current source in Fig. 9-14B operates on the basis of a relatively constant voltage being maintained across two forward-biased diodes in the base circuit. The unvarying voltage in the base circuit is transferred to the emitter circuit. A constant voltage is developed across the emitter resistor establishing constant emitter and collector currents.

There are many variations on the driver circuits shown here. Some of these have been provided as integrated circuits for use as low power amplifiers. Here, only the power supply, signal load, and feeback components must be added to the chip to complete the overall amplifier.

## CLASS-C AMPLIFIERS

Class-C amplifiers cannot be used to reproduce audio signals because the waveshape at its output is not even a reasonably magnified replica of the signal at its input. Here, collector current flows for less than half a cycle. With the exception of leakage current, no collector current flows when the transistor is idling nor when the signal applied to its input is of the wrong polarity, nor when the input signal level is below a predetermined amplitude. The transistor is normally biased beyond the point where current is cut off, when it is operating in the class-C mode. Collector current (or drain current in an FET) exists only when the signal is sufficiently high and of the proper polarity to turn on the transistor.

Radio frequency (rf) circuit requirements do exist that make good use of class-C amplifiers. A resonant or tuned circuit involving an inductor and capacitor is frequently placed at the output of the rf transistor circuit. This transistor can be biased to work in class-C. A single frequency is applied to the input of the class-C amplifier. The resonant circuit is usually turned to the frequency of the input signal or to a multiple of that frequency. Each time the amplitude of the input signal is sufficient to turn on the transistor, energy is supplied to the LC circuit. Because of a type of "electronic inertia" starting with the current flow in the output circuit, an entire cycle appears across the resonant circuit.

Because current flows for short periods of time during each cycle, efficiency of the class-C circuit can be as high as 85 percent. It varies with the percentage of the time during which current flows in the output circuit.

## BIAMPLIFICATION

A high-fidelity amplifier is required to reproduce frequencies between at least 20 and 20,000 Hz. This wide frequency range can severely tax any amplifier system.

Statistics for the usual audio intelligence (speech and music) show that much more output power is required at low frequencies than at high frequencies. Biamplification takes advantage of this statistical fact. In the biamplification audio system, the audio input signal is split by a filter so that high frequencies are fed to one audio amplifier circuit and low frequencies are fed to a second one. The crossover frequency or frequency above which the low frequencies are rolled off or reduced, and below which the high frequencies are rolled off, is usually between 400 and 800 Hz. Rolloff starts slowly at one of these frequencies. Output is reduced considerably at frequencies just a few octaves on each side of the crossover frequency. Low frequencies (which are primarily below the crossover frequency) are fed to a high power class-AB output stage, while the high frequencies (which are primarily above the crossover frequency) are fed to a class-AB stage that can deliver somewhat less power.

The primary advantage of the biamplification technique are realized when extremely high output levels and low distortion are simultaneous requirements. As a secondary factor, biamplification has a somewhat better efficiency than does the conventional single power amplifier system.

## CLASS-D AMPLIFIERS

Class-D amplifiers have been devised to run at well over the 90 percent efficiency figure and yet deliver good audio output. Devices here are used as switches. Audio signals applied to the input of the amplifier are converted into flat-top, high-frequency pulses.

There may be as many as 500,000 of these pulses each second. All pulses generated here are at the same voltage, but their duration or width depends upon the instantaneous amplitude of the audio signal being coded. Pulses are amplified by a pulse-power amplifier using a circuit similar to the one used in Class-AB and Class-B amplifiers shown in Fig. 9-12. But pulse amplifiers are not linear. The collector current of each transistor jumps from 0 to saturation. This jump depends upon the polarity of the voltage applied to the inputs of the transistors. If the polarity of the pulse is such as to turn on the top transistor, the voltage at the emitters jumps from 0 to $+V_{CC}/2$. Should the polarity be reversed and the bottom transistor is turned on, the emitter voltage swings from 0 to $-V_{CC}$. This is not the voltage across the loudspeaker load, only the voltage that is at the emitters of the transistors. Voltage exists at the emitter for the full period of time that the pulse exists.

The loudspeaker load is not connected directly to the emitter, as would normally be the case when using a class-AB or class-B amplifier. There is a low-pass filter across or in series with the loudspeaker load, so the high switching frequencies do not enter into the coil of this transducer. The low-pass filter is an integrator circuit; as such, it adds the instantaneous voltages present at each instant in the pulses to provide higher amplitudes across the loudspeaker when the pulses are longer than it does for shorter pulses. As the instantaneous voltage developed across the load is dependent upon the width of the particular pulse, a wider pulse produces a larger instantaneous voltage at the output. The form of the original signal is thereby recreated, since the pulse widths were initially determined by the instantaneous amplitudes in the cyclic variance of the audio signal applied to the input of the amplifier.

## CLASS-E, F, G, AND H AMPLIFIERS

In quest for highly efficient amplifiers, Hitachi developed what they originally referred to as the class-E audio power amplifier. Now they feel that they have advanced so much in the design that they skipped the class-F designation and renamed it the class-G circuit. It involves two sets of power amplifiers. One set delivers the output to the loudspeaker load while the input signal voltage is low. The second set of amplifiers is turned off during this period. When the input signal surpasses a predetermined voltage level, the second set of amplifying transistors is turned on. Efficiency is high, probably over 90 percent, because the high-power transistors used to deliver the higher power are off most of the time. This is the situation as average levels of music and speech are low. Only short-lived large amplitude peaks turn on the high-power transistors.

Class-H amplifiers differ from the class-E, F, or G types of circuits. Here, all output stages deliver the entire signal. However, logic circuits are made part of the power-supply circuit. Less voltage is supplied to the output stages when the signal is weak than on large peaks. Among the advantages of class-H operation are high efficiency and a cool-running power amplifier.

## VFETS

Up to this point, output devices shown in the various circuits were large bipolar transistors. Much research was done to develop power field-effect transistors. The newly developed devices were given the name VFETs for Vertical Power FETs. VFETs have also been developed for use in rf applications and have been applied successfully to such circuits.

VFETs were discussed in some detail back in Chapter 2. Their characteristics will be briefly reviewed here as they relate to power amplifiers.

Output impedances of the bipolar transistor and ordinary low-power field-effect transistor (HFET) are high compared to the impedance of the VFET. This can be seen by comparing the characteristic curves. While the output curves of bipolar and HFET devices are predominantly horizontal, the drain characteristic curves of the VFET may be considered vertical, as shown in Fig. 9-15A. As the output impedance of any device is $\Delta V_{DS}/\Delta I_D$, it becomes obvious that impedance is lower when changes in drain current are higher for a fixed change of drain voltage. The drain impedance of the VFET is usually less than 10 ohms, whereas the collector impedance of the bipolar power device is 20 or more times that value.

Loudspeaker loads at the output of a power amplifier circuit want to see low impedances when looking back into output transistors. This is important if the damping is to be good, keeping the loudspeaker from vibrating once the signal has been removed or the musical tone no longer exists. Feedback is usually provided around the amplifier to further reduce the output impedance. Less feedback is required, and hence better stability can be achieved, if the output impedances of the transistors themselves are low.

Characteristic curves of the VFET show another important quality of the device. Curves are evenly spaced, hence there is little distortion caused by greater amplification of one portion of the cycle than of a second portion. This is not the case with bipolar transistors. When using the VFET, fewer odd harmonics are generated than when bipolar

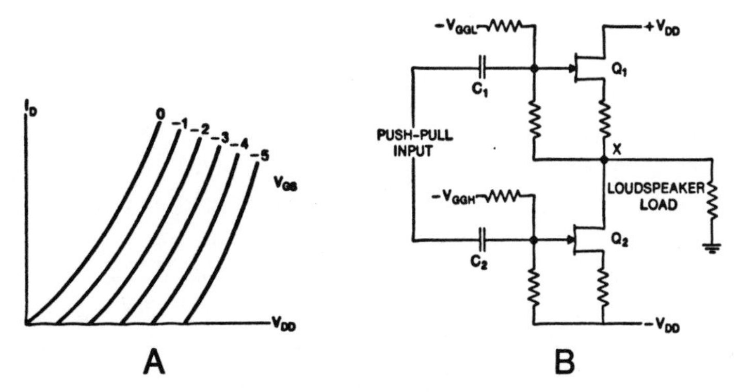


Fig. 9-15. The graph (A) shows the drain curves for a typical VFET, and a circuit using VFETs is shown at (B).

devices are used. While many higher harmonics are present in amplifiers using bipolar power devices, it has been found that harmonics above the third are insignificant where VFETs are used.

Although amplifiers are subject to different types of distortion, harmonic and intermodulation distortion are most important. *Harmonic* distortion can be noted when the input signal becomes misshapen by the time it reaches the output load. Thus, if a perfect sine wave is fed to the input of an amplifier, it should appear as a perfect sine wave at the output. If it is not perfect, harmonics have been added. The frequency of the original sine wave is known as the fundamental. Harmonics are frequencies that are multiples of the fundamental. A misshapen sine wave consists of the fundamental plus different harmonics. The size and number of these harmonics at the output depends upon the shape of the distorted sine wave.

On the other hand, intermodulation distortion can be noted when more than one frequency is fed to the amplifier at any one time. If the amplifier is not perfectly linear, these frequencies beat with each other producing sum and difference frequencies. Thus if, let us say, 60 and 7000 Hz are fed to an amplifier with poor intermodulation distortion characteristics, 7000 + 60 = 7060 Hz and 7000 − 60 = 6940 Hz in addition to the original 60 Hz and 7000 Hz would appear at the output of the amplifier. This is very similar to the heterodyne principle used in the ordinary radio to produce the 455 i-f. Frequency components owed to intermodulation distortion are more distasteful to the listener than are frequency components due to harmonic distortion.

Intermodulation distortion (IM) is high in amplifiers with bad crossover characteristics. That is, if one transistor in a push-pull pair stops conducting on one half-cycle before the other transistor starts to conduct for the second half-cycle, IM components are generated. If amplifiers are completely off until signals of the proper polarity are applied to the inputs, the input signal must be above 0.7 volt before collector current flows. This problem is alleviated in class-AB operation as both transistors are kept turned on by the bias voltage. Now there is always some minimal amount of collector current flowing at all times. Idling or quiescent current must be high if IM components are to be negligible. While collector idling currents of bipolar devices are usually limited to from 20 to 50 mA, the idling current can be as high as 500 mA where VFETs are used. Thus the IM distortion characteristics of VFET amplifiers are usually better than those of amplifiers using bipolar devices. High idling current can be used in circuits involving VFETs, because they do not exhibit thermal runaway problems. This is true because drain current drops as the temperature of the FET transistor rises. All FETs have negative temperature coefficients.

A circuit using VFETs in the output stage is shown in Fig. 9-15B. $+V_{DD}$ and $-V_{DD}$ are equal drain supply voltages of opposite polarities. Bias voltages $-V_{GGL}$ and $-V_{GGH}$ are adjusted to assure that there is 0 volt dc at point X. Hence, a loudspeaker load can be connected from X to ground without any flow of dc current through the load. As these are depletion types of FETs, the gate must be negative with respect to the source if the idling drain current is to be set at some value below $I_{DSS}$. Hence $-V_{GGL}$ is negative with respect to voltage at X, while $-V_{GGH}$ is negative with respect to $-V_{DD}$. These gate bias voltages are adjusted so that the quiescent drain current is about 500 mA in each transistor and the voltages across both transistors are equal. Only then is X at zero except when signal is applied to the inputs of the transistors.

A push-pull input (not shown) is applied to the gates through C1 and C2. Only half of the input cycle is applied to each gate. During one half-cycle, the gate of Q1 swings positive with respect to its source, and voltage at X rises toward $+V_{DD}$. On the alternate half-cycle, when the gate of Q2 turns more positive with respect to $-V_{DD}$, point X drops towards the $-V_{DD}$ level. During each sinusoidal cycle, voltage at X varies from almost $+V_{DD}$ to almost $-V_{DD}$. This voltage is a reconstituted and amplified version of the input signal. It appears across the loudspeaker load between junction X and ground.

VMOS power FETs or HEXFETs, are insulated gate versions of the VFET. Characteristic curves are similar to those of bipolar devices, but are more horizontal and more evenly spaced. These are enhancement-type devices so bias can be at zero without producing crossover distortion. Despite the horizontal curves, impedance at saturation can be considerably less than 1 ohm. These transistors are extremely useful in audio as well as in high power rf amplifier stages. Unlike bipolar devices, there are no thermal runaway nor second breakdown problems despite their ability to handle well over 100 amperes.

## PROTECTION CIRCUITS

When bipolar output transistors are used, they are subject to breakdown. This can occur even if they are only abused momentarily. While loaded with the specified impedance, output transistors can provide reliable service; but a short circuit at the output can force the output devices to deliver so much current that they will dissipate more power than they can handle safely. Despite the proper loudspeaker load at the output, an unusually inductive or capacitive impedance in the crossover network can, in some cases, upset the stability of an amplifier. Excess current can flow if the amplifier breaks into sustained or damped oscillations.

A number of circuits have been devised to provide protection for the output devices. In the circuit shown in Fig. 9-16, the magnitude of the output current causes a voltage to be developed across a sensing resistor. This voltage is fed to transistors Q1 and Q2 at the input of the driver stages. When the current at the output exceeds a predetermined level, the voltage across the sensing resistor is of the proper magnitude to turn on Q1 and Q2. In turn, these transistors short the input signal of the drivers. As this signal does not reach the output devices, they cannot be destroyed.

Output transistors in the circuits in Fig. 9-11 can be protected through the use of a sensing resistor, R, and two diodes connected in parallel with the cathode of one diode connected to the anode of the other diode. In this protection circuit, R is connected between $R_L$ and ground. The parallel diode combination is connected from the junction of

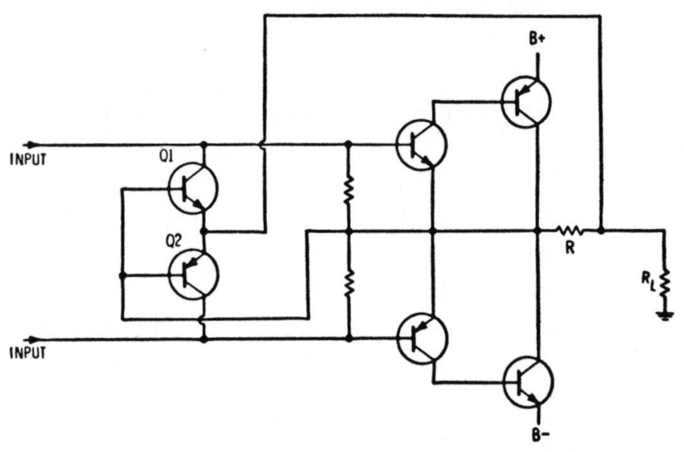


Output current sensed across resistor R in series with load.

Fig. 9-16. This protection circuit senses the output current.

R and $R_L$ to base of Q1. When more than a permissible quantity of current flows through R, the voltage developed across it should be sufficient to turn on one or both of the diodes. The signal voltage developed across R is fed back through the diodes to the base of Q1 very much like the signal across $R_L$ is fed back through $R_FC_F$ in the interest of reducing distortion. The signal fed back through the diodes reduces the output of the overall circuit.

In a more sophisticated circuit developed by RCA, information is fed back about the amount of current flowing through the output transistors along with data about the size of the load $R_L$. This can be explained with the help of the circuit in Fig. 9-17. When excess current flows through the output transistor Q3 while output load resistor $R_L$ is

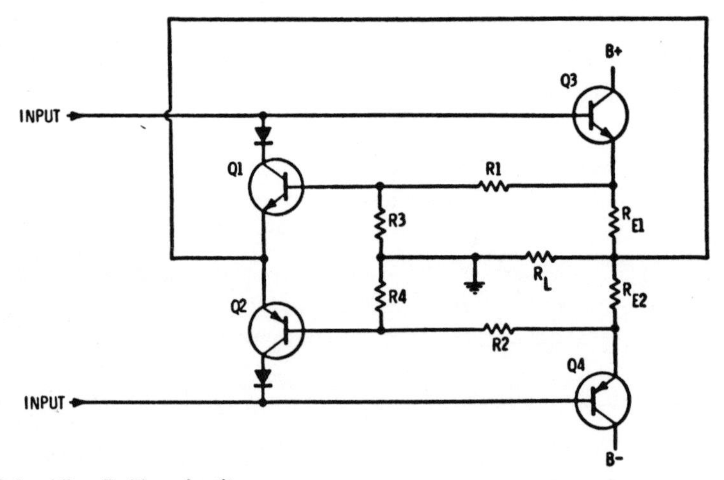


Fig. 9-17. A load-line limiting circuit.

shorted, the voltage developed across $R_{E1}$ is divided between R1 and R3. It is sufficient to turn on Q1 and shunt signal away from the base of Q3. The same relationship holds true for the lower half of the circuit involving $R_{E2}$, R2, R4, Q2, and Q4.

Now assume the load resistor is equal to or larger than a predetermined value. When signal is flowing through this resistor, the voltage developed across $R_L$ reverse biases Q1 and Q2. If they are to conduct, the voltage developed across $R_{E1}$ and $R_{E2}$ must be greater than the sum of the voltage required to overcome this reverse bias.

## THERMAL CONSIDERATIONS

Because drain current of the FET is reduced as the temperature of the device increases, thermal considerations are not as critical when applied to these devices as when applied to bipolar transistors. Procedures and mechanisms of removing heat from all types of devices, are essential considerations.

Power is dissipated at the collector junctions of bipolar transistors. The amount of power that may be dissipated is limited by the temperature at which the transistor operates. The maximum junction temperature is about 100° C for germanium devices and 200° C for silicon transistors.

Heat must be conducted away from the junction to the cooler case of the bipolar transistor if the junction is not to be overheated due to the continuous dissipation of power. This applies as well to the channel of the FET. There is a thermal resistance, $\theta_{JC}$, between the junction and case that limits the quantity of heat carried from the junction to the case. This is not unlike the ohmic resistance in an ordinary dc circuit. The case must be cooled by the surrounding air. There is a thermal resistance, $\theta_{CA}$, between the case and the surrounding air that cools it.

The case temperature is an indication of how well heat is transferred from the junction to the cooler air. The temperature of the case is a function of $\theta_{JC}$ and $\theta_{CA}$. A hot case means that little of the heat transferred from the junction to the case is radiated into the air. The transistor's ability to dissipate power decreases as the temperature of the case increases. A curve illustrating this for the 2N3055 is shown in Fig. 9-18. Other curves show the maximum permissible transistor dissipation as a function of the ambient temperature.

If this curve is not available, it can usually be drawn from other transistor data. Mark the axes on graph paper as in Fig. 9-18. Put a dot at the point where 25°C intersects the maximum power dissipation rating of the device. In this case, it is at 115 watts. Mark the 0 watts point at the maximum permissible operating temperature of the transistor. Connect these two points.

Some manufacturers supply the derating data in units of watts per degree Celsius. For the 2N3055, the derating factor is 0.66 watt/°C. The transistor's ability to dissipate power is reduced by 0.66 watt for every degree the case temperature is above 25°C. If the case temperature were at 125°C or a rise of 125°C − 25°C = 100°C, the transistor

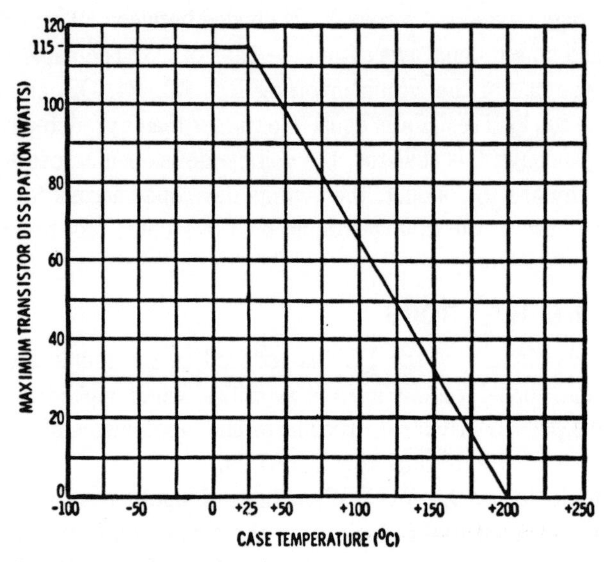

Fig. 9-18. Graph showing the $P_{diss}$ derating curve for a 2N3055 transistor (courtesy of RCA).

would be able to safely dissipate only 115 watts − (0.66) 100 watts = 49 watts. This checks with the curve in Fig. 9-18.

The power dissipated by the transistor is related to the thermal resistance from the junction to the air, $\theta_{JA}$, and the difference between the ambient temperature, $T_A$, and the junction temperature, $T_J$, by the equation:

$$P_{diss} = \frac{T_J - T_A}{\theta_{JA}} \tag{9-17}$$

where,

$\theta_{JA}$ is the thermal resistance from junction to air,
$T_A$ is the ambient temperature,
$T_J$ is the junction temperature.
$P_{diss}$ is measured in watts, T in °C, and $\theta$ in °C/watt.

The resistance, $\theta_{JA}$ is composed of two thermal resistances, $\theta_{JC}$ and $\theta_{CA}$, $\theta_{CQ}$ is frequently broken up into two smaller components—$\phi_{CS}$, the thermal resistance from the case to the heatsink, and $\theta_{SA}$, the thermal resistance from the heatsink to the surrounding air.

A heatsink is a large piece of metal, preferably black anodized aluminum. The hot transistor, mounted on the sink, heats the metal, which in turn radiates the heat into the air. The metal may be flat or it may have fins so that there is more surface in contact with the air. The resistance, $\theta_{CA}$, is reduced if there is a heatsink, and $\theta_{CS} + \theta_{SA}$ is smaller than $\theta_{CA}$ would be without the sink. Formula 9-17 can be extended to the form:

$$P_{diss} = \frac{T_J - T_A}{\theta_{JC} + \theta_{CS} + \theta_{SA}} \quad (9\text{-}18)$$

From Equations 9-17 and 9-18, it is obvious that the junction temperature, $T_J$, should not increase above the ambient temperature, $T_A$, by more than the product of the power the transistor dissipates with the total thermal resistance from the junction to the air. As for the JFET, heatsinking is required to keep the channel within its maximum rated temperature limits despite the quantity of power it dissipates.

## THERMAL CYCLING

It is well known that after sheet metal is bent back and forth several times, it will break. A similar phenomenon can be observed with power transistors. After the device has been heated and cooled a number of times, there will be a mechanical breakdown. In many applications, the number of temperature changes or cycles can be a limit on the life of the device. RCA developed a chart relating the number of such thermal cycles a device is capable of withstanding to the power the device can dissipate. This relationship differs with the magnitude of the changes in temperature of the case. A chart for the 2N6100 transistor is shown in Fig. 9-19. The case-temperature changes can frequently be held within desirable limits by using a proper heatsink.

Assume that a transistor does not change an equal amount in temperature during each cycle. For example, let us assume that the transistor will operate for 30,000 cycles with a 20°C case-temperature change while dissipating 1.5 watts, and that it will operate

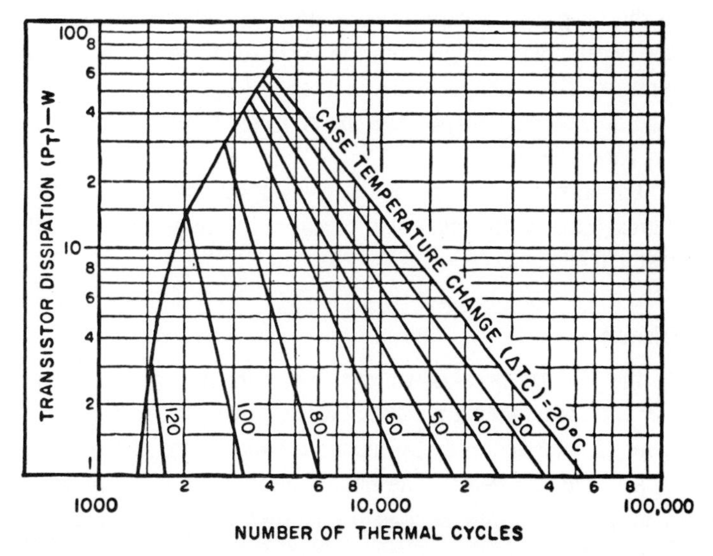


Fig. 9-19. A graph of the thermal-cycling curves for a RCA 2N6100 (courtesy of RCA).

for an additional 5,000 cycles with a 50° C case-temperature change while dissipating 10 watts. Is the device operating within its thermal limits?

In the first group of temperature changes, the maximum number of cycles permitted with a variation in case temperature of 20° C and 1.5 watts of dissipation is 40,000 cycles. The ratio of the actual number of cycles to the maximum permissible number of cycles for a 20° C case-temperature change is 30,000/40,000 = 0.75.

As for the second group of variations, when the case temperature changes by 50° C and 10 watts is dissipated by the device, the maximum number of permissible cycles is 6,500. The ratio of the actual number of cycles to the maximum permissible number of cycles for a 50° C temperature change while dissipating 10 watts is 5,000/6,500 = 0.77.

If the sum of the two fractions is less than 1, the transistor is operating within its thermal-cycling rating. In this case, the sum is 0.75 + 0.77 = 1.52. The transistor is not being used within its rating and will probably break down sooner than desired.

## RATING LIMITS

There are a number of symbols used to indicate the limits of the bipolar power transistor. The absolute maximum ratings should never be exceeded, even by transients of extremely short duration.

- $BV_{CBO}$ is the breakdown voltage from the collector to the base with the emitter open.
- $BV_{CEO}$ is the breakdown voltage from the collector to the emitter with the base open.
- $BV_{CER}$ is the breakdown voltage from the collector to the emitter with a specific resistor between the base and emitter. This voltage is higher than $BV_{CEO}$.
- $BV_{CES}$ is the breakdown voltage from the collector to the emitter with the base shorted to the emitter. This voltage is higher than $BV_{CER}$.
- $BV_{CEV}$, is the breakdown voltage from the collector to the emitter with the base-emitter junction reverse-biased.
- $BV_{CE}$ (sus), the sustaining voltage, is the breakdown voltage at elevated collector current. An additional letter in the subscript indicates the resistance or voltage placed in the base-emitter circuit.

### Second Breakdown

In addition to the primary breakdown voltages, currents, and powers, there is a second breakdown. This is a limiting factor in bipolar power transistors only. The second breakdown phenomenon seems to be caused by a large amount of current flowing through a tiny area in the transistor. The base does not exert an influence over the collector current in controlling this second breakdown current. $V_{CE}$ drops while the collector current rises, regardless of the applied current or voltage at the base. Because this is an irreversible phenomenon, the transistor is destroyed.

Transistors designed to operate at high frequencies and those with low thermal resistance, are good candidates for second breakdown. When designing a circuit in which the transistor is to be forward biased, first choose a transistor with the poorest high-frequency capabilities consistent with the design. Next, use the lowest possible collector-to-emitter voltage that will not adversely affect the performance of the circuit.

When the transistor is reverse biased, observe the following to minimize chances of second breakdown. The resistor in the base-emitter circuit should be large while the base-emitter voltage should be small. Do not let the transistor be turned off rapidly from the conducting mode.

Manufacturers supply a set of safe operating curves for the different transistors. The $V_{CE}/I_C$ swings of the transistor collector voltage and current, should be within limits set by these curves to minimize the chances of second breakdown.

### Rating Limits on FETs

Just as with bipolar transistors, various symbols are used to denote upper limits of voltages than may be applied to FETs. Specifications for these voltages should never be exceeded.

- $BV_{DSS} = V_{(BR)DSS}$ is the breakdown voltage from the drain to the source when the gate is connected to the source.
- $BV_{DSX} = V_{(BR)DSX}$ is the breakdown voltage from the drain to the source when the gate is biased at a particular voltage with respect to the source.
- $BV_{GSS} = V_{(BR)GSS}$ is the breakdown voltage from the gate to the source when the drain is connected to the source.
- $BV_{GSS} = V_{(BR)GSS}$ is the breakdown voltage from the gate to the source when the drain is connected to the source.
- $BV_{DGS} = V_{(BR)DGS}$ is the breakdown voltage from the drain to the gate when the drain is connected to the source.

These are just a few of the limits set on an FET. Maximum specifications for other factors such as drain current extremes, power dissipation maximums, operating temperature limits, and so on, should also never be exceeded.

# 10

# Transistor Switches

Besides being used as an amplifier of ac signals, the transistor can also be used as a switch. A small pulse in the base circuit can turn the large collector current on or off. Any device or load in the collector circuit is activated by the presence or absence of current in that circuit.

There are three basic methods of biasing a transistor for operation as a switch. These biasing methods are known as modes of operation, and each will be discussed in the following pages. The three switching modes for a transistor are:

- saturated mode
- current mode
- avalanche mode

Each mode has its own individual advantages, and is suited to different specific applications.

## SATURATED MODE SWITCHING

Many transistor switches are operated in the saturated mode.

If the transistor is off, the circuit is usually designed so that only $I_{CBO}$ flows through the collector. The supply voltage is across the transistor. It behaves as an open circuit. In the *on* state, the collector current is a function of the load in the collector circuit and the supply voltage. This can be illustrated with the help of Fig. 10-1.

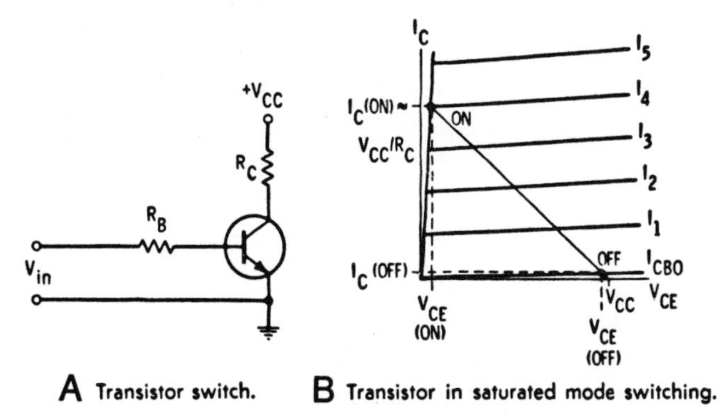


Fig. 10-1. A simple transistor switch circuit, and a graph of its operation.

In the saturated mode of operation, the voltage across the transistor drops to near zero when the transistor is turned on. Practically the entire supply voltage is across the collector load. The transistor is effectively a short circuit, putting $R_C$ between $+V_{CC}$ and ground. $I_C$ is then equal to $V_{CC}/R_C$.

A switching transistor is normally either in the *on* or *off* state. As there is a small voltage across the transistor in the *on* condition and a small current in the *off* state, little power is dissipated by the transistor. A reverse bias is usually placed on the base to keep the current at a minimum (about $I_{CBO}$) in the *off* condition.

Switching is not instantaneous, but must go through a number of steps. Assume the square wave shown in Fig. 10-2 is fed to the input terminal of the transistor in Fig. 10-1A. Initially, the transistor is turned on during the positive portion of the cycle. The base-emitter diode behaves like the diode switch described in Chapter 1. The base current starts without any delay, but there is a delay time, $t_d$, until the base-emitter voltage reaches zero and collector current can start to flow.

Delay time is defined as the period of time from when the turn-on pulse is initiated until the collector current reaches 10 percent of its peak value, $I_C$(max). It is directly proportional to the reverse bias and inversely proportional to the turn-on current.

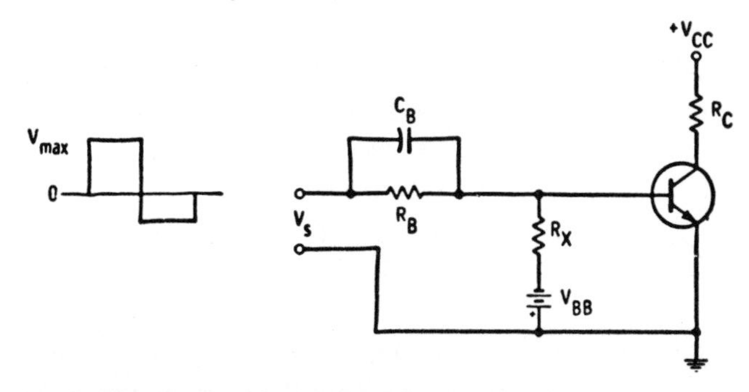


Fig. 10-2. Switching circuit featuring a speed-up network.

In the sections describing high-frequency amplifiers, two capacitances in the transistor were introduced. These have influence over the duration of $t_r$, the rise time. This is the time it takes the collector current to rise from 10 percent to 90 percent of its maximum. The sum of $t_d$ and $t_r$ is known as the turn-on time.

From there, all remains stable during $t_p$, the pulse duration time, until the input signal delivers the turn-off pulse. A transistor in saturation will not go out of saturation instantly, since a capacitor inside the transistor stores a charge. The storage capacitor must be discharged before the transistor will go out of saturation.

The base current turns negative without any delay and then gradually returns to zero. The base-emitter voltage remains positive for a while. The collector current does not change until that circuit comes out of saturation and the storage capacitor is discharged. The period of time from the turn-off pulse until the collector current comes out of saturation (drops to 90 percent of its maximum) is known as the storage time, $t_s$. The storage time increases when the transistor is driven deeper into saturation. It decreases if the reverse drive pulse is made larger.

The period of time that it takes the collector current to drop to zero is known as $t_f$, the fall time. The base current and base-emitter voltage then return to quiescent values and the transistor is turned off. The turn-off time is $t_s + t_f$.

In order to speed the switching action of the transistor, $R_B$ in Fig. 10-1A may be shunted by a capacitor, $C_B$. A negative voltage may be applied to the base to stop any base-to-emitter current when the transistor is off, so that only $I_{CBO}$ flows through the collector circuit. A complete drawing of this is shown in Fig. 10-2. The design procedure may be pursued as follows.

1. Determine the collector saturation current. It is $I_C(sat) = V_{CC}/R_C$. The base current must be capable of producing at least $I_C(sat)$.

2. Calculate $R_X$. It is equal to $V_{BB}/I_{CBO}(max)$, where $I_{CBO}(max)$ is the maximum $I_{CBO}$ rating of the transistor when operated at the highest temperature. $V_{BB}$ must establish a negative current equal to $I_{CBO}(max)$ to counter the $I_{CBO}(max)$ that is ordinarily flowing through the base-emitter junction of the transistor.

3. Plot the load line on the characteristic curve, as shown in Fig. 10-1B. Determine the base current needed to drive the transistor into saturation. It is $I_4$. The $R_B - C_B$ circuit must conduct $I_4$ as well as $I_{CBO}(max)$ to the base of the transistor, to overcome the current supplied by $V_{BB}$. Determine $R_B$ from Equation 10-1.

$$R_B = \frac{(V_{max} - V_{EE})}{I_4 + I_{CBO}(max)} \tag{10-1}$$

4. The rise time, which is determined from the speed required in the switching operation, can be reduced by overdriving the base circuit of the transistor. If $f_\beta$ is the

$\beta$ cutoff frequency of the transistor, the required base current for a rise time of $t_r$ can be calculated from:

$$I_B = \frac{I_C(\text{sat})}{\beta(D)} \tag{10-2}$$

where, D can be determined from the curve in Fig. 10-3.

5. Calculate current $I_B - I_4$. This is $I_S$, the current that must pass through the capacitor.

6. The capacitor, $C_B$ is:

$$C_B = (I_s)\,\frac{t_r}{V_{max}} \tag{10-3}$$

where,

$C_B$ is the capacitance in $\mu$F,
$I_s$ is the current in amps,
$V_{max}$ is the switching voltage in volts,
$t_r$ is the rise time $\mu$s.

In Fig. 10-2, current flows through the load, $R_C$, only when the transistor is turned on. In Fig. 10-4, current flows through the load, $R_L$, when the transistor is turned off. Ignoring the diode, the voltage $V_{CC}$ divides between $R_C$ and $R_L$ when the transistor is off. When it is on, the transistor is a short across $R_L$.

When the diode, D, is in the collector circuit, the voltage across $R_L$ is clamped to the voltage across the diode plus $V_{DD}$. This protective circuit takes effect when the volt-

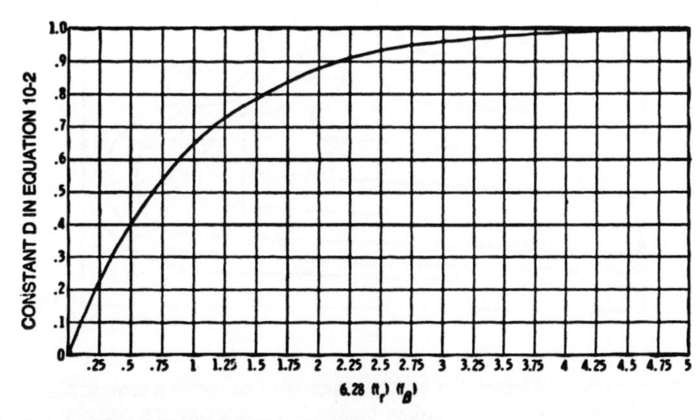


Fig. 10-3. Curve used for determining D in Equation 10-2.

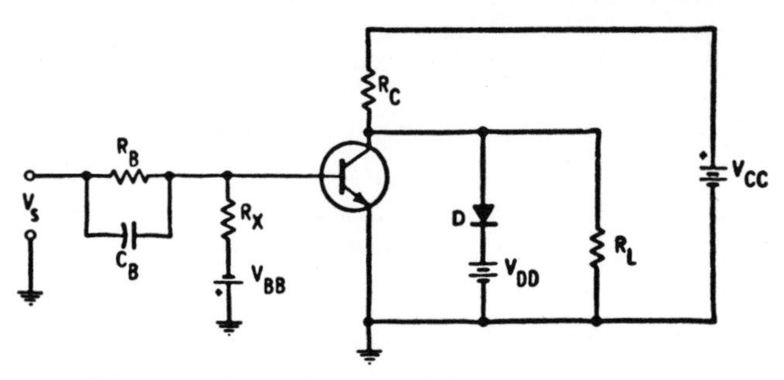


Fig. 10-4. In this switching circuit the load is grounded.

age across $R_L$ exceeds $V_{DD}$, and the diode conducts. The major advantage of this circuit is that the load is returned to ground.

In designing either type of switching circuit, latch-up must be avoided. This condition can be described with the help of Fig. 10-5.

A number of rated breakdown voltages have been noted on the drawing. The largest, $BV_{CBO}$, is the reverse breakdown voltage of the collector-base junction, with the emitter open. It is the maximum voltage that can be applied to the transistor. At this voltage, the collector current rises rapidly with but a minute increase in collector voltage.

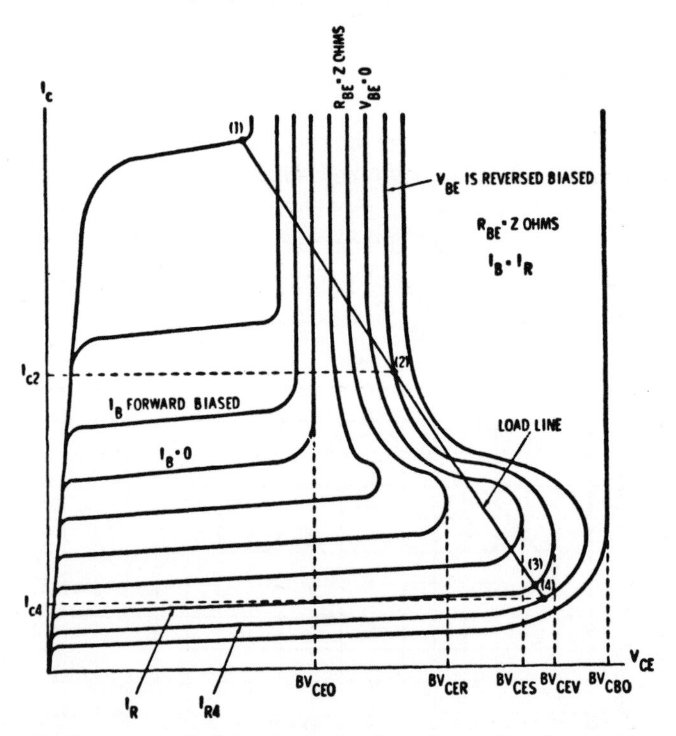


Fig. 10-5. Graph of collector characteristics showing latch-up. Curves for a forward biased base-junction above the $I_B$ = O curve. Curves for a reverse biased base-junction are below the $I_B$ = O curve.

$BV_{CEO}$ is the collector-to-emitter breakdown voltage, with the base junction open. It is the maximum collector voltage on the $I_B = 0$ curve. The collector-to-emitter breakdown voltage is increased when there is a resistance, R, between the base and emitter junctions of the transistor. The symbol for this is $BV_{CER}$. R may be indicated on the specification sheet although the data is frequently stated for a 10-ohm resistor. Should the base be shorted to the emitter, the breakdown voltage increases to $BV_{CES}$. In this case, the base-to-emitter voltage is zero. A reverse-biased base-emitter junction shunted by a resistor of R ohms will raise the breakdown voltage to $BV_{CEV}$.

In the avalanche region just below the $BV_{CBO}$ level, the collector-to-emitter voltage decreases as the current rises. If the load line in the switching circuits is as shown, the transistor will be on when the swing of the input pulse forces the collector current to (1). When turned off, let us say to a reverse base current $I_R$, it is desired that the quiescent point drop to (3). At this point, little collector current flows when the voltage is at a maximum. Actually, as the base current is switched off to $I_R$ from point (1), the collector current drops only to $I_{C2}$, because (2) is the first point where the load line crosses the $I_B = I_R$ curve. The transistor is in a latch-up condition. To get out of latch-up, the reverse base current must be increased so that the collector current reaches $I_{C4}$ at (4) on the $I_{R4}$ curve. Only after being at $I_{R4}$ can the quiescent point be returned to (3) on the $I_R$ base current curve.

## CURRENT MODE SWITCHING

In current mode switching, the transistor is not driven into the saturated region when switched on. This is only true of saturated mode switching. Here, when the transistor is on, it is usually near, but not quite in, saturation.

The curve showing current mode operation is drawn in Fig. 10-6A. Although there is more power dissipated here in the *on* condition than in the saturated mode of operation, current mode is highly desirable where fast switching speeds are required. Storage time problems are eliminated. The speed is increased with the distance that the *on* state is

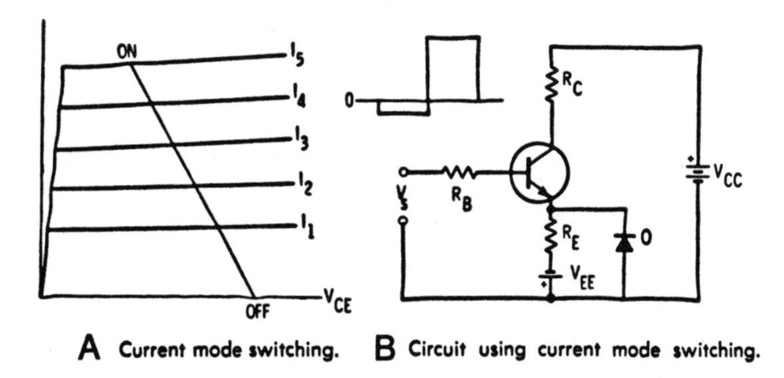


Fig. 10-6. A current mode-switching circuit.

away from saturation. A typical current mode switching circuit is shown in Fig. 10-6B.

When the transistor is switched off, a complete circuit is formed with the emitter supply, $V_{EE}$, the emitter resistor, $R_E$, and the diode. In this condition, the voltage across the diode appears between the emitter and ground. Since this voltage will forward-bias the transistor into condition, the input voltage must counter this to keep the transistor off. Hence, for the transistor to be off, the input voltage must be negative. When the input pulse is positive, the transistor is turned on. The diode will be turned off since the emitter current flowing through $R_E$ will bias the diode in the reverse direction.

The transistor can never be in saturation if the saturated collector current is greater than the current flowing through the emitter. For current mode operation, $V_{CC}/R_C$ must be greater than $V_{EE}/R_E$.

## AVALANCHE MODE SWITCHING

Avalanche mode switching exceeds the speed of current mode switching. The curves and load line showing this mode of switching and the basic circuit are shown in Fig. 10-7. $V_{BB}$ reverse biases the transistor so that operation is on the $I_{R1}$ curve. Assume that this transistor is idling at A in the off condition. A positive pulse turns the transistor on, let us say to the $I_B = 0$ curve. Operation jumps to D on the curve. The pulse is removed in a short time and the operation must revert to the $I_{R1}$ curve at C.

This portion of the curve is a negative resistance, which is characterized by instability. It can be shown that if the slope of the transistor curve is greater than the slope of the load line, $1/R_C$, the operation is stable. If the reverse is true, operation is unstable. Hence, C is a stable point on the curve.

The reverse pulse can force operation from C back to a point E on the $I_{R2}$ curve. When the pulse is removed, the transistor will then revert to B on the $I_{R1}$ curve. Since the slope at this portion of the curve is less than the slope of the load line, operation

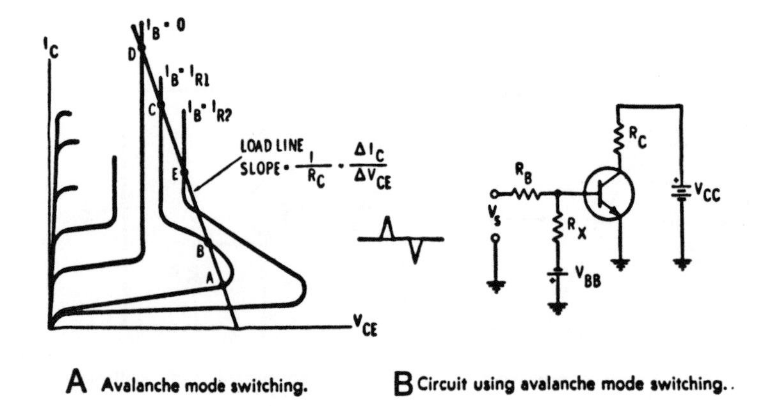


A Avalanche mode switching. B Circuit using avalanche mode switching..

Fig. 10-7. An avalanche mode-switching circuit.

is unstable. It jumps to point A, a stable point on the positive resistance portion of the curve.

The switching action in the avalanche mode can be initiated by pulsing the base current or collector supply voltage. In the latter case, the load line is moved parallel to the load in its quiescent state, shifting the operating point from one base current curve to the other.

The avalanche mode is used when high switching speeds are needed. In this mode the transistor essentially reaches its maximum switching speed. If higher switching speeds are required, some other specialized semiconductor device, such as a tunnel diode or a PIN diode, must be used.

## SINGLE TRANSISTOR SWITCHING CIRCUITS

In relatively non-critical, low to medium power applications, the switching function can be performed by a single transistor.

A simple transistor switching circuit is illustrated in Fig. 10-8. In addition to the transistor itself, only a single resistor ($R_B$) is required. Resistance $R_L$ in the collector circuit represents the load being switched.

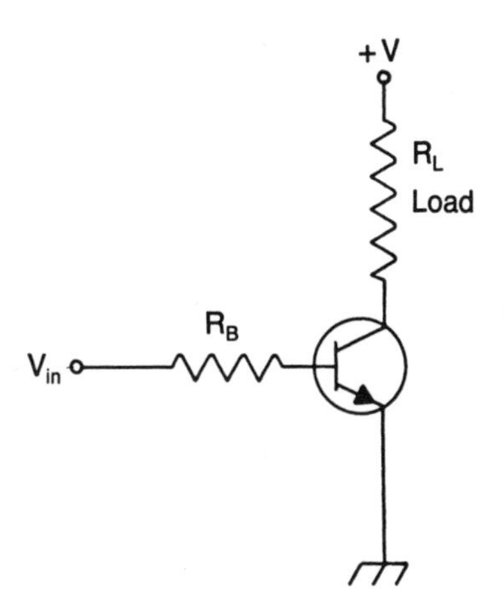


Fig. 10-8. The basic transistor switching circuit in its simplest form.

Calculating the value of $R_B$ is not difficult. The first step is to determine the collector current, $I_C$. This is the maximum current drawn by the load ($R_L$).

Next, you estimate the current gain ($h_{fe}$) for the specific transistor being used. This value is included in the manufacturer's specification sheet for the device. Be aware that the $h_{fe}$ value given in the spec sheet is an approximation. Current gain values are usually listed as typical or minimum on specification sheets. The exact value will vary from unit to unit. For optimum performance, it is advisable to breadboard the circuit, and experiment with various values for $R_B$. The current gain figure just gives you a useful starting point.

If you do not have the manufacturer's spec sheet handy, assume that $h_{fe}$ is approximately 40 for low-power transistors, and about 20 for high-power units. These values should put you more or less in the ballpark.

The important factor in choosing a transistor to be used in this type of switching circuit is that it will handle the signals to be switched. Obviously, it must be able to supply a collector current at least equal to the load current ($I_C$). To be safe, allow a considerable amount of leeway. Select a transistor with a maximum current rating that is at least 150 percent to 300 percent the maximum estimated value of $I_C$.

Similarly, the maximum collector-to-emitter voltage ($V_{ceo}$) of the transistor used is important. When the transistor is cut-off, so that no current (except a small leakage value) reaches the load, the full supply voltage of the circuit will be applied across the emitter and collector leads. To avoid possible disaster, select a transistor with a maximum $V_{ceo}$ rating that is at least double the supply voltage for the circuit.

When in doubt, it never hurts to use a larger (higher power) transistor. However, there is no sense in going overboard. Heavy duty transistors are more expensive (and bulkier) than lower power devices. It would be silly to use a two or three dollar power transistor when a low power unit will do the job just as well for less than a dollar.

The next step in designing the circuit is to determine the base current ($I_B$). This current value can be derived from the collector current ($I_C$) and the transistor's current gain ($h_{fe}$), according to this formula:

$$I_B = I_C/h_{fe} \qquad (10\text{-}4)$$

To allow a sufficient margin of error, add about 20 percent to the calculated value. Now, the approximate value for $R_B$ can be found simply by using Ohm's law;

$$R_B = V_{in}/I_B$$

Because of differences between individual transistors, all of the calculated values should be considered approximations. For the best results, breadboard the circuit, and experiment until the switching function is optimum.

The functioning of this circuit is illustrated in Fig. 10-9. The collector current is normally near zero. There will be some slight leakage current, of course, but it will almost always be negligible. The base input ($V_{in}$) is normally held close to ground potential. The voltage across the load ($R_L$) will be close to the supply voltage.

When a sufficiently positive voltage appears at the input ($V_{in}$), the collector current ($I_C$) suddenly shoots up to a fairly high level, and the voltage across the load drops close to zero.

If the circuit has a negative power supply, a pnp transistor should be used instead of the npn transistor shown in Fig. 10-8.

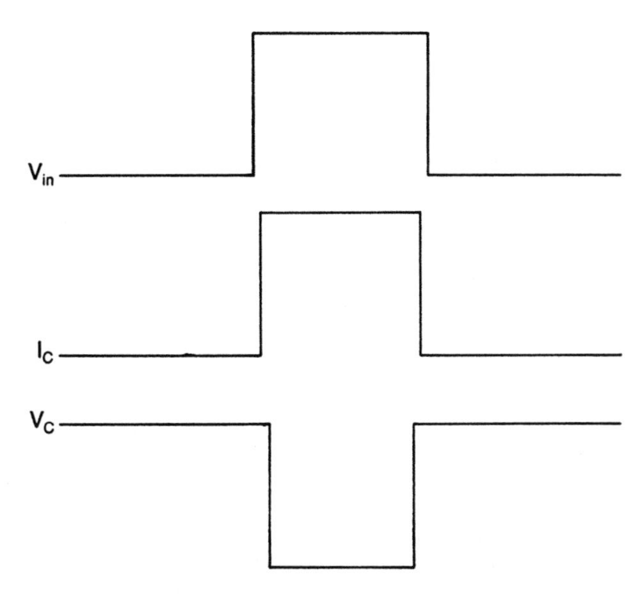


Fig. 10-9. The signals occurring in the circuit of Fig. 10-8.

In some applications, it might be necessary to reverse the switching polarity. That is, the collector current ($I_C$) should remain high except when a high positive pulse appears at the input. This reverse polarity operation is illustrated in Fig. 10-10. The circuit shown in Fig. 10-11 will accomplish this. Actually, this is very similar to the circuit of Fig. 10-8, except the supply voltage (V+ and ground) have been reversed.

In using this circuit, it is important that the driving signal becomes high enough to cut the transistor completely off.

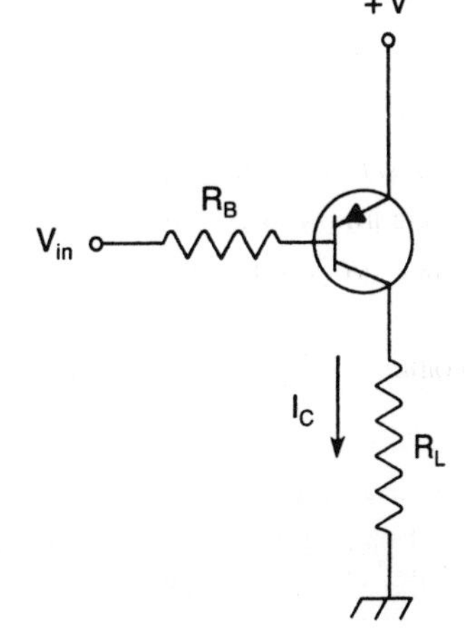


Fig. 10-10. This switching circuit inverts polarity.

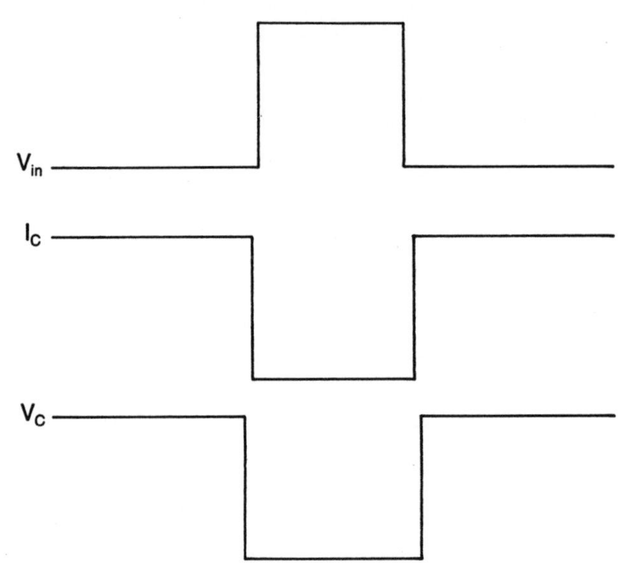


Fig. 10-11. The operating signals for the circuit of Fig. 10-10.

## MULTIPLE TRANSISTOR SWITCHING CIRCUITS

In many practical applications, a single transistor will not be adequate for switching the desired load. This will usually be because the load's required current exceeds the capability of a single transistor, or the switching signal source is loaded down by the switch circuit.

The solution is to add a second transistor. Transistor 1 acts as a buffer, switching transistor 2 on and off, while transistor 2 controls the load.

The circuit shown in Fig. 10-12 is a basic two transistor switch. The design is fundamentally similar to the single transistor version shown back in Fig. 10-8. The operating signals for this circuit are illustrated in Fig. 10-13.

The controlling input voltage ($V_{in}$) must be close to the supply voltage (V+) to cut transistor Q1 off completely.

Figure 10-4 shows a variation of this basic circuit. As the operating signals, shown in Fig. 10-15, indicate, this circuit is an inverting switch.

A moderately large load can also be controlled by a switching circuit built around a Darlington pair, as illustrated in Fig. 10-16.

## FET SWITCHES

In the low voltage region, the FET acts as a resistor. Special switching transistors have been designed with very low $r_{DS}$(on) characteristics. Should a transistor be set to a $V_{DS}$ voltage in this region, the FET is a low resistance. This is ideal if it is in a switching circuit and the switch is supposed to be closed and conducting current.

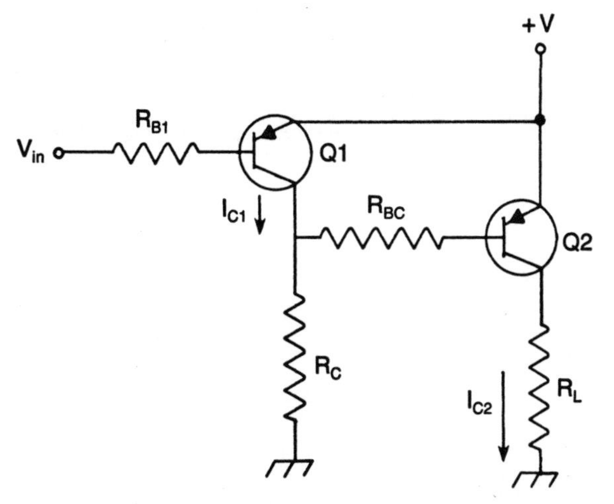


Fig. 10-12. To increase the load handling capacity, a second transistor can be added to the switching circuit.

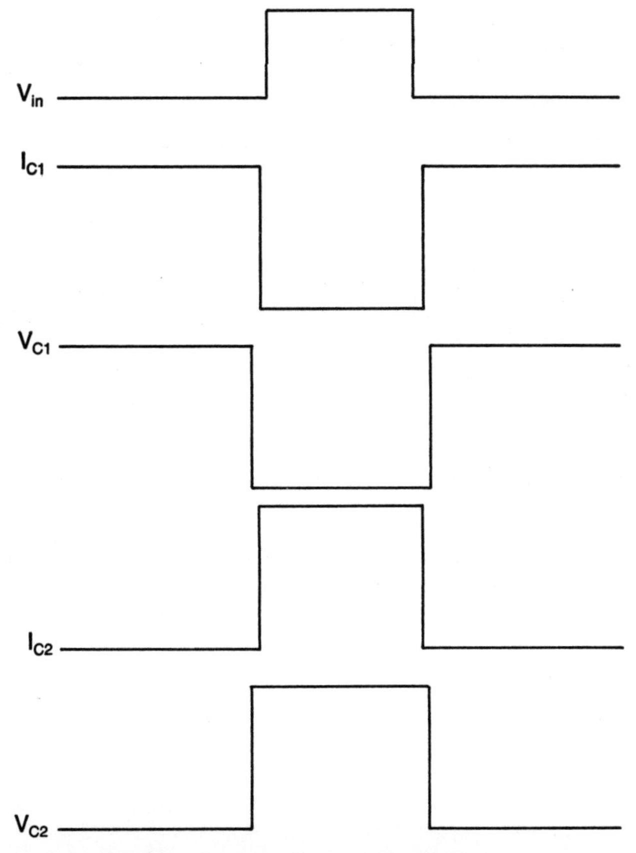


Fig. 10-13. The operating signals within the circuit of Fig. 10-12.

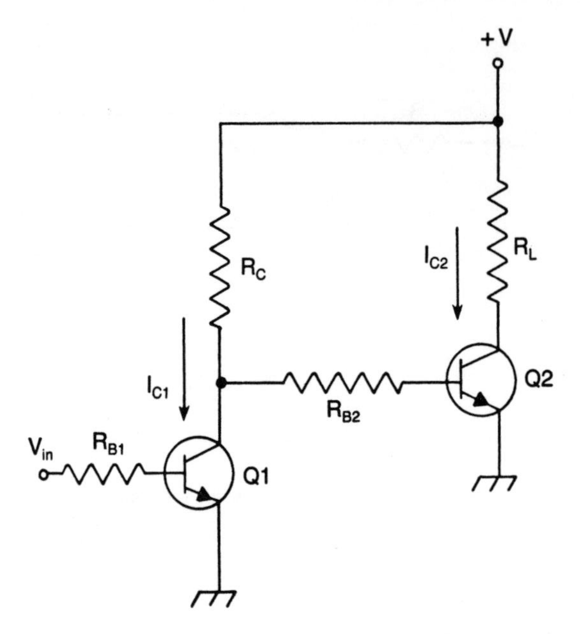


Fig. 10-14. This two transistor switching circuit features polarity inversion.

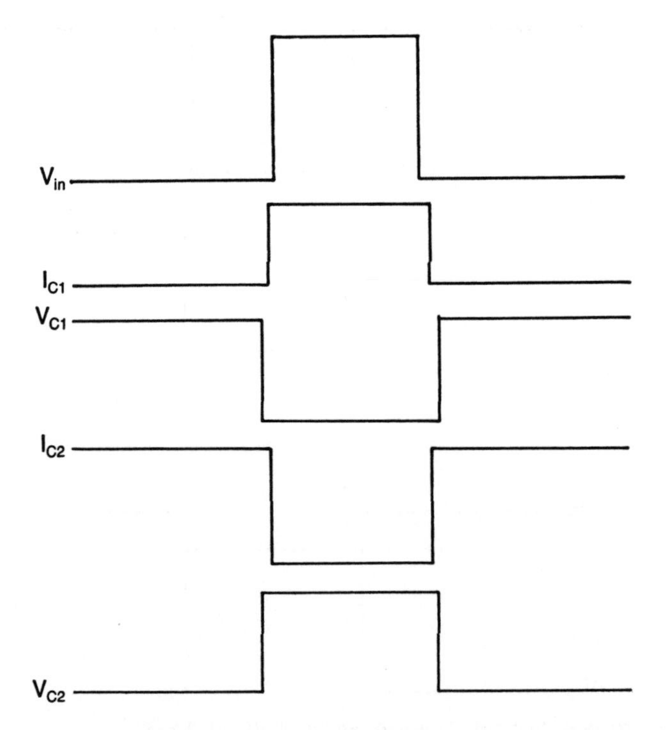


Fig. 10-15. The operating signals within the circuit of Fig. 10-14.

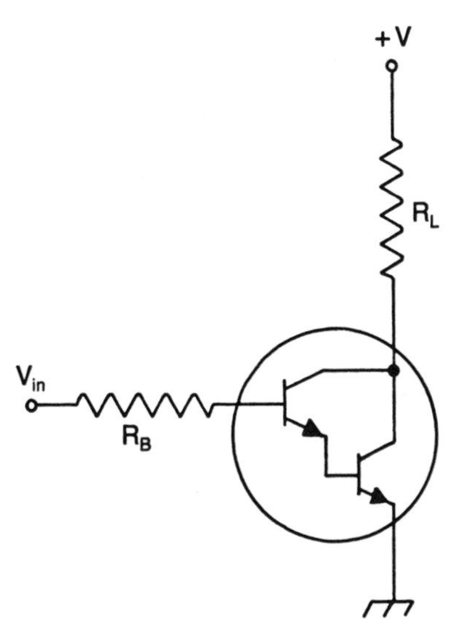


Fig. 10-16. Darlington pair transistors can also be used for moderately large load switching applications.

On the other hand, when the transistor is operated in the pinch-off region, resistance is in the thousands of megohms. The transistor behaves as an open circuit—the ideal switching characteristic when the switch is opened or set to *off*. Thus, if the gate voltage were to swing the drain-to-source current (and hence the voltage) from a large value to a very small value, the resistance presented by the FET can vary from several ohms to many megohms. This is an ideal switch.

The main drawbacks of the FET switch are its input and output capacitances. Because of these, the switching speed of the FET is slower than that of the bipolar device. Devices with low capacities should be chosen when switching speed is a factor.

## TRANSISTOR VOLTAGES IN THE VARIOUS STATES

When a transistor is just at cutoff, the voltage from the base to the emitter of a silicon transistor can be assumed to be zero, while the voltage across the terminals of a germanium device is about −0.1. As the silicon transistor starts to conduct, however slightly, the base-emitter voltage reading will be about 0.5. In the case of a germanium device, it is about 0.1 volt. In saturation, the voltage across the base-emitter silicon junction is about 0.7 while across the germanium junction it is about 0.3 volt. The base-emitter voltage decreases 2 mV for each 1° C rise in temperature.

In saturation, the collector-emitter voltage for silicon devices is about 0.3 and for germanium transistors is about 0.1. When the collector current is less than 10 mA, the collector-emitter saturation voltage varies only minutely with temperature. At high collector currents, the voltage will *increase* at about 0.2 mV for each 1° C rise in temperature. The collector-emitter voltage increases at about 2 mV/° C.

## MULTIVIBRATORS

A multivibrator circuit usually consists of two transistors in a switching circuit arrangement. In the quiescent state, one device is on and the other is off. They can be switched to reverse their states. During switching, both devices can be on simultaneously. In the conventional multivibrator circuit, regenerative or positive feedback is placed around two transistors. Should both transistors conduct and the feedback is greater than 1, the circuit is unstable. The circuit must be unstable while switching. Circuit components can be so specified that the *on* transistor is operating either in the saturation or current mode.

Three different types of multivibrators are commonly used in electronic equipment. The astable or free-running multivibrator is the equivalent of a feedback oscillator. It is stable for short periods of time, or quasi-stable. Both transistors interchange their *on* and *off* states continuously. During one portion of the cycle, one transistor is on while the second is off. The states are reversed during the second portion of the cycle.

On the other hand, the monostable or one-shot multivibrator has one stable state. Assuming that switching transistor No. 1 is on in the stable state and switching transistor No. 2 is off, a pulse can be applied to the circuit, turning transistor No. 1 off and transistor No. 2 on. After a period of time predetermined by the circuit components, the circuit will revert to the original condition. The transistors will remain in their original states until a second switching pulse is applied, once again reversing the states. After a period of time, the transistors will again revert to their original states awaiting other switching pulses.

The bistable multivibrator or flip-flop circuit has two stable states. Either transistor can be on or off. They will reverse states only on the application of a switching pulse. They will retain the new states until a second pulse forces the transistors to revert to the original states.

### The Bistable Multivibrator

The bistable multivibrator is a switching circuit that can be held in either an on or off condition indefinitely. Each time an input pulse is received, the output reverses state.

To understand how this type of circuit works, look at the typical circuit shown in Fig. 10-17. Assume that Q1 is cut off, while Q2 conducts. Q1 is held in cutoff by virtue of the negative $-V_{BB}$ supply. The collector voltage of Q1 is at $+V_{CC}$. This voltage, applied to the base of Q2 through $R_{C1}$, counters $-V_{BB}$ to keep transistor Q2 turned on. Should a positive pulse be applied to the base of Q1, this transistor will be turned on and the collector voltage will drop to zero. Q2 will be turned off because there will be no positive voltage to counter $-V_{BB}$.

A similar chain of events will be triggered by applying a negative pulse to the base Q2. It will be turned off. The collector voltage will rise to $+V_{CC}$. Applied to the base of Q1 through $R_{X1}$, it will counter $-V_{BB}$ and turn on Q1. During the changing cycle from

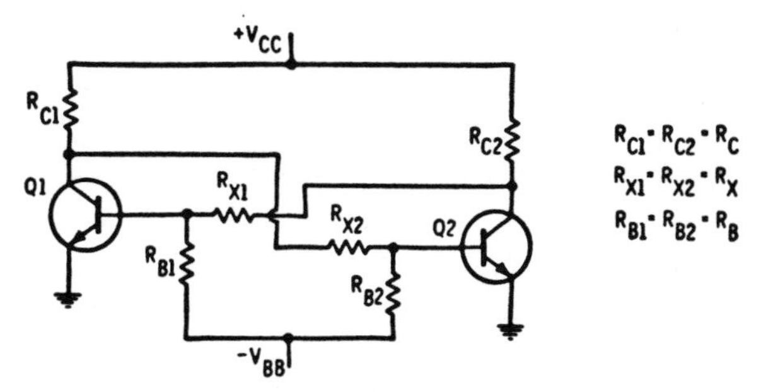


Fig. 10-17. This basic bistable multivibrator circuit uses a fixed bias supply.

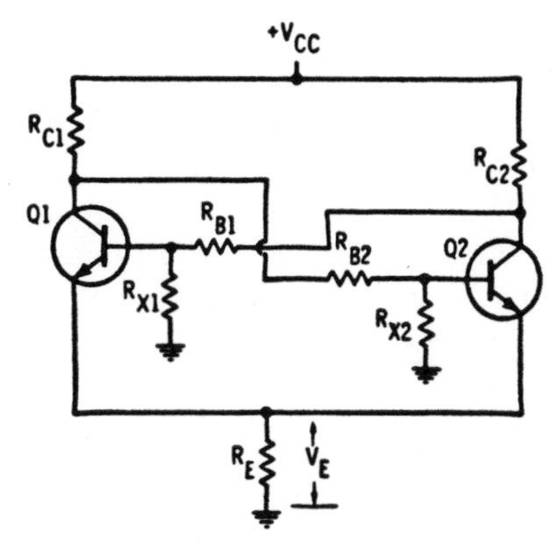


Fig. 10-18. This bistable multivibrator circuit operates from a single-ended power supply.

one state to the other, the circuit must be unstable. This is achieved when $\beta R_C$ is greater than $R_X$. The loop gain will then be greater than 1.

The output voltage swing at either collector is approximately $V_{CC}$ volts, because it is the difference of the voltages across either transistor in the *on* and *off* states.

A bistable circuit requiring only one power supply is shown in Fig. 10-18. The voltage required to bias the transistor off, replacing $-V_{BB}$, is developed across the emitter resistor, $R_E$. Since there is current flowing through one transistor while the other transistor is turned off, the voltage across $R_E$ is equal to the saturation current of one transistor multiplied by the resistance of the emitter resistor.

## The Monostable Multivibrator

The monostable multivibrator has one stable state and one quasi-stable state. In Fig. 10-19, while the negative base supply voltage, $-V_{BB}$, acting through $R_X$ keeps Q2 turned off, $V_{CC}$ and $R_{B1}$ supply the proper current to the base of Q1 to keep that transistor in saturation. C1 is used to couple any pulse that may appear at the collector of Q2 to the base of Q1. C2 is the speed-up or commutating capacitor. The states are switched by applying a negative pulse to the base of Q1 or collector of Q2, thus turning off Q1.

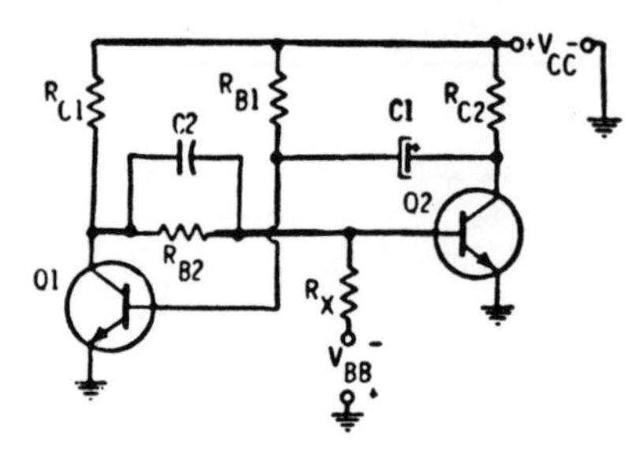


Fig. 10-19. A basic monostable multivibrator circuit.

$R_{B1}$ must be connected to a large voltage, preferably $+V_{CC}$ as shown in the drawing, if C1 is to discharge readily. The discharge time of C1 is equal to $0.69R_{B1}(C1)$. It increases somewhat with an increase in temperature, but the effect of the temperature on the discharge time is diminished if $V_{CC}$ is large.

In the stable state, the collector current of the normally *on* transistor, Q1, is approximately $V_{CC}/R_{C1}$ when the transistor is in saturation. The base current of this transistor is about equal to $V_{CC}/R_{B1}$. The transistor will be in saturation if the minimum collector current required for saturation, $V_{CC}/R_{C1}$, is equal to or less than beta multiplied by the base current, $V_{CC}/R_{B1}$.

As for the *off* transistor, the voltage present from the base to emitter of Q2 involves the saturation voltage, $V_{CE}(sat)$, of Q1. The base-emitter voltage is $-\{V_{BB}R_{B2} - V_{CE}(sat)\ (R_X)\}/(R_{B2} + R_X)$. The transistor is usually off when the base-emitter voltage is more negative than $-0.1$ for a germanium transistor and more negative than zero for a silicon device.

## The Astable Multivibrator

Neither of the two transistors remains indefinitely in any one state when they are arranged in an astable multivibrator circuit similar to that shown in Fig. 10-20. They keep alternating between *on* and *off* states despite the absence of an applied signal or pulse. Effectively, this circuit is the basis of a square-wave oscillator with feedback from the collector of one transistor to the base of the other.

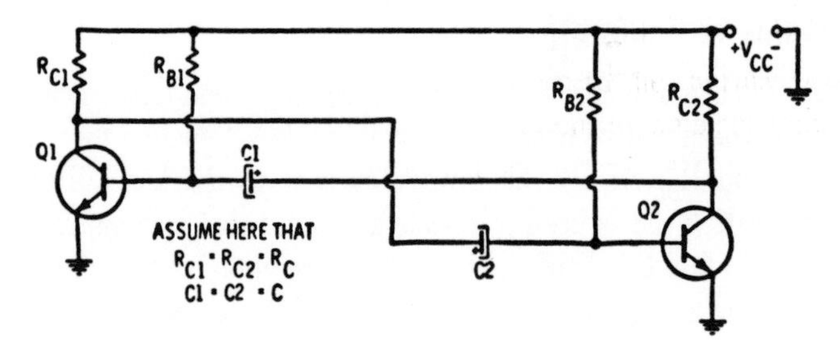


Fig. 10-20. A basic astable multivibrator circuit.

Assume for the moment that Q1 is on and Q2 is off. The collector of Q2 is at the supply potential, $V_{CC}$. The collector of Q1 is near ground potential. (It is at $V_{CE}$(sat) with respect to ground.) Since C2 has been charged during the previous half-cycle, there is a voltage across the capacitor in the direction noted in the drawing. The positive end is at ground potential because the collector of Q1 is at ground. The negative end makes the base of Q2 negative with respect to ground, cutting it off.

While C1 is charging (through the base circuit of Q1 and through $R_{C2}$) to $V_{CC}$ with the polarity as shown, C2 is discharging through $R_{B2}$. When the process has been completed, Q1 will be turned off and Q2 will be turned on. The time for one half-cycle of the square wave is $0.69R_{B1}(C1)$, and the time for the second half-cycle is $0.69R_{B2}(C2)$.

## SOLID-STATE RELAYS

Devices such as optical couplers (discussed in Chapter 15) and silicon-controlled rectifiers (discussed in Chapter 11) can serve as relays. Each device has its advantages and disadvantages in this application.

Semiconductor relays have the advantage of greater reliability than their mechanical counterparts. Besides being smaller in size, the semiconductor relay has no contact bounce and does not deteriorate through use. However, there are disadvantages. With the exception of the optical coupler, most solid-state relays suffer from a lack of complete isolation between the tripping and switching circuits. Other disadvantages are that complexities must be added to the circuit if several poles must be switched simultaneously. The semiconductor switch may be damaged by transients and an undesirable voltage drop is always present across a semiconductor device.

Two transistor relay circuits are shown in Fig. 10-21. When S1 is open in Fig. 10-21A, no current (or only leakage current) flows through Q1 or Q2 and the load. The base of Q2 is essentially at ground potential, so this transistor is turned off. When S1 is closed,

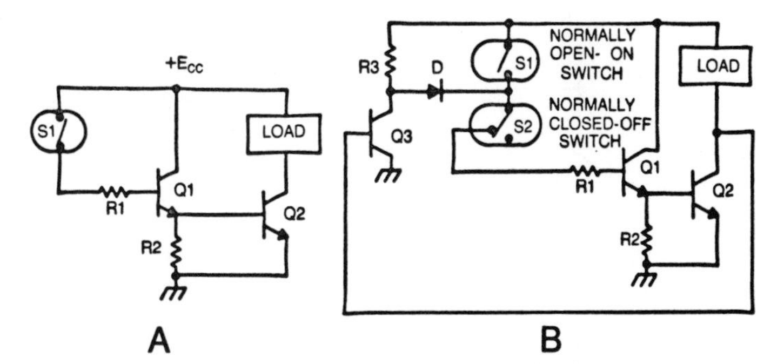


Fig. 10-21. In many applications a solid-state relay can be very useful. (A) This circuit operates a solid-state relay using a Darlington pair. (B) This circuit uses a Darlington pair to create a latching relay function. S1 is a normally open switch, and S2 is a normally closed switch.

current flows through Q1 putting its emitter, and hence the base of Q2, at a voltage above ground. Q2 is turned on so that current flows through its load.

Latching relays use two momentary contact switches. One switch is used to trip the relay circuit, while the second switch is used to open the relay circuit. The relay *on* switch is a normally open type and the relay *off* switch is a normally closed type. Transistor Q3 is driven by the voltage at the output of Q2. When Q2 is off, its collector voltage is high, turning on Q3. The collector current of Q3 is high, so its collector voltage is very low. When this voltage is applied to the base of Q1, it keeps Q1 as well as Q2 turned off. Should switch S1 be set to the position where it conducts momentarily, Q1 is turned on. As the voltage at the collector of Q2 becomes low when this happens, Q3 is turned off. Its collector voltage jumps up, thereby keeping Q1 and Q2 turned on until S2 is pressed. When S2 is pressed, it breaks the continuity of the base circuit of transistor Q1, turning it and Q2 off, and Q3 is turned on to fulfill its function in the latching circuit to keep Q1 and Q2 turned off. The diode is an ordinary silicon device used to keep voltage at the base of Q1 from being fed back to the collector of Q3. Resistors can be added in series with the various bases to keep all base and collector currents at safe levels.

## UJT SWITCHING CIRCUITS

The UJT (UniJunction Transistor) is essentially a switching device. Most of its applications are variations on electronic switches. The UJT was described in Chapter 2.

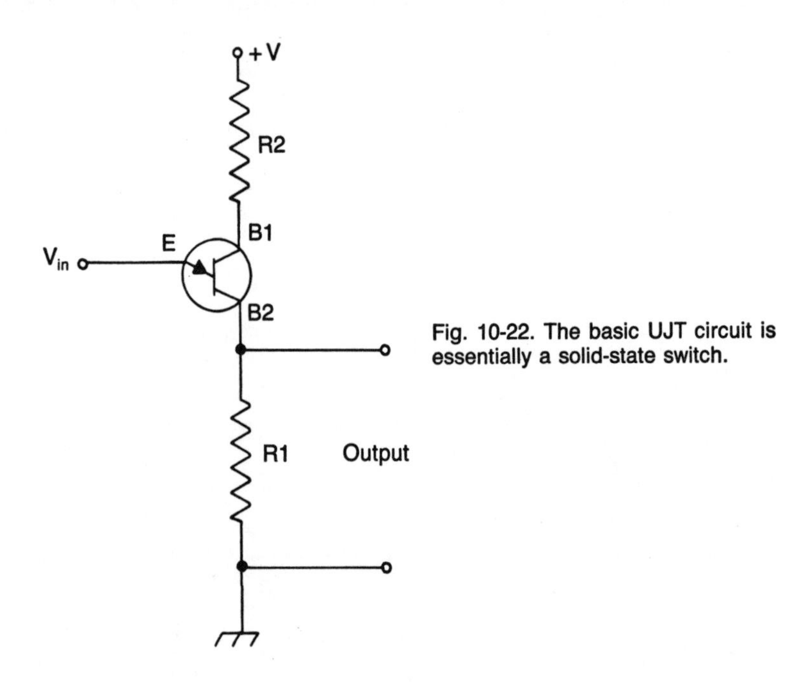


Fig. 10-22. The basic UJT circuit is essentially a solid-state switch.

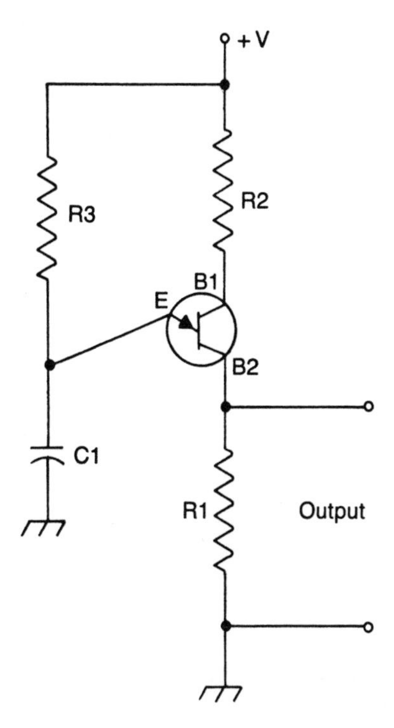


Fig. 10-23. Adding an RC network to the basic UJT switching circuit of Fig. 10-22 causes it to oscillate.

The basic UJT circuit is shown in Fig. 10-22. The voltage applied to the emitter (referenced to ground) must exceed a specific trigger value. When this happens, the UJT opens up a conducting path between its two base connections. Current flows from V+, through R2, into base B1, out of base B2, through R1 and on to ground. When the voltage at the emitter is less than the trigger voltage, the current flow through the bases will be blocked.

If a series of pulses are fed into the emitter, the output (tapped off across resistor R1) will be a series of pulses in step with the input.

In some applications, resistor R2 might not be needed. Its primary function is to improve the circuit's thermal stability.

A UJT switching circuit can be forced into oscillation by adding a simple RC network, as shown in Fig. 10-23. The output frequency will be approximately equal to;

$$F = 1/(C1 \times R3)$$

# 11

# THYRISTORS

A large number of semiconductor devices have been considered throughout this text. Many other types are also being produced. This chapter will be devoted to the various types of thyristors currently available.

A thyristor is a four-layer semiconductor device. To some extent, it is similar to an ordinary junction diode, but there are some important differences. For one thing, a thyristor has three leads instead of just two. In addition to the usual anode and cathode, there is also a control electrode or gate. The gate determines whether or not the device conducts current.

Thyristors are used primarily in switching applications. The first thyristor device to be considered here is probably the most common—the silicon-controlled rectifier, or SCR.

## The SILICON CONTROLLED RECTIFIER (SCR)

The SCR is basically a silicon diode. As with all other diodes, there is no conduction when it is reverse-biased, except when the reverse breakdown voltage is exceeded. Furthermore, the SCR does not conduct even when forward-biased, except when the forward breakdown or breakover voltage is exceeded. The forward and reverse breakdown voltages are about equal.

## Basics of the SCR

The SCR can best be explained with the help of the curve in Fig. 11-1A and the circuit in Fig. 11-1B. When the diode is reverse-biased so that the anode is negative with respect to the cathode, leakage current is small and less than the reverse blocking current, $I_{RRM}$, until the voltage $V_{RRM}$ is exceeded. $V_{RRM}$ is the peak-inverse voltage that may be repetitively applied across the diode. A higher voltage, $V_{RSM}$, is a transient or surge voltage that may be applied only on a nonrepetitive basis. The maximum permissible duration of the reverse surge current, $I_{RSM}$, when rated for a half-cycle surge pulse, is 8.5 ms.

If the diode is forward biased, it will conduct only a small quantity of current until the repetitive forward-blocking voltage, $V_{DRM}$, is exceeded. The forward-blocking current is $I_{DRM}$. When the forward voltage becomes as large as the breakover voltage, $V_{(BO)}$, it will drop instantly to $V_{TM1}$. (The difference voltage between $V_{(BO)}$ and $V_{TM1}$ is developed across $R_L$ in Fig. 11-1B). SCR current will then rise rapidly as the voltage across the diode is increased. In this portion of its operating characteristic, the device behaves much as any ordinary silicon diode. The SCR will stop conducting only when the diode current drops below the holding current, $L_H$.

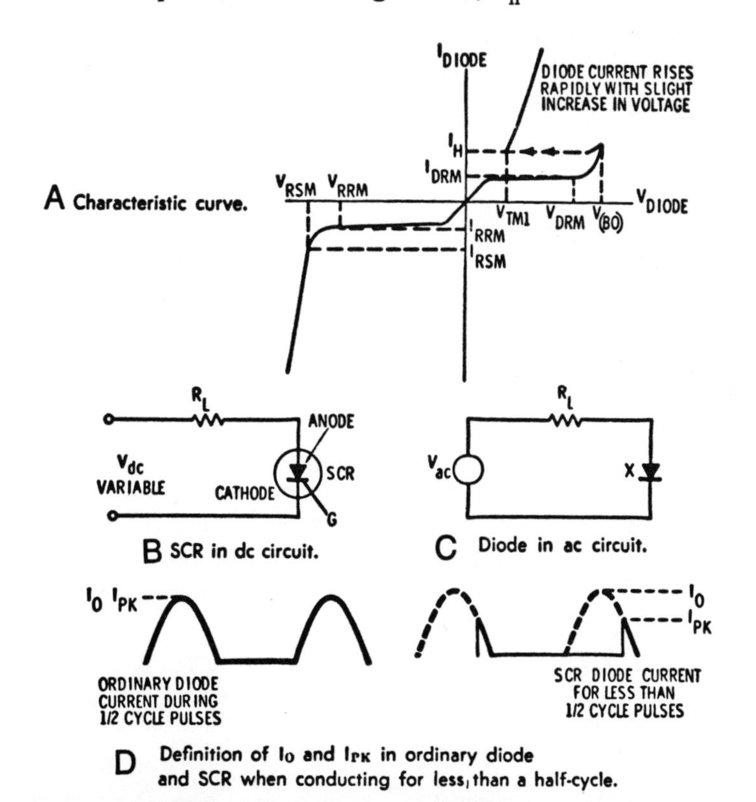


Fig. 11-1. Diagrams illustrating the characteristics of the SCR.

## Triggering the SCR

In Fig. 11-1B, the SCR is shown with a third terminal, marked G, for gate. This terminal forms a diode with the cathode of the SCR. If this terminal is made positive with respect to the cathode, even for a short period of time, the SCR will conduct even if the applied voltage is less than $V_{(BO)}$. The more positive the gate pulse is, the lower will be the anode voltage required to put the SCR into the conducting mode. $I_{GD}$ is the maximum gate current and $V_{GD}$ is the maximum gate voltage that will not cause the SCR to be turned on. Undesirable triggering can occur if the voltage from the anode to the cathode is much larger than its rated value, a voltage may be suddenly applied across the SCR, or the device is subjected to high temperatures.

SCRs are rated for the maximum transient voltage that may be applied between the anode and cathode without causing the device to turn on. This is specified by a dv/dt rating, where dv is a change or increment in voltage, while dt, is a change or increment in time. Using the rating for the Westinghouse 2N690 as an example, dv/dt is specified at 100 volts per microsecond. If the slope of any pulse applied to the SCR is greater than this, the SCR will probably turn on. To avoid undesirable triggering from a sudden transient voltage, a 0.05-$\mu$F capacitor may be connected from the gate to the cathode.

Leakage current may be of sufficient magnitude to trigger the SCR. The device may be stabilized against such objectionable triggering by connecting a small resistor from the cathode to the gate, as specified by the manufacturer of the device. Similar results can be achieved by reverse-biasing the gate-cathode diode.

Even though large gate current pulses, $I_G$, can be used to trigger the SCR when voltages less than $V_{(BO)}$ are applied, the gate cannot turn the SCR off. It will stop conducting only when the current drops below $I_H$ or a reverse voltage is applied across the SCR.

Gate current is limited by the maximum power, $P_{GM}$, it can dissipate, as well as by its maximum gate current, $I_{GFM}$, rating. Care must be exercised not to exceed the maximum reverse voltage, $V_{GRM}$, that may be applied between the gate and cathode, as well as the maximum current, $I_{GR}$, the gate may carry in the reverse direction.

Should the SCR be reverse-biased while the gate is positive, the reverse leakage current of the diode will increase considerably. Under these conditions, the SCR will dissipate extra power. The situation can be alleviated by connecting a series resistor-diode combination from the anode to the gate, with the cathode end of the diode connected to the anode of the SCR. The gate is thus clamped to the negative anode, reducing conduction from the gate to the cathode.

The gate may be fired from a low-impedance voltage source or a high-impedance current source. (The unijunction transistor, discussed in an earlier chapter, is frequently used for this purpose.) The circuit in Fig. 11-2 uses the ac supply applied to the SCR diode for triggering purposes. During the portion of the cycle when the upper power-supply terminal is positive with respect to the lower one, diode D conducts, and the

Fig. 11-2. An SCR gate can be fired from an ac power supply.

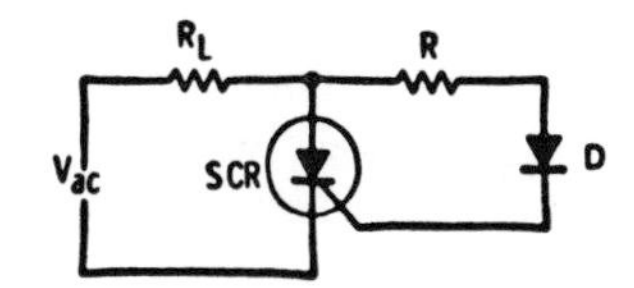


SCR is turned on. As for the second portion of the cycle, the reverse anode-cathode voltage on the SCR turns it off, and there is no voltage at the gate due to the diode D.

There is a delay time, $t_d$, that extends from the instant the gate is fired until the SCR diode current reaches 10 percent of maximum value. The time during which the diode current rises from 10 percent to 90 percent of its final value, is the rise time, $t_r$. The turn-on time, $t_{on}$,is $t_d + t_r$, and it depends primarily on the size of the gate pulse. The period of time from when the diode current turns negative until the gate once again regains control is $t_o$, the turn-off time.

## Designing SCR Circuits

Triggering can be arranged so that the SCR will conduct for less than a half-cycle. Conduction can start after the cycle has passed through its peak. However, had there been conduction for a full half of the cycle, the peak current, $I_O$, would be $V_{PK}/R_L$. ($V_{PK}$ is the peak voltage applied across the diode and load, $R_L$, as shown in Fig. 11-1C.) The peak current while the diode is in the conducting mode is $I_{PK}$. It is equal to the peak voltage across the circuit while the diode conducts divided by $R_L$. The relationship between the four currents, $I_o$,$I_{PK}$,$I_{AV}$ and $I_{RMS}$, when diode conduction is less than for half a cycle, can be determined from Fig. 11-3. For a full half cycle, $I_{AV} = I_{PK}/\pi$ and $I_{RMS} = I_{PK}/2$.

Forward-current transients can destroy diodes. While triggering the SCR for a high dv/dt will not damage the device, the SCR cannot withstand, without damage, current pulses sharper than its maximum di/dt rating, at any time in the cycle. The increment or change in current is noted by di. A resistor placed in series with the capacitor, connected from the anode to the cathode, will alleviate the transient problem. The resistor may be about twice the size of the load resistance.

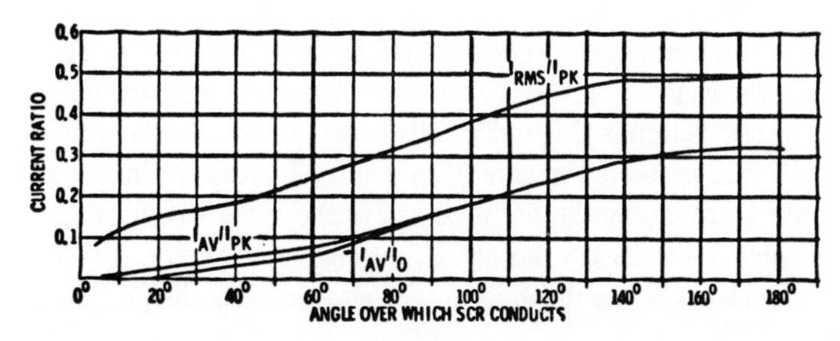


Fig. 11-3. The ratio of current through an SCR for various conduction angles.

Should several SCRs be connected in series, each diode should be shunted with a parallel resistor-capacitor combination. The capacitor value ranges from 0.01 μF to 0.05 μF. The maximum size of the resistor can be determined as follows:

1. Subtract the maximum peak voltage, $V_{PK}$, that appears across all the SCRs connected in series, from the maximum voltage that may, from the ratings, be placed across the series string of diodes.
2. Divide this by the number of diodes in the series circuit, minus one.
3. Multiply the result by 50 if the average forward-current rating of the diode is 5 amps or less; multiply by 30 if the average current rating is between 5 and 15 amps; multiply by 10 if the average current rating is between 15 and 50 amps.
4. The resistor determined above is a maximum value. Smaller resistors are most desirable. Resistors used in the circuit should be within a few percent of each other.

If SCRs are connected in parallel, a resistor should be inserted in each gate lead to equalize the various gate impedances. The gates must be fired simultaneously. SCR diode currents should be divided equally between them. Matched pairs may be used. Another method is to place identical resistors in series with all diodes. The total current flowing through all SCRs should be about 80 percent of the total rated current of all parallel diodes in the circuit.

### Applications of SCRs

SCRs can be triggered by some device or sensor included in the circuit to do a specific job. Suppose you want a light to go on when it gets dark. A circuit that can be used to do this is shown in Fig. 11-4A. The resistance of the photosensitive resistor in the drawing increases as the intensity of light striking it is reduced. The portion of the $V_{BB}$ voltage across the photosensitive resistor (in the voltage divider formed by the resistor and Rl) is sufficient to turn on the emitter-base 1 circuit of the UJT. It fires. A voltage is developed across R2. It is applied to the gate-cathode circuit of the SCR through resistor R3. R3 is used to limit gate current to a safe value. This current turns on the SCR so that current from battery B will flow through bulb L, lighting it.

When ac is applied to an SCR circuit, it can be arranged to limit the interval during which the SCR is conducting to a predetermined amount of time. One circuit that can accomplish this is shown in Fig. 11-4B. Capacitor C charges through the variable control. When voltage across capacitor C is sufficiently large, it triggers the gate of the SCR through diode D2. D2 protects the gate from voltage breakdown by not allowing any voltage to appear there during the negative half of the cycle when the ground end of the circuit is positive. C discharges through D1 during this half cycle. A variable resistor (potentiometer) is shown in the drawing. It is used to control the portion of the cycle

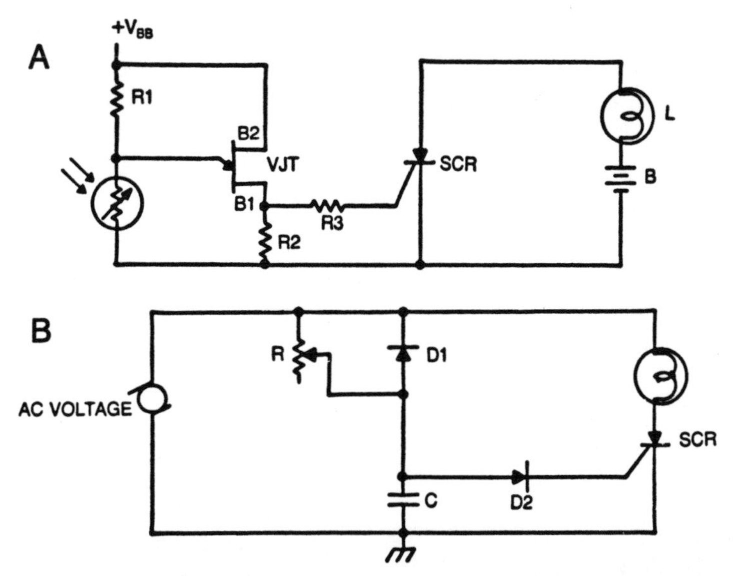


Fig. 11-4. Two typical SCR applications. (A) Bulb L lights when the ambient light is low in intensity. (B) This circuit is used to select the on time of the SCR.

during which the SCR conducts. It can be used to vary the conduction period from 0° to 180°.

## THE TRIAC

Unlike the SCR, the triac conducts current in both directions if an ac signal is applied to its main terminals. It will also be triggered regardless of the polarity of the voltage applied to its gate, as long as there is sufficient voltage available across the main terminals and there is sufficient current available to flow through these terminals.

### Triac Behavior

The symbol of a triac and its characteristic curve are in Fig. 11-5. Triacs are essentially two paralleled SCRs with the anode of one connected to the cathode of the other. Gates of both SCRs are connected together. Leads are referred to as main terminal 1($MT_1$) and main terminal 2 ($MT_2$).

In quadrant I in Fig. 11-5B, the curve looks exactly like the curve of the SCR. Here, $MT_2$ acts as the anode and is positive with respect to $MT_1$. The gate can be positive or negative with respect to $MT_1$ and still turn on the triac. In quadrant III, the curve looks like a redrawn version of the curve in quadrant I, because it retains its general shape. All voltages and currents are generally the same. When the triac is working i quadrant III, $MT_2$ is negative with respect to $MT_1$. Once again, a positive or negative gate with respect to the $MT_2$ will turn on the triac.

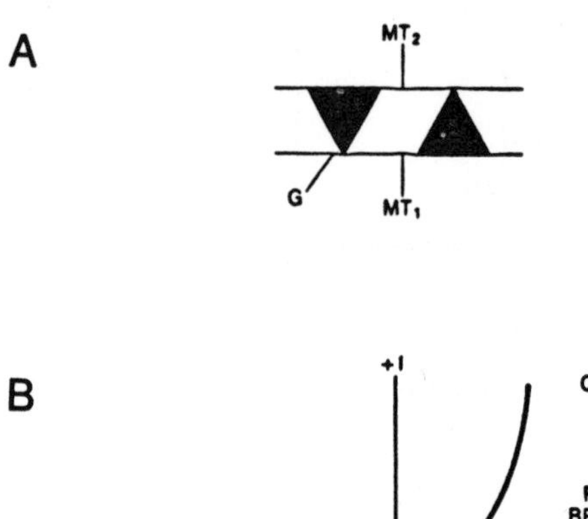


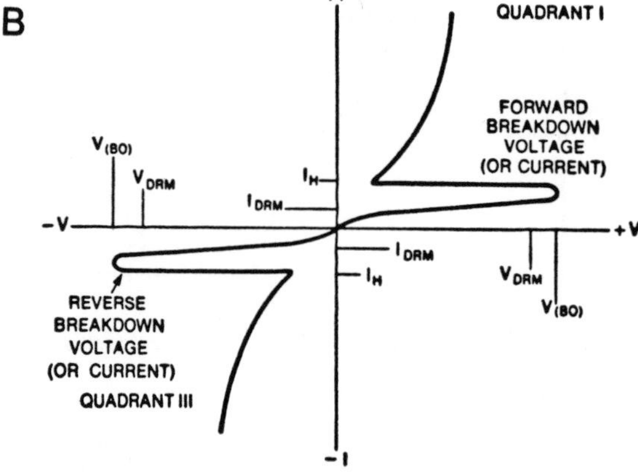


Fig. 11-5. The Triac is like a pair of back-to-back SCRs. (A) Schematic symbol. (B) Characteristic curve.

The behavior of the triac in the first and third quadrants is very similar to the behavior of the SCR in quadrant I. The triac conducts if the applied voltage exceeds $V_{DRM}$. It conducts at a lower voltage if a pulse of either polarity is applied to the gate. Conduction ceases when the triac main terminal current drops below $I_H$, even if a gate current is present. Operation is usually limited to input voltages or currents at the 60 Hz power-line frequency or less.

Inductive loads pose problems when used in triac or SCR circuits. After the current flowing through the triac drops below $I_H$, a voltage is still present across the thyristor if an inductor is the load. The voltage due to the change of current in the inductive load is equal to L(di/dt). Here, L is the inductance in henrys, di is a change in current, and dt is a change in time. This voltage may keep the triac in a conducting state so that current keeps flowing. A cure for this situation is to connect a series R-C circuit across the $MT_1$ and $MT_2$ terminals. This should limit the rate of rise of voltage (dv/dt, where dv is a change of voltage). The final voltage should be held to below $V_{DRM}$, dv/dt is referred to as the critical rate of voltage rise.

The four methods of triggering the triac from a gate voltage are: (1) The gate and $MT_2$ are both positive with respect to $MT_1$, (2) the gate and $MT_2$ are both negative with respect to $MT_1$, (3) the gate is negative and $MT_2$ is positive with respect to $MT_1$, and (4) the gate is positive and $MT_2$ is negative with respect to $MT_1$. Obviously, situations

1 and 3 exist in quadrant I because $MT_2$ is positive in both instances, while situations 2 and 4 are in quadrant III because $MT_2$ is negative in both instances. Sensitivity to the size of the gate pulse varies with the trigger method. Maximum sensitivity is achieved when either method 1 or 2 is used. Sensitivity is slightly reduced when using method 3 and is poor when method 4 is applied.

### Practical Circuits

Triacs are used in circuits to limit the conduction time during the positive and negative portions of the ac cycle, whereas the SCR can be used to limit current only during the portion of half-cycle through which it conducts. Because of this, the triac can be used in a circuit to dim lights, control the speed of a motor, and so on.

In Fig. 11-6A, VR limits the amount of current that can be fed to the gate of the triac. It starts to conduct only after the gate current exceeds the specific value necessary to turn on the device. Conduction can range anywhere between 90° and 180° of each half-cycle. This situation can be expanded through use of the circuit in Fig. 11-6B, where capacitor C must be charged to a voltage sufficient to turn on the triac. This voltage must be large enough to exceed the breakdown voltage of the intervening semiconductor device, the diac.

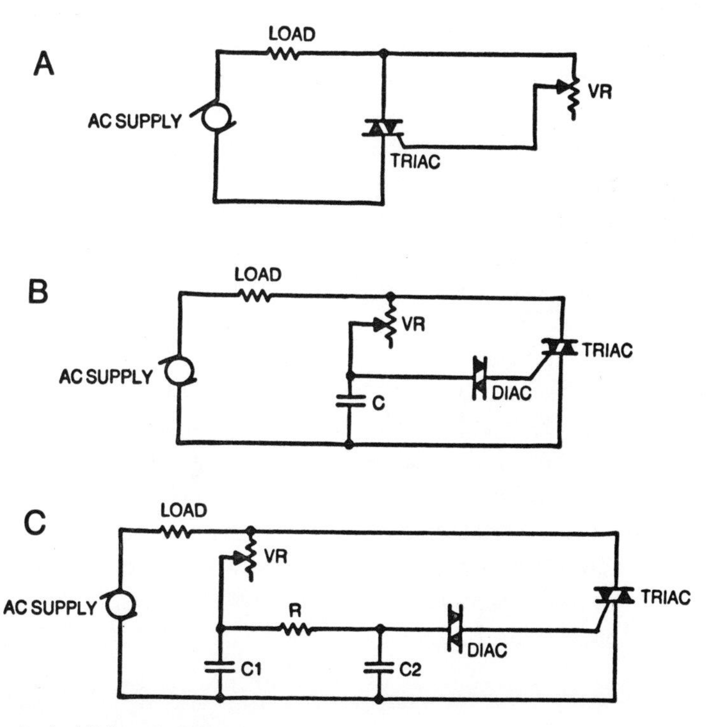


Fig. 11-6. Some typical Triac circuits.

Circuit B is not ideal, as it does not allow the triac to fire when the conduction angle is relatively small. The circuit in Fig. 11-6C was developed to alleviate this situation through use of the second resistor-capacitor network, R-C2. When the voltage across C2 is greater than the breakover voltage of the diac, the voltage across the diac drops more than 5 volts from its breakover value. C2 is discharged through the gate. No new trigger pulse is available until after C2 has been recharged during the next half-cycle.

The load in all circuits shown here is drawn as resistors. This load may, however, take one of many different forms. It may be a bulb or a group of bulbs when the triac circuit is used as a dimmer. A motor in a speed-control application can be used instead of the resistor load shown in the drawing. Any number of different devices may be used as a load where limiting the conducting period in the cycle limits some of the load's activity.

## THE DIAC

It is best to trigger a thyristor by applying a large pulse to the gate for a short period of time to instantly and fully turn on the triggered device. Should the gate current be small, there is conduction initially in only a portion of the main junction area of the triac or SCR. Damaging hot spots may be formed in the device. A large gate pulse would prevent this situation from occurring, but the time duration of the pulse should be limited so as not to exceed the average gate power ratings.

A neon bulb can be placed in series with the gate circuit to limit the conduction time while letting a good solid pulse be applied to the gate. It will light and conduct current after the applied peak voltage exceeds about 65 volts. In a triac circuit, this voltage is usually across a charged capacitor. Voltage across the bulb drops instantly to about 50 volts. Current flows through the bulb and gate circuit from a charged capacitor (such as C in Fig. 11-6B), until the capacitor is partially discharged and the voltage across it drops below the sustaining voltage of the bulb. This voltage reduction across the neon bulb after the breakdown voltage has been exceeded permits a large current pulse to flow through the bulb and through the gate connected in series with it. Current flows until the voltage across the capacitor drops below the sustaining voltage. Hence, there is a short-lived current pulse.

Diac behavior is similar to that of the neon bulb. The diac conducts in both directions. It exhibits a negative resistance as did the neon bulb where the voltage drops, while the current flowing through the diac increases. Similar to the triac, the diac operates in the first and third quadrants where it can be turned on by positive and negative voltage peaks. The primary advantage of the diac over the neon bulb is that the diac breaks down at a lower switching voltage, $V_S$. $V_S$, the positive or negative voltage necessary to turn the diac on from a nonconducting state ranges between 20 and 40 volts. There is a small switching current, $I_S$, flowing through the diac when it is at its switching voltage, $V_S$.

Two factors must be considered when specifying a diac. One factor is the leakage or *peak blocking current*, $I_B$, that flows through the device before the switching voltage has been reached. $I_B$ is specified at a particular voltage. A second important factor describes the amount the voltage across the diac dropped after $V_S$ has been reached, when a specific amount of current (usually a short pulse) is flowing through the device. This *differential voltage*, $\Delta V$, is that voltage change across a particular diac. The reduced voltage is equal to $V_S - \Delta V$.

## THE PUT

A thyristor with only one gate like the SCR is the *programmable unijunction transistor* (PUT). Unlike the SCR, the PUT's gate is at the anode rather than at the cathode. Its applications are similar to those of the UJT. The basic circuit in which a PUT is used is drawn in Fig. 11-7. Two biasing resistors, R1 and R2 are shown. They replace the interbase resistance, $R_{BB}$, in the UJT. The intrinsic standoff ratio, $\eta$, of the circuit is the ratio of R1 to the total resistance in the bias circuit, R1 + R2. Because these resistors can be chosen to establish different values for $\eta$, the PUT is referred to as "programmable."

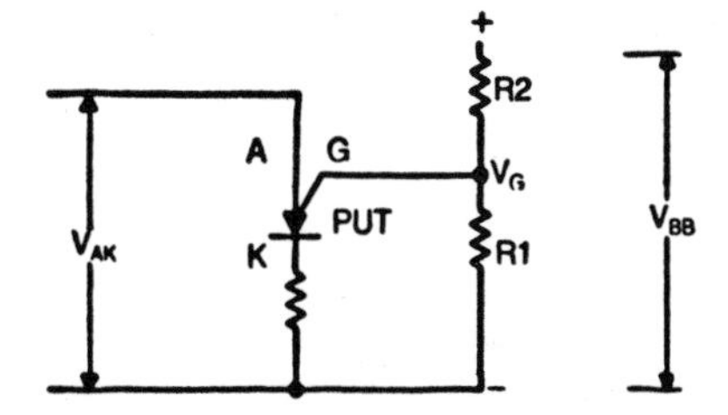


Fig. 11-7. A typical bias arrangement for the PUT.

The device is off as long as $V_{AK}$ is lower than $V_G$. It will turn on when the voltage $V_{AK}$ exceeds $V_G$ by about 0.6 volt, the voltage drop across a forward biased diode. When used in this type of circuit, the PUT surpasses the UJT in a number of ways:

1. $\eta$ can be made anywhere between 0 and 1.
2. Because $I_P$ is considerably lower for the PUT than for the UJT, leakage current can be very low—in the order of 0.1 mA.
3. The PUT is a very efficient switch because the forward voltage drop between the anode and cathode is low.

Two applications of the PUT are shown in Fig. 11-8. In A, the PUT is used as a relaxation oscillator in an arrangement which is very reminiscent of the equivalent UJT circuit. Because $I_V$ is low, this oscillator can operate from a low voltage source. It can also have long time delays between pulses because of the low leakage current. Should

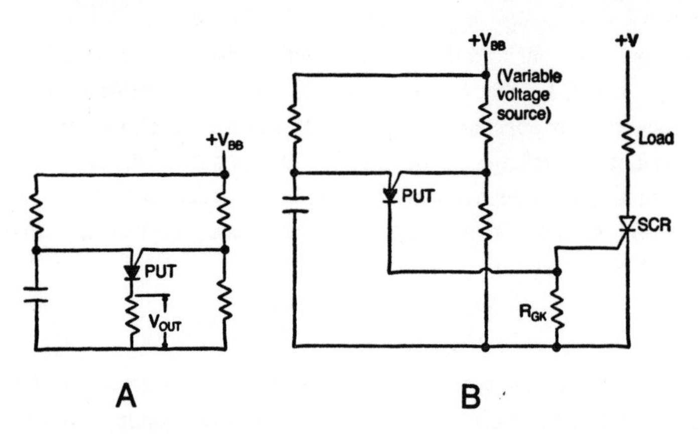


Fig. 11-8. Some typical PUT circuits. (A) Relaxation oscillator. (B) Gate-trigger circuit.

voltage pulses be applied across one of the bias resistors, the oscillator can be synchronized with the frequency of these pulses. A similar PUT circuit can be used to trigger the gates of an SCR.

Another application of the PUT is the threshold detector. Should a variable voltage supply be across the bias resistors, the PUT is turned on when the voltage exceeds a threshold value.

# 12

# Power Supplies

The power supply is conventionally thought of as a device to convert ac power into smooth dc power. The p-n junction rectifier diode serves well in the transformation process. Resistor-capacitor filters are frequently used to smooth ripple produced by the ac fed to the input of the circuit. Zener diodes also perform as filters while at the same time stabilizing the output voltage across the load against the line voltage and load current variations.

## HALF-WAVE POWER SUPPLY

The basic circuit of a half-wave power supply is shown in Fig. 12-1. At the secondary of the transformer, the sinusoidal voltage is $V_{rms}$. It varies repeatedly between the positive peak $+V_p$ and the negative crest of $-V_p$ volts. The complete cycle is divided into 360°.

The average voltage over a complete ac cycle is zero since all portions of the positive half of the cycle from 0° to 180° are cancelled by all portions of the negative half of the cycle from 180° to 360°. Despite this, the voltage during each half of the cycle, in conjunction with the current, can supply power to a load. The effective voltage over a half-cycle is $V_p/2$ and over the complete cycle is $V_p/\sqrt{2}$. These are rms voltages.

In Fig. 12-1, during the portion of the cycle when the upper half of the secondary winding of the transformer is positive with respect to the lower half, the diode conducts,

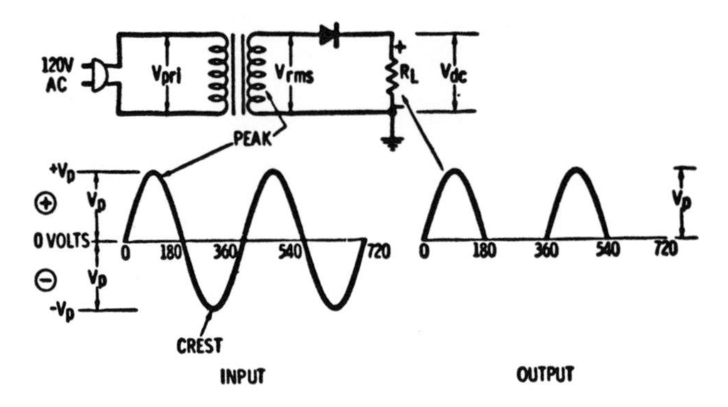


Fig. 12-1. The half-wave power supply is one of the simplest power supply circuits.

and a voltage is developed across $R_L$, with the polarity shown. During the second half of the cycle, the voltage across the diode is in the reverse direction, and there is no conduction. The shape of the output voltage is shown in the figure. Note the half-cycle pulses with peak voltages, $V_p$. It is dc because the sinusoidal half-cycles are only in one direction—positive with respect to ground in this instance. The output is a pulsating dc that has an rms value of $V_p/2$. Because the negative half of the cycle has been eliminated, there is also an average dc voltage across the resistor. It is $V_p/\pi$.

The total equivalent representation of this circuit, Fig. 12-2, includes the impedance of the transformer and diode, in series with the rest of the components. The impedance of the diode is usually so small as to be negligible. The transformer impedance, $R_T$, can be calculated from the formula:

$$R_T = R_s + \left(\frac{V_{rms}}{V_{pri}}\right)^2 R_p \tag{12-1}$$

where,

$R_T$ is the transformer impedance,

$R_p$ is the primary winding resistance,

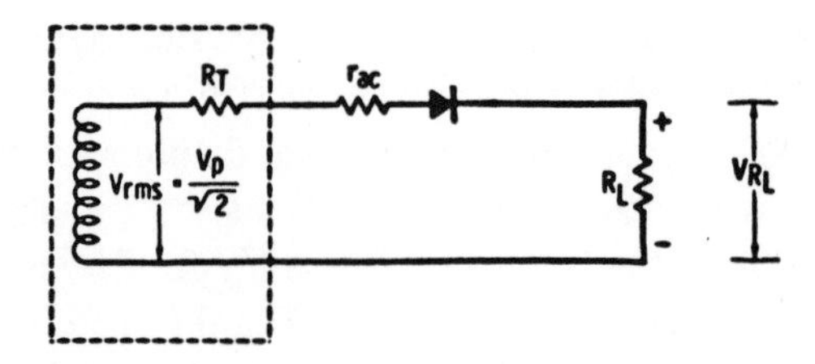


Fig. 12-2. An equivalent circuit showing the impedances in Fig. 12-1.

$R_s$ is the secondary winding resistance,

$V_{rms}$ is the unloaded voltage across the secondary for a specific rms supply voltage $V_{pri}$.

The peak voltage, $V_{PR_L}$ across $R_L$ is:

$$V_{PR_L} = \frac{R_L}{R_T + R_L} V_P \qquad (12\text{-}2)$$

where,

$R_L$ is the load resistance,

$R_T$ is the transformer resistance,

$V_{PR_L}$ is the peak voltage across the load,

$V_P$ is the peak supply voltage.

However, $R_T$ is usually so small that it is negligible when compared to $R_L$.

## Characteristics

Regulation is the amount the output voltage varies with changes in current through the circuit. The percent of regulation is defined by the formula:

$$\%\ \text{Regulation} = \frac{V_2 - V_1}{V_1} 100 \qquad (12\text{-}3)$$

where,

$V_2$ is the dc voltage across the load when $R_L$ is open circuited,

$V_1$ is the dc voltage across the load when $R_L$ is at its rated value.

Smaller numbers for percent of regulation are most desirable. Obviously, regulation is best when $R_T$ is at a minimum.

Another important characteristic of a power supply is the ripple content in the output. There is usually some ac riding on the dc at the load. This ac is undesirable ripple. The ripple factor, r, is determined from the formula:

$$r = \frac{\text{rms voltage components across the load}}{\text{average dc voltage across the load}} = \sqrt{\left(\frac{I_{rms}}{I_{dc}}\right)^2 - 1} \qquad (12\text{-}4)$$

The numerator is the difference between the dc and the rms ripple components.

In the example in Fig. 12-1, the rms voltage across the load is $V_p/2$; the average dc voltage is $V_p/\pi$. These numbers, divided by $R_L$, are the rms and average currents, respectively.

$$\frac{I_{rms}}{I_{dc}} = \frac{V_p/2R_L}{V_p/\pi R_L} = \frac{\pi}{2} = 1.57$$

Substituting this into Equation 12-4, the ripple factor is 1.21. This is quite high.

One other important factor in power-supply design is efficiency; the ratio of the dc power at the load to the ac power input to the system.

$$\text{efficiency} = \frac{P_{Load}}{P_{ac\text{-}input}} \times 100\% = \frac{40.6}{1 + [(R_T + r_{ac})/R_L]} \qquad (12\text{-}5)$$

for the circuit shown in Fig. 12-1.

### Specifying the Diode Rectifier

The diode can be specified from circuit requirements. The peak inverse voltage, or the peak voltage across the diode when it is not conducting is $V_p$ or $\sqrt{2}V_{rms}$ = $1.41_{rms}$. The peak current, $L_p$, is equal to the peak supply voltage divided by the total resistance in the circuit, or $I_p = V_p/(R_T + R_L)$. This is the minimum recurring peak current for which the diode need be rated. The average diode current is:

$$I_{AV} = \frac{I_p}{\pi} = \left(\frac{1}{\pi}\right)\frac{V_p}{R_T + R_L}$$

There is no large initial surge current, and thus this type of rating is not a factor in choosing a diode for use in the circuit under discussion.

If the power dissipated by the diode causes the temperature of the diode to greatly exceed 25° C, a heatsink may be required. An estimation of the power dissipation can be made if it is assumed that the current flowing through the diode is sinusoidal. Multiply the average diode current ($I_{AV} = I_p/\pi$) by the voltage across the diode. For safety, assume that this voltage is 0.5 volt for germanium units and 1 volt for silicon. Power is the product of the voltage across the diode and the average current.

### Specifying the Power Transformer

The power transformer can be specified from the accumulated data. For an average output voltage of $V_p/\pi$, the transformer secondary must supply an rms input voltage of $V_p/\sqrt{2}$. Setting up the ratio of rms to output voltage, the secondary rms voltage is $\pi/\sqrt{2}$

or 2.22 multiplied by the average output voltage. Specify the desired load current at the average output voltage. The minimum acceptable regulation of the circuit may be controlled by indicating a maximum or minimum voltage at a specific circuit current in addition to the current and voltage used as the center design values.

The permissible operating temperature is an important factor. Specify the maximum ambient temperature around the transformer. Also, specify the maximum winding temperature you require. Most transformers can be operated safely with temperatures up to 105° C. Temperature rise, ΔT, can be determined from the equation

$$\Delta T = \frac{R_H - R_C}{R_C(0.00393)} \tag{12-6}$$

where,

$R_C$ is the cold resistance of a winding of the transformer,
$R_H$ is the hot resistance of the same winding.

### Determine Ripple From a Curve

The supply just described has too much ripple but a capacitor across the load resistor will smooth much of the ripple. The effect of this capacitor is seen in Fig. 12-3.

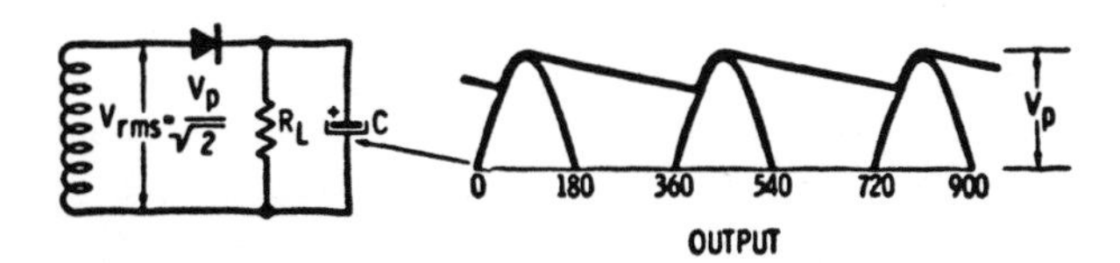


Fig. 12-3. Filter capacitor C smoothes out ripple.

The half-cycle output of Fig. 12-1 is redrawn here. The capacitor across the load is charged to the peak voltage by the half-cycle pulses. The capacitor begins losing its charge the moment the pulse drops from its peak. The voltage across the capacitor will discharge only until the next pulse recharges it to the peak value. The thick curve on top of the signal indicates the shape of the voltage across the capacitor and load resistor.

Ripple factor is a function of the size of the capacitor across the load, as well as the size of the load. The variation of ripple factor with the size of these components is shown in Fig. 12-4. A power-line frequency of 60 Hz is assumed here as well as throughout the remainder of the discussion. C is the capacitance in farads across the load, $R_L$ is in ohms.

Output voltage is related to $R_L$, C and the ratio of the impedance of the transformer, $R_T$, to the resistance of the load, $R_L$. Curves showing this information are shown in Fig. 12-5A.

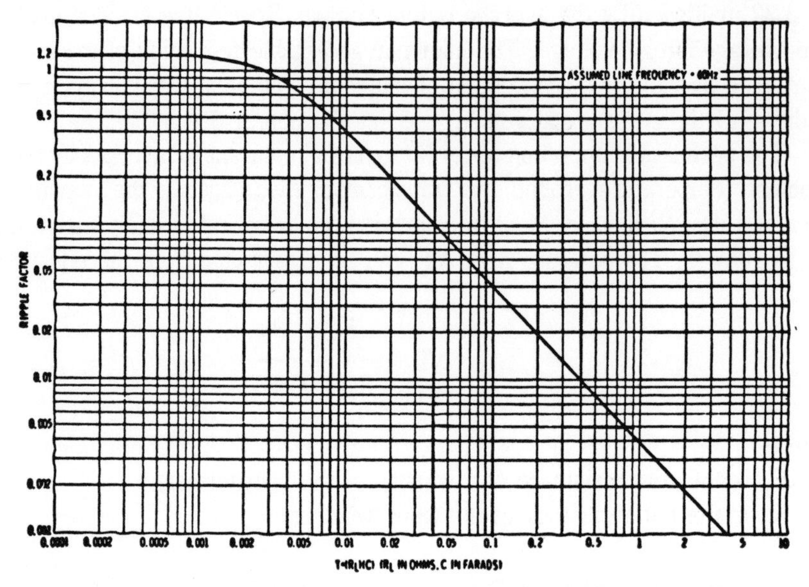


Fig. 12-4. The ripple factor for different values of $R_LC$ in a half-wave circuit.

## Power Supply Currents

An input filter capacitor causes a very large *initial surge current* to flow through the diode. This is because the capacitor is effectively a short when the power is first applied. This short is across the load, $R_L$. The only component that remains in the circuit is $R_T$, The peak surge current, $I_{ps}$, that the diode must be able to support is the peak of the supply voltage divided by $R_T$. It must withstand this for a period of time (time constant) equal to $\tau = CR_T$.

Manufacturers specify the permissible surge current in several different ways. One method indicates the maximum permissible current that can be sustained over one cycle of supply voltage. If the power-supply frequency is 60 Hz, the time for the cycle is $1/60$ second or $16.7 \times 10^{-3}$ secs. If the specification for a particular diode states, for example, a nonrepetitive permissible surge current of 20 amps over one cycle, it means that 20 amps is permitted on an initial surge if the time constant, $\tau$, is less than 16.7 $\times 10^{-3}$ secs. If the current is more than 20 amps or the period of time more than 16.7 $\times 10^{-3}$ secs, the life of the diode under the stress of these surges would be limited.

Other manufacturers indicate maximum surge current when $\tau$ is equal to or less than ½ cycle or $8.35 \times 10^{-3}$ secs. In this case, the surge current must not exceed its specified value if the time constant of the circuit is equal to or less than $8.35 \times 10^{-3}$.

Many manufacturers supply curves indicating the permissible surge current for different values of $\tau$.

Another significant factor in power supplies is the *repeated forward peak current*. The amount of this current flowing through the diode can be determined from curves in Fig. 12-5B.

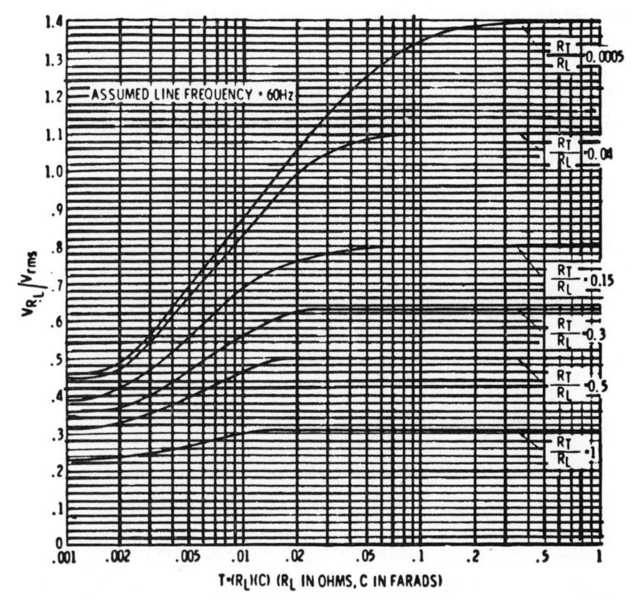


A Plot to determine dc voltage across $R_L$ from rms voltage across transformer.

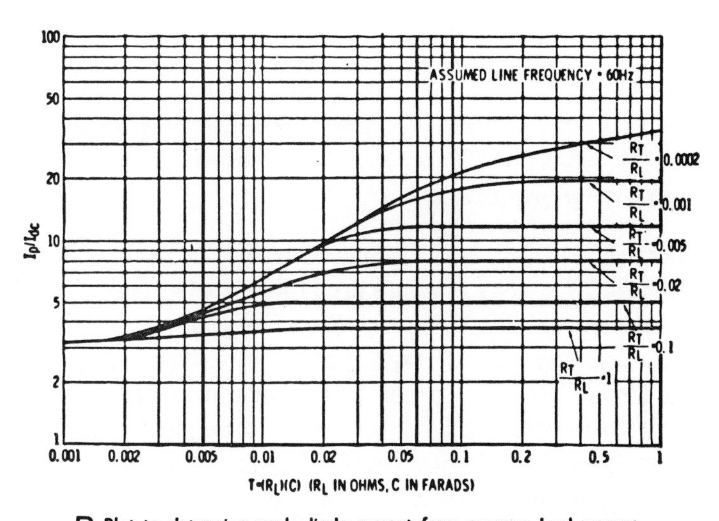


B Plot to determine peak diode current from average load current.

Fig. 12-5. Curves used to determine peak diode current and dc voltage for a half-wave circuit.

Despite all the problems caused by a capacitor, it is an important item in filter circuits. Zener diodes are sometimes used in addition to the capacitor, to help reduce ripple. But the capacitor can seldom be omitted.

The action of the zener diode becomes obvious when it is recalled that the variable input voltages used to derive the zener circuit equations in Chapter 1 may change at a random rate with power-line excursions. The ripple on the dc produced by a filtered power supply may be thought of as a variation of the dc input voltage. This periodic variation may frequently be eliminated by a zener diode rather than a filter capacitor.

The ripple across the diode and load is less than the raw power-supply ripple because of the low ac diode impedance across the load. The circuit in Fig. 12-3 consists of a large resistor, $R_s$, in series with the low ac impedance of the zener diode. Because of the voltage-divider action, ripple in the power supply is primarily developed across the large series resistor, while very little ripple remains for the diode and load. Ripple is reduced when the series resistor is large and the zener diode impedance is small.

## FULL-WAVE POWER SUPPLY

The half-wave circuits discussed thus far can be extended and improved by use of the full-wave power supply shown in Fig. 12-6. Here, each diode conducts for an alternate half-cycle and the two half-cycles add across the load, $R_L$, as shown.

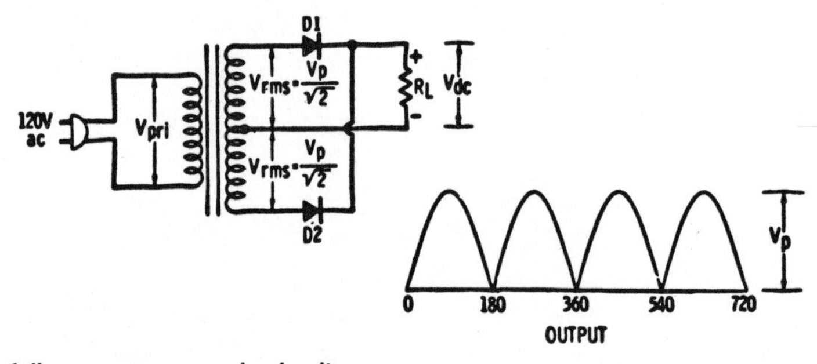


Fig. 12-6. A full-wave power supply circuit.

The transformer secondary is divided into two parts by means of a lead from the center of the winding. $V_{rms}$ volts is across each half of the secondary. The peak voltage across the load resistor is $V_p$. The rms voltage across the load is $V_p/\sqrt{2}$, the same as for an ordinary sine wave. The average or dc output voltage is $2V_p/\pi$, or double the average voltage for the half-wave circuit. Transformer and diode calculations are not unlike those for the half-wave rectifier circuit. In this case, the average current flowing through each diode is only half of the current flowing through the load. The ripple factor is 0.48. The major ripple component in this example is 120 Hz; in the half-wave circuit it was 60 Hz. Although the situation is better than in the half-wave circuit, this ripple is still much too high, and must be reduced. Once again, a capacitor, C, must be added across the load resistor, $R_L$. As before, the ripple can be determined from the curves.

Use Fig. 12-7 to find the ripple factor if $R_T$ is 1/10 of $R_L$ or less. Figure 12-8 may be used to determine the output voltage. When using either curve, it should be noted

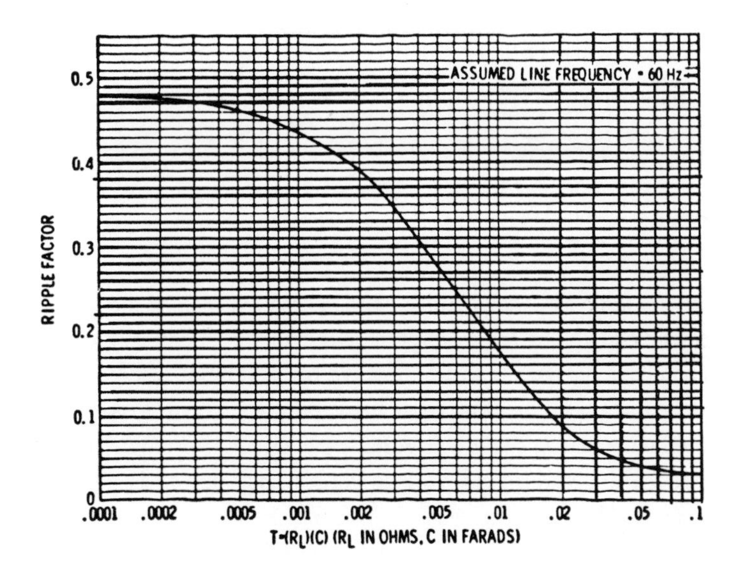


Fig. 12-7. The ripple factor for different values of $R_LC$ in a full-wave circuit.

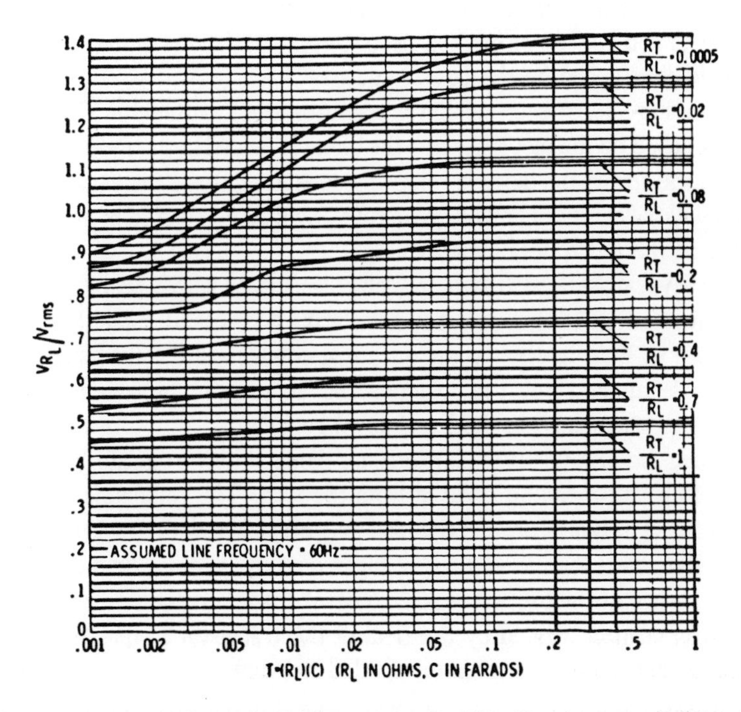


Fig. 12-8. Curves used to determine dc voltage across $R_L$ when $V_{rms}$ is across half the secondary winding. This is for a full-wave circuit.

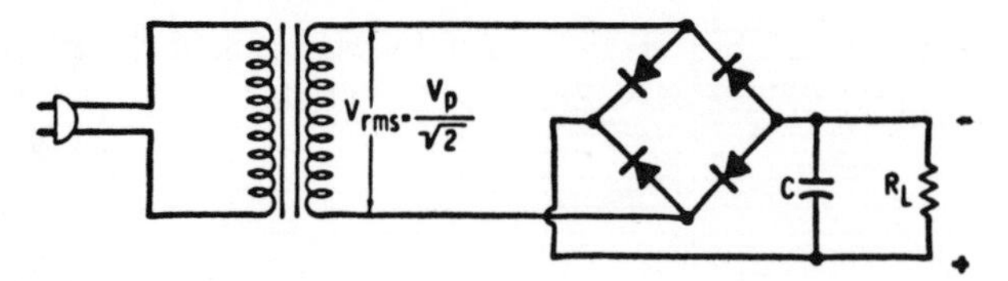


Fig. 12-9. A full-wave bridge circuit.

that $R_T$ refers to the equivalent impedance seen across one-half of the secondary winding of the transformer. $V_{rms}$ is the voltage across half the secondary winding.

A more economical design uses the full-wave bridge circuit of Fig. 12-9. A smaller transformer can be used because, for the same dc output current, less rms transformer current is required here than in the previous full-wave circuit. Although twice the number of diodes are used, the inverse peak voltage across each device is equal to $V_p$ of the supply voltage, or (1.41) $V_{rms}$.

## VOLTAGE REGULATORS

A modern power supply is designed to provide a load at its output with either a constant voltage or a constant current, or both, although not at the same time. Regulator circuits supply a fixed voltage or current despite some variations in the size of the load and/or the voltage supplied by the power source.

Zener diode regulator circuit design procedures were outlined in Chapter 1. The discussion was based on the fact that once the diode is biased in the reverse breakdown region, the voltage across the diode is relatively constant despite the variation in applied voltage to the overall circuit. The voltage across any load connected in parallel with the diode is equal to the diode breakdown voltage.

There are several practical design problems associated with a zener regulator such as the one shown in Fig. 12-3. First, it is not always possible to obtain low-impedance zeners. The size of $R_S$ is limited because it must conduct all the load current as well as the zener diode current. Yet, the voltage across $R_S$ must be of such dimensions that the zener diode is maintained in the breakdown region.

To overcome the drawbacks of simple zener diode voltage regulators, the $R_s$-zener diode circuit of Fig. 12-3 can be isolated from the load. The regulating circuit can be in the base circuit of a transistor, and the load may be placed in the emitter circuit. The zener diode will regulate the base circuit voltages directly, and, consequently, the voltage across the load in the emitter will remain constant. Base current here is approximately equal to the load current divided by the beta of the transistor in question.

Regulator circuits can be conveniently divided into three groups—series, shunt, and feedback. The placement of the output load with respect to the transistor in the regulating circuit is the primary factor in determining if the arrangement is a shunt or series regulator. Feedback may be applied to elements of either circuit. In the feedback regulator circuit, usually of the series type, a portion of the output voltage is sensed and fed back to the regulating circuit for proper compensation.

### Series Regulator

The series regulator in Fig. 12-10 is essentially an emitter follower with the potential from the base to ground maintained at the zener breakdown voltage. Since the emitter is a fixed voltage away from the base, the voltage across the load is also constant with respect to ground.

The dc portion of the unstable power supply is a voltage, $V_{sdc}$, in series with the internal impedance, $R_i$. Similarly, the ripple voltage in the supply is $V_{RIP}$. The path of the load and zener current is through $R_i$ with the zener current being negligibly small. $R_s$, the resistor in series with the zener diode, limits zener current to safe values. Zener and base currents both flow through $R_s$. The voltage across the load resistor, $R_L$, is equal to the breakdown voltage, $V_B$, across the zener diode minus the base-emitter voltage of the transistor. The transistor should be kept out of saturation at all times.

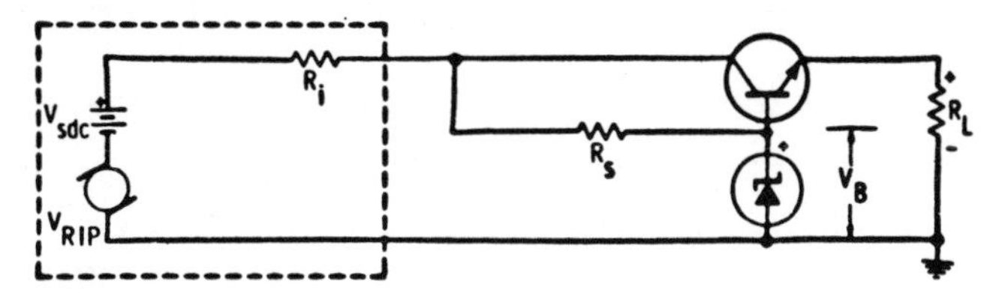


Fig. 12-10. A series regulator circuit.

Regulation is improved if the output impedance of the transistor is small. The emitter resistance, $r_e$, is a very important factor in setting the output impedance. It is equal to $26/I_E$. $I_E$ can be increased to make $r_e$ and the output impedance smaller by shunting the load resistor, $R_L$, with a second resistor, $R_E$.

The regulated output voltage can be made variable if the circuit in Fig. 12-11 is used. A potentiometer of about 1000 ohms is placed across the zener diode. The diode is maintained in the breakdown region by the $R_s$-diode circuit. The regulated voltage is the portion of the zener voltage at the wiper arm of the control minus the base-emitter voltage of the transistor. This is all accomplished at the expense of some of the regulating ability of the circuit.

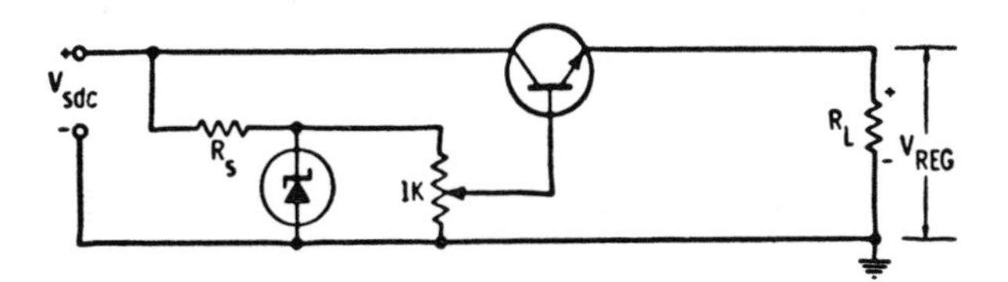


Fig. 12-11. Series regulator circuit with an adjustable output.

## Design Considerations

It is advantageous to ripple reduction and regulation for the zener diode to see a high impedance. Hence, $R_s$ as well as the impedance the zener sees at the base of the transistor, should both be large. To make the latter impedance large, two emitter-followers arranged in the Darlington pair, as in Fig. 12-12, can be used. The beta of the combination is the product of the betas of the two transistors. As an alternative, any of the complementary beta-multiplying circuits discussed in Chapter 5 may be used. As an additional advantage, the output impedance seen by the load, $R_L$, is greatly reduced.

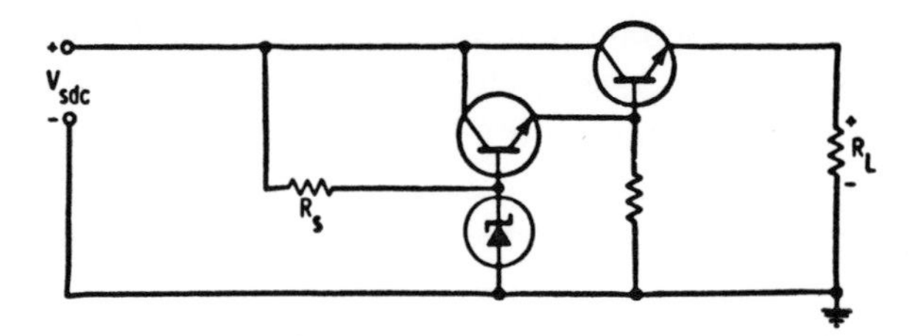


Fig. 12-12. A Darlington pair can be used for better regulation. Any beta multiplier circuit can be used.

As for $R_s$, it should be, ideally, a constant-current source. In Fig. 12-13, it is shown as it may be used with the circuit of Fig. 12-10. As before, the regulated output voltage is equal to the voltage across the zener diode less the base-emitter junction voltage of Q2.

The base current of Q2 is equal to the emitter or load current divided by the beta of that transistor. The collector current of Q1 is the sum of the base current required by the zener diode. $R_{E1}$ is chosen so that the current flowing through $R_{E1}$ and hence the emitter and collector current of Q1 are all equal to the sum of the base current of the Q2 and the current flowing through the zener diode: $R_{E1} = 0.6/(I_Z + I_B)$.

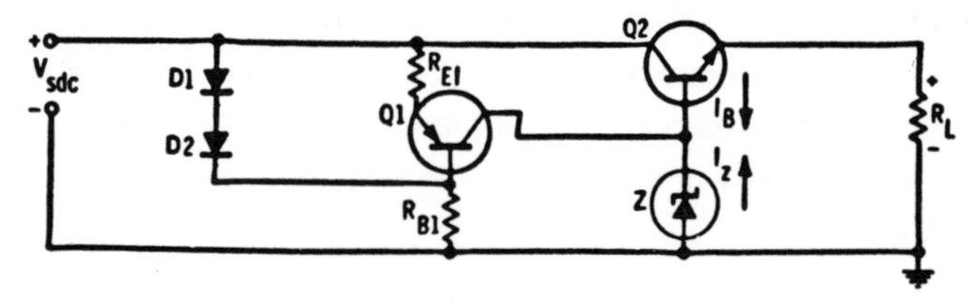


Fig. 12-13. A solid-state constant-current source has replaced $R_s$.

The current flowing through diodes D1 and D2, as well as the current flowing into the base of Q1, must pass through $R_{B1}$. The base current of Q1 is equal to the emitter current just determined, divided by $\beta$. The current flowing through the diodes must be sufficient to keep them conducting, with a voltage drop of about 0.6 volt across each device. The voltage across the resistor, $R_{B1}$, is the supply voltage less than 1.2-volt drop across the two diodes. Therefore, the resistor is equal to the voltage just determined divided by the sum of the diode current and base current of Q1.

### Protecting the Series Transistor

The series transistor is subject to breakdown by an accidental short that may be placed across the regulated output. Circuits have been devised to protect the series transistor from such failure. One method, used in low-current supplies, employs a current limiter. Should $R_L$ be made equal to zero, a high impedance will appear in series with the shorted load. The circuit returns to normal operation when the short is removed. Two arrangements are shown in Fig. 12-14. In both circuits, Q1, $R_s$, and Z1 are the elements ordinarily used in the regulator. The other components comprise the overload protection circuit.

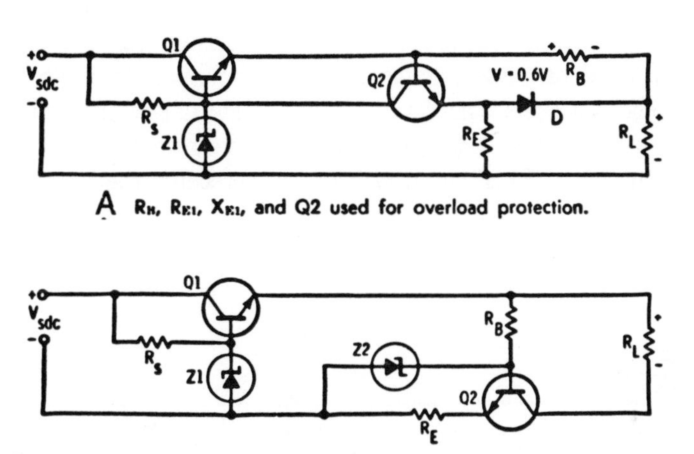


A $R_H$, $R_{E1}$, $X_{E1}$, and Q2 used for overload protection.

B Series regulator with overload-protection components Z2, $R_E$, $R_H$, and Q2.

Fig. 12-14. Some typical overload protection circuits.

In Fig. 12-14A, Q2 is normally off. When the diode is on, there is 0.6 volt across it. The transistor is just about turned on when there is 0.5 volt across its base-emitter junction. Hence, the transistor will be turned on when 0.5 + 0.6 = 1.1 volts is developed from its base to the cathode of the diode. $R_B$ is connected across the two devices. The size of the resistor is set to such value that when there is a predetermined current flowing through the load, 1.1 volts will be developed across $R_B$ to turn on Q2. Q2 is on when there is any current equal to or greater than this predetermined amount flowing through the load.

When Q2 is on, it draws current from $R_s$, so that less current remains for the base of Q1. Q1 is partially or entirely turned off, limiting the current for that transistor. Hence there will be no excess current flowing through Q1. Should the short or partial short be removed from the output, Q2 will be turned off and the regulator will operate normally.

In the system in Fig. 12-14B, $R_B$ keeps Q2 in saturation. $Z_2$ is not in the breakdown region. If the current demanded by the load increases beyond a predetermined value, a relatively large voltage will develop across $R_E$. This voltage, added to that across the base-emitter junction of Q2, will drive Z2 into the breakdown region. Q2 will come out of saturation providing a high impedance in series with the load. Conditions will be restored to normal when the load current demand is reduced. The disadvantage of the second circuit is the high power that transistor Q2 must dissipate. The entire load current flows through this transistor.

Current can also be limited by connecting an FET in series with the pass transistor. If the circuit in Fig. 12-10 were used with this modification, the source and gate of a p-channel FET are both connected to the emitter of the series-pass transistor and the drain is connected to the load. With zero volts between the gate and source, load current is limited to the $I_{DSS}$ of the FET. Because $I_{DSS}$ is in the milliamperes range, this method of limiting current is applicable only to supplies providing small amounts of power.

Another method of protecting the series-pass transistor uses a means of interrupting the circuit upon overload. A fast-acting fuse will only race the transistor to destruction. Circuits have been designed that force a large current to flow through the fuse at the moment of overload. One such device, using an SCR is shown in Fig. 12-15. It is normally in a nonconducting state. R1 is chosen so that when a predetermined current flows through the load, the voltage developed across R1 will be sufficient to cause the zener diode, Z, to break down. The breakdown voltage of the diode is low so that R1 can be made small. R1 is equal to the zener breakdown voltage divided by the minimum overload current.

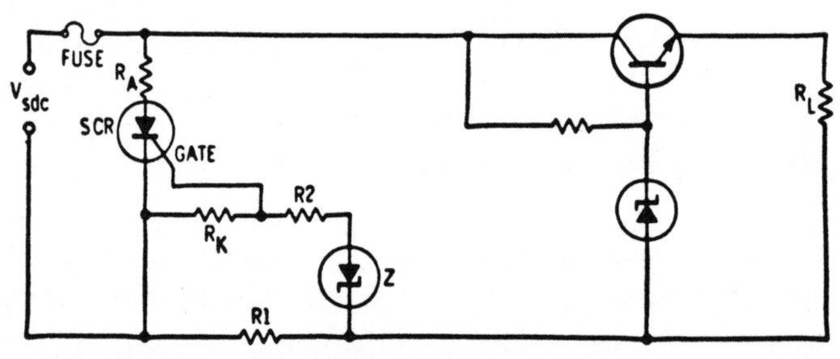

Fig. 12-15. An interruption-type overload protection circuit.

The gate current for the SCR is supplied through the zener diode. When the gate current is equal to or greater than a predetermined value, the SCR will conduct. This high surge of current passing through the SCR will cause the fuse to open (blow) rapidly. The fuse is chosen so that it will not blow under normal load current, but will blow rapidly once the normal load current is exceeded by a considerable factor. This considerable factor is supplemented by the current flowing through the SCR. $R_A$ limits the SCR surge current to safe values. $R_A$ is equal to the maximum unregulated voltage. $V_{sdc}$(max), divided by the maximum surge current that is allowed to flow through the SCR (or divided by the maximum surge current that is permitted to flow through the rectifier in the unregulated power supply, if the surge current rating of these rectifiers is less than the surge current of the SCR).

The gate current is set by $R_K$ and R2. $R_K$ is usually about 1000 ohms, whereas R2 is a small resistor limiting the gate current to safe values.

## Protecting the Load

When the pass transistor breaks down, high voltage will appear across the load. Some loads can be destroyed if this voltage is excessive. To protect the load, a crowbar circuit is frequently used. An arrangement is shown in Fig. 12-16. The basic element of this circuit is an oscillator built around a unijunction transistor (UJT).

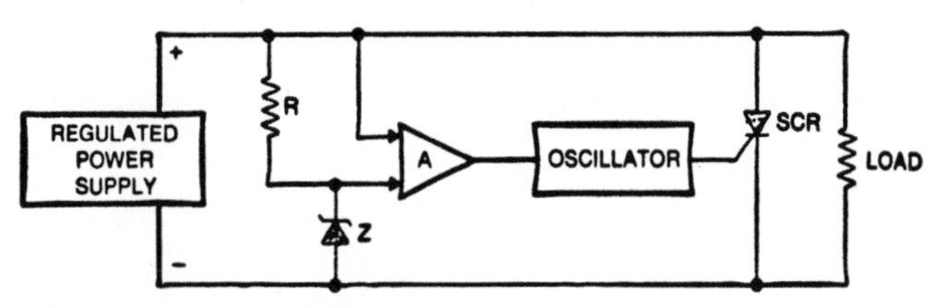


Fig. 12-16. A basic crowbar load-protection circuit.

In the *crowbar protection circuit*, the UJT gets turned on to trigger an SCR. The chain of events is as follows. The output from the regulated power supply is fed to one input of a differential comparison amplifier, A. A can also be an open-loop op amp. A fixed voltage supplied by zener diode Z is at the other input. The two voltages are compared. If the power-supply voltage is greater than the zener voltage, the voltage at the output of the comparison amplifier is sufficient to trigger an oscillator, turning it on. (The oscillator is frequently built around a UJT as described.) Voltage at the output of the oscillator is, in turn, applied to the gate of the SCR, turning on that thyristor. When it is turned on, only the 0.6 forward voltage usually developed across an ordinary silicon diode is across the SCR and across the load. This low-voltage SCR shunt protects the load from a high voltage that would have been at the output of the supply had the SCR not been in the circuit.

A zener diode connected across the output rather than the crowbar circuit shown can also protect the load. The diode acts to limit the load voltage to the zener breakdown voltage and must be capable of carrying the normal load current without being damaged. In this arrangement, the cathode of the zener must be connected to the more positive terminal of the supply.

## Shunt Regulator

The zener regulators discussed in Chapter 1 were used in shunt regulator circuits.

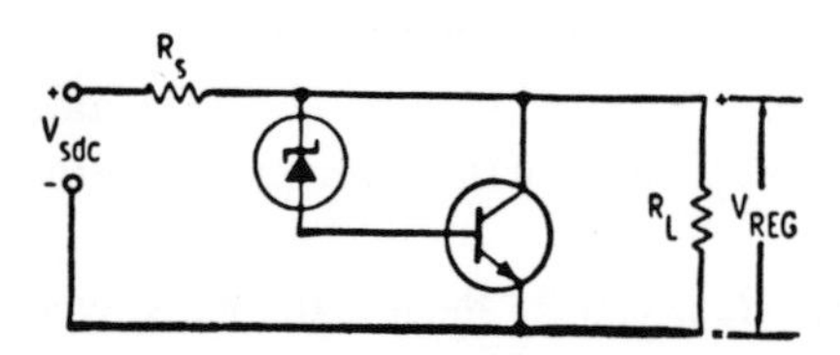


Fig. 12-17. A basic shunt regulator circuit.

The output load was connected across the diode. Similarly, in the following regulators to be discussed, the load shunts the transistor. In the basic circuit in Fig. 12-17, the voltage from the base to the collector is identical to the breakdown voltage across the zener diode. The voltage from the base to the emitter is the usual voltage developed across a forward-biased diode. As both voltages are relatively constant, a regulated voltage is developed across the series combination of zener diode and the base-emitter junction of the transistor. The load, $R_L$, is connected across the series circuit.

There are several advantages to this circuit over the zener shunt regulator. Here, only base current flows through the diode. A small diode can therefore be used to regulate a large amount of current at the output of the transistor. Furthermore, the regulating capabilities of the circuit are improved by a factor of beta.

Current flowing through $R_S$ should remain constant regardless of load-current variations. If the output is shorted, no current flows through the transistor, while all the current flows through $R_S$ and the short. If the load, $R_L$, is infinite, all current will flow through $R_S$ and the transistor. In both conditions, the current flowing through $R_S$ is identical. The current that flows through $R_S$ remains constant even when it is divided between the transistor and the load.

For good voltage regulation, the base and emitter resistances of the transistor, as well as the zener diode resistance, must all be small. The beta of the transistor and the series resistor, $R_S$, should be large. Since the current flowing through $R_S$ must be constant, the magnitude criterion can be satisfied by replacing $R_S$ with a constant-current source.

Good regulation requires the use of high beta transistors. Extremely high beta can be achieved using a Darlington pair arrangement rather than a single zener/shunting-transistor combination. The complementary circuit discussed in Chapter 5 may be used as an alternate. In either case, the beta of the combination is the product of the betas of all transistors in the circuit. The base and zener current is equal to the current flowing through the final shunting transistor divided by this product.

The output voltage of the circuit in Fig. 12-17 is slightly greater than the voltage across the zener diode. Should less regulated voltage be required, a voltage divider can be placed across the output. The circuit in Fig. 12-18 can be used if voltages considerably greater than the zener diode voltage are necessary. In this circuit, the output voltage is approximately:

$$V_{REG} = \frac{V_B(R_X + R_B)}{R_X} \tag{12-7}$$

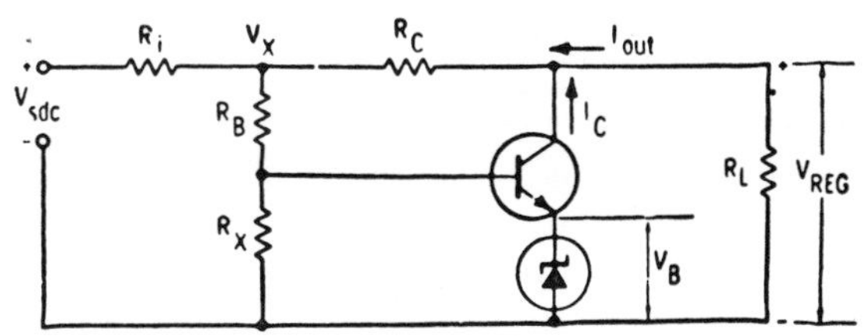

Fig. 12-18. The regulated output is greater than $V_Z$.

$V_{REG}$ is slightly smaller than $V_X$. Neglecting the base current, the voltage at $V_X$ is

$$V_X = V_{sdc} - R_i(I_{out} + I_C) \tag{12-8}$$

$R_C$ can be determined experimentally. Vary the unregulated input voltage, $V_{sdc}$, and note the output at $V_{REG}$. Adjust $R_C$ for the minimum variation of $V_{REG}$ as $V_{sdc}$ is varied. If $R_C$ is made too big, it will be found that $V_{REG}$ will drop as $V_{sdc}$ is increased.

The ripple factor can be improved by connecting a large electrolytic capacitor from the collector to the base of the transistor. A capacitor at this point in the circuit appears as a much larger capacitor when reflected across the output load because of the current gain of the transistor. The output capacitance is approximately equal to $C(R_X/R_Z)$, where $R_X$ is the resistor noted in Fig. 12-18, $R_Z$ is the resistance of the zener diode or a resistor placed in the emitter circuit, and C is a capacitor connected from the collector to the base of the transistor.

## Feedback Regulators

A voltage regulator can serve one or more functions. It can deliver a regulated output voltage despite line, supply voltage, load resistance, or load current variations. All this was properly accomplished with shunt and series regulators discussed previously. The feedback regulator in Fig. 12-19 is an improvement over the types of regulating circuits described in the earlier sections of this chapter.

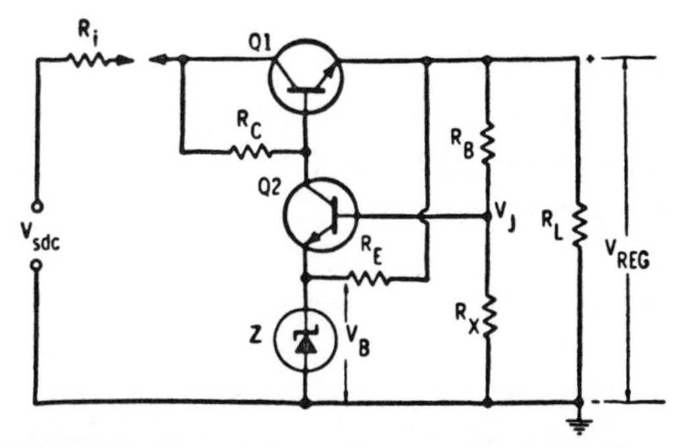

Fig. 12-19. A feedback regulator circuit.

The basic circuit of a feedback voltage regulator is shown in Fig. 12-19. Transistor Q1 responds to changes in relative voltages between the base and emitter of Q2. The regulated output voltage is sampled and applied to the base of Q2. The emitter voltage is maintained constant by virtue of the zener diode. Should the regulated voltage rise, the voltage at the base of Q2 will rise, as will the collector current. Both the base of Q1 and collector of Q2 derive their current through $R_C$. Ideally, the current through $R_C$ is fixed. As current increases through Q2, there is less remaining for Q1. Consequently, the collector and emitter current of Q1, as well as the load current, is reduced, forcing down the voltage across the load resistor. The reverse will be the case should the voltage at the base of Q2 drop. Consequently, the voltage at the output is regulated.

Regulated voltage is developed across the series combination of $R_B$ and $R_X$. Voltage at the junction of the resistors is fed back to the base of Q2. This voltage is a sampling of the regulated voltage, $V_{REG}$, as well as any variation of this regulated voltage. The sum of $R_B$ and $R_X$ should be much larger than the maximum value of the load resistance, $R_L$. If this requirement would force the Thevenin resistance to be high, $R_B$ can be replaced with a zener diode. In this case, the Thevenin voltage would be $V_{REG} - V_Z$, where $V_Z$ is the breakdown voltage of the diode replacing $R_B$, and the Thevenin resistance is slightly less than the resistance of that zener diode.

The voltage at the emitter of Q2, is the breakdown voltage, $V_B$, of the reference zener diode, Z, shown in the drawing. $V_B$ plus the voltage from the base to the emitter of Q2 is the voltage $V_J$ at its base, and hence at the junction of $R_B$ and $R_X$. Assuming that $R_B$ and $R_X$ are small, the regulated voltage is a function of the relative sizes of $R_B$, $R_X$, and $V_J$. It is approximately equal to $\{(R_X + R_B)/R_X\}\ \{V_J\} = \{(R_X + R_B)/R_X\}\ \{V_B + V_{BE}\}$, where $V_{BE}$ is the base-emitter voltage of Q2. Information is fed to Q1 about the variation of $V_{BE}$ owed to the change of $V_J$ with respect to the reference voltage, $V_B$. Q1 reacts to the information by controlling the current fed to the output, maintaining $V_{REG}$ constant. If the beta of Q1 is insufficient to do its job properly, a Darlington pair can be substituted for the individual device.

The zener diode, Z, should be fed from a high-impedance source. Current flows through $R_E$ and through the diode. Here, $R_E$ behaves as the constant-current source.

## Switching-Type Voltage Regulator

In the circuit shown in Fig. 12-19, transistor Q1 is on, conducting all the time. Should a square wave be fed to the base of the transistor, it could be arranged for it to conduct for one-half of the time. The square wave could turn the device off for half of its cycle so that pulses flow through $R_L$ for only half the time. Because of this, the average current is half what it would be if conduction were continuous. Consequently, the average voltage across the load is cut to half. If the period of time the current flows is somewhere between one-half of the cycle and the full cycle of the square wave, the voltage developed across the load increases proportionally above what it would be when a perfect square wave is applied. Similarly, if the period of time the current flows is less than one-half the square-wave cycle, the output voltage is reduced accordingly. It follows the formula:

$$E_O = E_{in}\left(\frac{T_{on}}{T_{on} + T_{off}}\right) \tag{12-9}$$

where,

$E_O$ is the output voltage at the emitter of Q1 in Fig. 12-19.

$E_{in}$ is the input voltage at the collector of Q1.

$T_{on}$ is the time the transistor is turned on during the cycle.

$T_{off}$ is the time the transistor is turned off during the cycle.

Should the on and off times of the pass transistor be controlled by a square wave or a pulse oscillator, the output voltage from a circuit such as the one in Fig. 12-19 can be adjusted by simply changing $T_{on}$ and $T_{off}$ of the pulses. Furthermore, if information about the output voltage is fed back to this oscillator, and its on and off times are controlled by this output voltage information, the oscillator can be used in a feedback circuit to turn the pass transistor on and off for appropriate periods of time. In this manner, it is used to regulate the output voltage. This type of regulator works on what is referred to as *pulse-width modulation.*

Two types of switching regulators are in use. Examples of each are shown in Fig. 12-20. Voltage at the output of the series regulator in Fig. 12-20A is lower than that at the input of the circuit while voltage at the output of the circuit in Fig. 12-10B is higher. In Fig. 12-20A, dc voltage, $V_{IN}$, passes through transistor Q and appears across the load, $R_L$. Output from the pulse generator (usually a free-running or astable multivibrator) operating normally above 12,000 Hz, is applied to the base of Q to turn it on and off. Voltages at the output of the comparator (usually an op amp without feedback loops) are fed to the pulse generator. It informs that generator how long the $T_{on}$ and $T_{off}$ pulses should be. Relative voltages at the inputs of the comparator determine the output voltage it is to apply to the pulse generator. One of these input voltages is held constant by a zener diode. The other is determined by the voltage at the junction of R1 and R2, which changes as the output voltage across the load changes. Thus the complete feedback loop from Q to $R_L$ and back to Q through the comparator and pulse generator is designed to keep the voltage constant. L and C smooth the ripple because of the pulses of current from transistor Q. D is a *free-wheeling diode,* which keeps current from L flowing into the filter capacitor C while Q is turned off.

In the shunt regulator in Fig. 12-20B, voltage developed across the load is higher than $V_{in}$ because of the inductive kick caused by the presence of inductor L. When the transistor is on, a ramp current flows through it. None of this current can reach the load or capacitor C. When Q is off, a voltage is developed across the capacitor and load

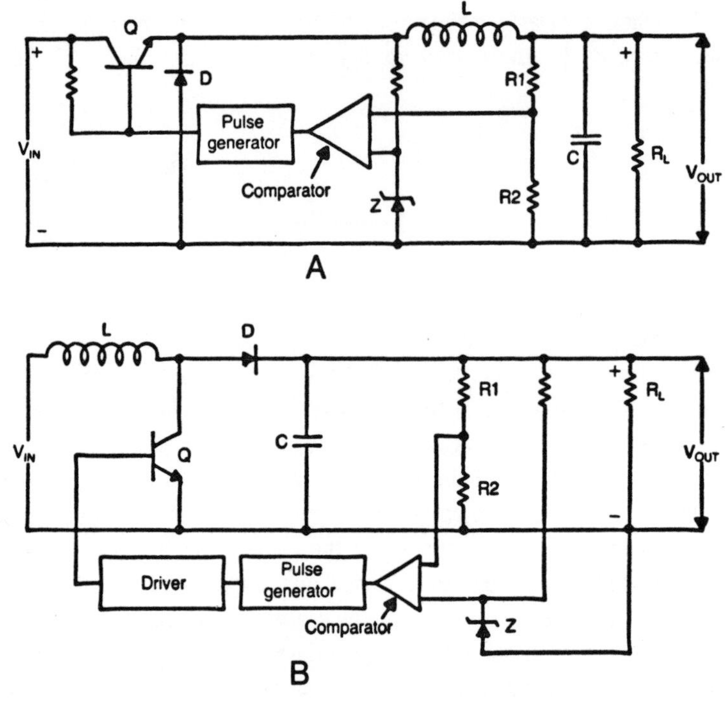


Fig. 12-20. Two versions of the switching-type voltage regulator circuit.

for the current remains constant because of the action of inductor L. The comparator and pulse-signal generator behave in a fashion similar to their behavior in the series-regulator circuit. A driver stage is added to increase the amplitudes of the pulses fed to the base of the transistor so that it can be either fully in saturation or fully cut off.

Output voltage is determined by the pulse width of the signal fed to transistor Q as well as by the frequency applied to the transistor. The output frequency from a monostable multivibrator can be set by the voltage from the comparator and designed to control and stabilize the output voltage. Some switching regulators use only frequency changes to help set the output voltage. Others use both frequency and pulse width.

## CURRENT REGULATORS

Up to this point, the discussion has been on the various types of voltage-regulator circuits. Current regulators have been ignored. Current regulators present a high resistance to a load so that current flowing through the overall circuit depends upon the resistance of the regulator rather than upon the resistance of the load. Voltage developed across the load is equal to the regulated current, I, multiplied by the resistance of the load, $R_L$, or $IR_L$. Any constant-current source can be used as a current regulator.

# 13

# Digital Systems

We have been concentrating entirely on linear (analog) circuits, at one time the vast majority of electronics circuits were of this type.

In an analog circuit, input and output signals can take on any value within a specific range. Between any two specific output levels, for example, there can be an intermediate value. If an analog signal is graphed, it will be a more or less smooth line, which may or may not be curved.

There is another branch of electronics which is becoming increasingly important today. Of course, this is digital circuitry. All signals can be represented by numbers (digitally). Intermediate values are not allowed. A signal might have a value of three or four, but not three and a half.

Digital circuits use the binary numbering system. Many people find binary numbers very intimidating. They are awkward for a human being to use, but they are ideal for use in electronic circuits.

We are all familiar with the decimal numbering system that we use in day-to-day life. In the decimal system there are ten possible digits:

0 - 1 - 2 - 3 - 4 - 5 - 6 - 7 - 8 - 9

If we have to express a number larger than nine, we simply add another digit. For example, 27, or 52.

The binary system works in exactly the same way, except there are only two digits, instead of ten:

$$0 - 1$$

To express a value greater than one, we have to add more digits. For example, two is written in the binary system as 10.

This two digit system is easy to set up electronically. In the simplest possible arrangement, each binary digit is represented by a simple switch. If the switch is off, it means "0." If the switch is on, it stands for a "1."

In practical digital circuits, two signal levels are used to unambiguously indicate the two possible states—high and low. Usually, the circuit is turned on when the applied signal is high and turned off when it is low. In some cases, the reverse may be true. In either case, when the applied signal is high, it is usually assigned the digit 1. When it is low, the digit 0 is usually used. Because only two digits are required for digital circuits, the binary number system is applied, rather than the commonly used decimal system with ten digits from "0" to "9."

## BINARY NUMBERS

Before turning to a discussion of the binary system, consider the significance of the placement of digits in a decimal number. Use the four-digit number 8692 as an example. It can be considered as the sum of four numbers: 8,000 + 600 + 90 + 2. Another way of stating this sum is $(8 \times 10^3) + (6 \times 10^2) + (9 \times 10^1) + (2 \times 10^0)$. Number 10 is the base or *radix* of the system. The digit at the extreme right in the decimal number is multiplied by the radix to the exponent zero, or $10^0$. Proceeding from right to left, the next digit is multiplied by $10^1$. The exponents of 10 increase by one number as we progress.

The least significant digit (LSD) of the whole decimal number 8692 is multiplied by $10^0$. In the example cited above, the least significant digit is 2. If the number had several decimal places with digits to the right of 2, such as in the number 8692.34, the 3 is multiplied by $10^{-1}$ and the 4 by $10^{-2}$. Both these numbers are then added to the other quantities to form the entire number. In this case, 4 would be the least-significant digit rather than 2. Using a similar concept, the most significant digit (MSD) in a number is the one at the extreme left. It is the digit to be multiplied by the 10 with the largest exponent. In this decimal number, the most significant digit is 8.

As was the case with numbers in the decimal system, numbers in the binary system also have least- and most-significant digits. Each digit in the binary system is referred to as a *bit*. The least significant bit is abbreviated LSB. Similarly, the most significant bit is abbreviated MSB.

Similar to the practice with decimal numbers, the MSB in the binary system is the first digit at the left of the number and the LSB is the first digit at the right. Here, how-

Table 13-1. Binary Numbers Zero Through Fifteen and Their Decimal Equivalents.

| $2^3$ | $2^2$ | $2^1$ | $2^0$ | Decimal Equivalent |
|---|---|---|---|---|
| Binary Number | | | | |
| 0 | 0 | 0 | 0 | 0 |
| 0 | 0 | 0 | 1 | 1 |
| 0 | 0 | 1 | 0 | 2 |
| 0 | 0 | 1 | 1 | 3 |
| 0 | 1 | 0 | 0 | 4 |
| 0 | 1 | 0 | 1 | 5 |
| 0 | 1 | 1 | 0 | 6 |
| 0 | 1 | 1 | 1 | 7 |
| 1 | 0 | 0 | 0 | 8 |
| 1 | 0 | 0 | 1 | 9 |
| 1 | 0 | 1 | 0 | 10 |
| 1 | 0 | 1 | 1 | 11 |
| 1 | 1 | 0 | 0 | 12 |
| 1 | 1 | 0 | 1 | 13 |
| 1 | 1 | 1 | 0 | 14 |
| 1 | 1 | 1 | 1 | 15 |

ever, the radix is 2 rather than 10. As only a 1 or 0 can be used to form numbers, a typical number can look something like 11010. As was the case with the number in the decimal system, the number stated here is the sum of several numbers, namely 10,000 + 1,000 + 000 + 10 + 0. Adding these binary numbers provides us with the originally stated number in the example, 11010. The alternate way of stating this number as a sum uses the radix 2, and is $(1 \times 2^4) + (1 \times 2^3) + (0 \times 2^2) + (1 \times 2^1) + (0 \times 2^0)$. The decimal equivalent of this number is then $(1 \times 16) + (1 \times 8) + (0 \times 4) + (1 \times 2) + (0 \times 1) = 26$. Table 13-1 shows binary numbers from 0 to 15 and their equivalents in the decimal system. The multiplier is listed at the top of each column.

Binary numbers are frequently referred to as *bit patterns* or *words*. Words in the table have four bits. Four-bit words have the name *nibble*. Numbers or words can be composed of many more than the four bits shown here. A special designation has been assigned to eight-bit words—namely *bytes*.

### Binary Coded Decimals

Various codes are frequently used in computer work to transform decimal to binary numbers and binary to decimal. The *binary coded decimal* system, or BCD, is quite commonly used. Here, each digit of the decimal system is coded with its equivalent four-bit binary number from Table 13-1.

The binary number is split up into four bit nibbles, each representing a decimal digit. An example using number 49813 is:

| Decimal: | 4 | 9 | 8 | 1 | 3 |
|---|---|---|---|---|---|
| BCD: | 0100 | 1001 | 1000 | 0001 | 0011 |

In the BDC system, the following combinations are disallowed and not used:

| | |
|---|---|
| 1010 | 10 |
| 1011 | 11 |
| 1100 | 12 |
| 1101 | 13 |
| 1110 | 14 |
| 1111 | 15 |

These values cannot be expressed as single digit decimal values, so they do not fit into the BCD system. Avoiding these disallowed values can sometimes complicate the design of some BCD systems.

### Octal And Hexadecimal Numbering

Because the binary numbering system is so awkward for humans to use efficiently, several other numbering systems have been designed to make things a little easier.

If we split a large binary number into groups of three binary digits, each grouping can have any of eight possible values. For this reason, this is known as the *octal* (base eight) numbering system.

A single octal digit, or three binary digits can express any value from zero to seven:

| binary | octal |
|---|---|
| 000 | 0 |
| 001 | 1 |
| 010 | 2 |
| 011 | 3 |
| 100 | 4 |
| 101 | 5 |
| 110 | 6 |
| 111 | 7 |

To express a value higher than seven in the octal numbering system, we must use two or more digits. For example, eight is written as 10.

The octal system is used primarily to make binary numbers easier to remember.

For example, a binary value like 100110011 is awkward, and it is easy to make a mistake in copying it. Converting it to the octal numbering system makes life easier:

| | | | |
|---|---|---|---|
| binary | 100 | 110 | 011 |
| octal | 4 | 6 | 3 |

Another numbering system is the *hexadecimal* system which uses a base of sixteen. Four binary digits equal one hexadecimal digit with no disallowed values.

Since a single digit can represent any value from zero to fifteen, we need a few extra numerical symbols. The letters A through F are used to represent digits from ten to fifteen:

| binary | hexadecimal | decimal | octal |
|---|---|---|---|
| 0000 | 0 | 0 | 0 |
| 0001 | 1 | 1 | 1 |
| 0010 | 2 | 2 | 2 |
| 0011 | 3 | 3 | 3 |
| 0100 | 4 | 4 | 4 |
| 0101 | 5 | 5 | 5 |
| 0110 | 6 | 6 | 6 |
| 0111 | 7 | 7 | 7 |
| 1000 | 8 | 8 | 10 |
| 1001 | 9 | 9 | 11 |
| 1010 | A | 10 | 12 |
| 1011 | B | 11 | 13 |
| 1100 | C | 12 | 14 |
| 1101 | D | 13 | 15 |
| 1110 | E | 14 | 16 |
| 1111 | F | 15 | 17 |
| 1 0000 | 10 | 16 | 20 |
| 1 0001 | 11 | 17 | 21 |
| 1 0010 | 12 | 18 | 22 |

and so forth.

These various numbering systems are just different methods of expressing the same values. Each system will be convenient in certain circumstances, and awkward in others. To work well in digital electronics, it is very useful to have at least a passing familiarity with all four numbering systems. Other numbering systems also exist, but they are not normally used in electronics work.

The octal and hexadecimal systems are primarily used as memory aids for working with binary numbers. You will seldom, if ever, have to perform mathematical operations in these numbering systems. Converting between either the octal or hexadecimal system and the binary system is straightforward in either direction.

In working with digital circuits, it is sometimes necessary to convert between the binary and decimal numbering systems, or to perform mathematical operations on binary numbers. These tasks will be discussed in the next few pages.

### Converting Between Binary and Decimal Numbers

The procedure for converting from ordinary binary numbers (not BCD) to decimal numbers is simple enough. For every one in the binary number, add the appropriate power of two. Here is an example, converting binary 1001101 to its decimal equivalent;

| $2^6$ | $2^5$ | $2^4$ | $2^3$ | $2^2$ | $2^1$ | $2^0$ | |
|---|---|---|---|---|---|---|---|
| 1 | 0 | 0 | 1 | 1 | 0 | 1 | binary |

equals;

$$2^6 + 2^3 + 2^2 + 2^0 =$$
$$64 + 8 + 4 + 1 =$$
$$77 \quad \text{decimal}$$

It is just a little more complicated to convert from decimal to binary numbers. The procedure starts by writing in sequence all powers of radix 2 and its decimal values. Start with $2^0 = 1$ and continue with $2^1 = 2$, $2^2 = 4$, $2^3 = 8$, and so on until you find the first number equal to 2 to an exponent that is larger than the number to be converted to binary form. Discard that number. List all the smaller remaining numbers in a column. Put a 1 on the same line on which you wrote the largest of the exponential numbers. This 1 will be the MSB of the binary number. Subtract the exponential number on the top line from the number to be converted and write the difference on the next line. If the difference is equal to or larger than the exponential number on that line, write a 1 on that line and make a note of the difference between the number on that line and the exponential number. Write the difference on the next line. If the exponential number is smaller than the number on that line, write a 0 and rewrite that number on the next line. Continue this process until there is a 1 or a 0 on all lines that have numbers with an exponent of radix 2. Using the top bit as the MSB, list all bits in sequence as they appear on succeeding lines. The resulting number is the binary equivalent of the number converted. An example should help clarify the procedure. See Fig. 13-1.

Suppose you wish to convert the number 47 to binary form. Write 47 on the first line in column 1. The numbers that are powers of 2 are $2^0 = 1$, $2^1 = 2$, $2^2 = 4$, $2^3$

Fig. 13-1. How decimal number 47 is converted to binary form.

| 1 | 2 | 3 | 4 |
|---|---|---|---|
| 47 | $2^5 = 32$ | $47 - 32 = 15$ | 1 |
| | $2^4 = 16$ | $15 - 16 = X$ | 0 |
| | $2^3 = 8$ | $15 - 8 = 7$ | 1 |
| | $2^2 = 4$ | $7 - 4 = 3$ | 1 |
| | $2^1 = 2$ | $3 - 2 = 1$ | 1 |
| | $2^0 = 1$ | $1 - 1 = 0$ | 1 |

= 8, $2^4 = 16$, $2^5 = 32$, $2^6 = 64$, $2^7 = 128$, and so on. The smallest number equal to a power of 2 and larger than 47 is 64. Write the next smaller number, $2^5 = 32$, in column 2 on the same line as the 47. Write a 1 in column 4. Now list $2^4 = 16$, $2^3 = 8$, and so on in sequence, in column 2 headed by $2^5 = 32$. There are six lines involved beginning at the number $2^5$ down to the number to be multiplied by the LSB, namely $2^0 = 0$. Hence there will be six bits in the binary equivalent number of 47.

Now subtract the 32 from the original number 47. Write the difference, 15, on the next line. $2^4 = 16$ is shown in column 2 on the next line. It is larger than the 15 and thus cannot be subtracted from it. Because the 16 cannot be subtracted from the 15, put a 0 in column 4 under the 1 and rewrite the 15 on the next line—the third line from the top. On this line, the number with a power of 2 is $2^3 = 8$. This number is smaller than 15. Subtract it from 15 and place a 1 in column 4 under the 0. Write the difference, $15 - 8 = 7$, under the 15. On this line, the radix 2 with its exponent 2 is equal to 4. As it is smaller than 7, subtract it from 7 and place a 1 in column 4 on the fourth line. Write the difference, $7 - 4 = 3$, under the 7. $2^1$ on this line is equal to 2. As this number is smaller than 3, place a 1 in column 4 and subtract the 2 from the 3. Write the difference, $3 - 2 = 1$, under the 2. On the bottom line, $2^0$ is equal to 1. This is equal to the 1 on that line. Subtract it from the 1 and place a 1 on that line in the column 4. Start with the uppermost digit in the column at the left, and write all digits in sequence. The binary equivalent of the decimal number 47, is 101111.

### Binary Point

Fractions that are converted to decimals are written as digits to the right of a decimal point. Thus a number in the decimal system such as 5.3 is 5 3/10, where the one digit to the right of the decimal point is equal to that digit divided by 10 or multiplied by $10^{-1}$. Similarly, if there were a second digit to the right of the decimal point, it is multiplied by $10^{-2} = 1/100$. All fractions are then added to the whole numbers discussed earlier, to determine the total number.

Similar to the decimal point, a binary point is used in the binary system. The first digit to the right of the binary point is multiplied by $2^{-1} = ½$, the second digit is multiplied by $2^{-2} = ¼$,the third by $2^{-3} = ⅛$, and so on. Thus binary number 11.1011 is equal to the sum of $(1 \times 2^1) + (1 \times 2^0) + (1 \times 2^{-1}) + (0 \times 2^{-2}) + (1 \times 2^{-3}) + (1 \times 2^{-4})$.

### Binary Addition

Binary numbers are added using methods similar to those employed when adding decimal numbers.

1. 0 + 0 = 0. This is as it was in the decimal system.
2. 0 + 1 = 1. Once again, the decimal system rule applies.
3. 1 + 1 = 10. This too is the same type of sum used in the decimal system.

When determining that 5 + 5 = 10 in the decimal system, you put down a 0 and carried a 1. Considering this, 1 + 1 + 1 = 11 because 1 + 1 = 10 and 10 + 01 = 11 from rule 2.

Let us try an example adding two binary numbers. Proceed from right to left as usual, starting with the LSB. The sum of the LSBs are 1 + 1 = 10.

```
Bit # 7 6 5 4 3 2 1
-------------------
        1 1 0 1 1 1
      + 0 1 1 1 0 1
      -------------
      1 0 1 0 1 0 0
```

Note that in the LSB column you must write a 0 and carry a 1 to be added to the bits #2. In the bit #2 column, 1 + 0 = 1. Now add the carry bit 1 determined from the sum of the bits in the LSB column #1 so the sum of the 1 in column #2 with the carry bit from the LSB column becomes 0 with a carry bit of 1 to be added to bits in column #3. Here, 1 + 1 = 10. When this is added to the carry bit 1 from column #2, the sum is 11. Put down a 1 carry a 1 to bit column #4. By the time you are finished, the sum is as shown below the numbers being added.

### Binary Subtraction

Subtraction in the binary system does not differ from subtraction in the decimal system. The larger number is referred to as the *minuend*. The number to be subtracted from the minuend is the *subtrahend*. In the examples shown, the borrow system is used. For clarity, borrowed numbers in the examples are circled. They are written between the minuend and subtrahend.

```
Example 1:  1 1 0     Example 2:  1 0 1 1     Example 3:  1 0 0 0
              ①                     ①                       ①①①
          -     1               -     1 0 1               -   1 1 1
            -----                 ---------                 -------
            1 0 1                 0 1 1 0                   0 0 0 1
```

In Example 1, subtrahend 1 is subtracted from minuend 110. Starting with the LSB at the right, the 1 in the subtrahend cannot be subtracted from the 0 in the minuend, for the difference is negative. So borrow the 1 from the middle bit in the minuend. Note this by writing a 1 in a circle just below this borrowed middle bit. The borrowed 1 is added to the 0 in the LSB column of the minuend to form a 10. As it is bigger than the LSB 1 in the subtrahend, the 1 in the subtrahend can now be subtracted from the newly formed 10 in the minuend. The difference is 1. Now subtract the borrowed 1 in the circle, from the center bit 1 in the minuend. The difference is 0. As for the MSB, there is no bit in the subtrahend below the MSB 1 in the minuend. That spot can be assumed to be a 0. Subtract this 0 from the MSB 1 in the minuend. The difference is shown as a 1 just below the imaginary 0. Similar procedures can be applied to Examples 2 and 3.

### Binary Multiplication

Rules for multiplying two binary numbers are identical to the rules for multiplying two decimal numbers, as noted in the example.

1. The product of any number with a 0 is a 0.
2. The product of a 1 with a 1 is a 1.

```
       1 1 0 1
     ×   1 1 0
     ---------
       0 0 0 0
     1 1 0 1
   1 1 0 1
   -----------
   1 0 0 1 1 1 0
```

### Binary Division

Being the reverse of multiplication, binary division procedures follow the same rules and similar methods employed when dividing one decimal number by another. For example,

```
         1001
    101 )101111
         101
         ----
          0111
           101
          ----
            10  (Remainder)
```

The solution is $1001 = \frac{10}{101}$

## BOOLEAN ALGEBRA

Boolean algebra is a type of mathematics that can be applied to logic circuits. There are three basic functions in Boolean algebra. They are AND, OR, and NOT. Even though the Boolean symbols are the same as in conventional algebra, the indicated operations are different. For example, the symbol "+" between two quantities does not indicate a sum of two numbers or factors, but rather that the presence of a significant magnitude of either quantity at the input of a gate is sufficient to turn it on. It is the OR symbol.

$$A + B$$

means

$$A \text{ OR } B$$

An AND symbol is a dot located between two quantities indicating that both quantities must be of sufficient magnitude if a gate fed by both quantities is to be turned on.

$$A \bullet B$$

means

$$A \text{ AND } B$$

Often the dot is omitted so that the proximity of one number to another indicates the AND operation.

$$AB$$

is the same as

$$A \bullet B$$

The NOT operation is indicated by a bar over the quantity to be affected.

$$\overline{A}$$

means

$$\text{NOT } A$$

A NOT operation inverts the binary value being affected. A 1 becomes a 0, or vice versa. This is roughly analogous to a minus sign in conventional algebra.

The basic unit of digital electronics is a gate. A gate is a circuit (usually in IC form) that accepts one or more inputs and performs a Boolean operation to create one (occasionally more than one) output.

Signal levels applied to a gate are represented by a 1 or a 0. In some instances, the 1 represents a large input voltage or current, while a 0 represents a small input voltage or current. Depending upon the convention used, the reverse may be true. In a similar fashion, the 1 or 0 may represent the condition when a switch is closed or is in the *on* state. The alternate digit represents the condition when a switch is open or is in the *off* state. Unless otherwise stated below, the 1 will represent the larger of two quantities and a closed switch in the *on* state. The 0 will represent the alternate condition.

## AND Gates

For an AND gate to be on with a 1 at the output, all inputs must be in the 1 state. Output is 0 if any one of the inputs is 0. The symbol for a two-input AND gate and its *truth table* are in Fig. 13-2. Note that the truth table shows what happens at the output for every possible combination of inputs. The inputs and outputs are denoted by letters at the top of each column. The state that the output is in because of the states of the input is represented by the 1 and 0. Thus A AND B, written as A • B or simply AB, are two inputs applied to an AND circuit. They determine F, the output. Different states or combinations of A, B, and F are listed in the various rows.

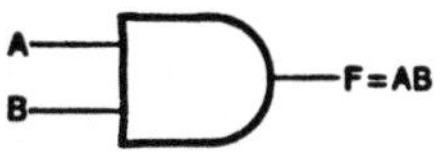

Fig. 13-2. An AND gate with its truth table.

A, B — F=AB

| A | B | F |
|---|---|---|
| 0 | 0 | 0 |
| 0 | 1 | 0 |
| 1 | 0 | 0 |
| 1 | 1 | 1 |

Inputs are listed in numerical order in the binary numbering system. For A = 0 and B = 0, the binary number is 00. For A = 0 and B = 1, the binary number is 01. This continues until all combinations of two-bit numbers are shown in the table. There are four variations or combinations when two input gates are involved. Note that F = 1 only when both A and B are 1, so that F = AB.

This system can be expanded to cases where there are more than just two inputs to the gate. A four-input gate is shown in Fig. 13-3 along with an expanded truth table to cover all possible combinations. Note the binary number sequence used to indicate the states of the inputs. As AND gates are used in both examples, outputs are 1 only when all inputs to the gates are 1, so F = ABCD.

Three rules may be stated for a two input AND gate.

1. If A is one input (we do not know if it is a 1 or a 0) to an AND gate and the second input is a 0, the output must be a 0 because both gates must be 1 for the output to be 1. Written simply, A • 0 = 0.

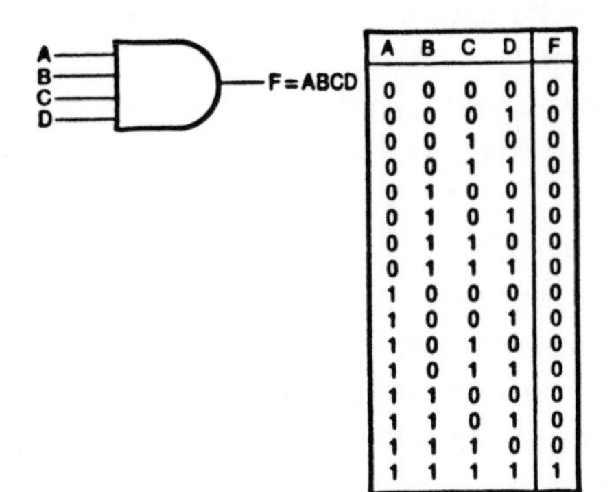


| A | B | C | D | F |
|---|---|---|---|---|
| 0 | 0 | 0 | 0 | 0 |
| 0 | 0 | 0 | 1 | 0 |
| 0 | 0 | 1 | 0 | 0 |
| 0 | 0 | 1 | 1 | 0 |
| 0 | 1 | 0 | 0 | 0 |
| 0 | 1 | 0 | 1 | 0 |
| 0 | 1 | 1 | 0 | 0 |
| 0 | 1 | 1 | 1 | 0 |
| 1 | 0 | 0 | 0 | 0 |
| 1 | 0 | 0 | 1 | 0 |
| 1 | 0 | 1 | 0 | 0 |
| 1 | 0 | 1 | 1 | 0 |
| 1 | 1 | 0 | 0 | 0 |
| 1 | 1 | 0 | 1 | 0 |
| 1 | 1 | 1 | 0 | 0 |
| 1 | 1 | 1 | 1 | 1 |

Fig. 13-3. An AND gate can be expanded to any number of inputs. This one has four.

2. If one input to an AND gate is the unknown A and the second input is a 1, the output must be whatever the A is. $A \bullet 1 = A$

3. If both inputs to an AND gate are A, the output must be whatever A is. Hence, $A \bullet A = A$.

## NOT Gates

Whatever input is fed to an inverting circuit, the output is opposite to that of the input. If a 0 is fed to the input, a 1 is at the output. The symbol for the NOT gate is a small circle at the input to the gate if the input signal fed to the gate is inverted, or at the output from the gate if inversion is to take place after the gate has done its job. A bar over the letter representing a particular input or output is the symbol used to denote that that input or output has been inverted. Two bars over a letter—one on top of the other, indicates two inversions have taken place, or that the input to the inverters has reverted to its original form. A triangular symbol similar to that used for the op amp, along with a circle at its input or output, designates two items. The triangle designates amplification, while the circle designates inversion.

The laws of inversion are obvious.

1. $\overline{1} = 0$. If $A = 1$, $\overline{A} = 0$
2. $\overline{0} = 1$. If $A = 0$, $\overline{A} = 1$
3. $\overline{\overline{A}} = A$

If A is the input to an inverter and B is the output, the truth table for the NOT circuit is:

| A | B |
|---|---|
| 0 | 1 |
| 1 | 0 |

It can be concluded from the truth table that $B = \overline{A}$ and $A = \overline{B}$.

## OR Gates

The OR gate is on (output = 1) when any or all of the inputs to that gate are in the 1 state. OR gates are illustrated in Fig. 13-4. Figure 13-4A shows the basic two input version, and Fig. 13-4B shows a three input OR gate.

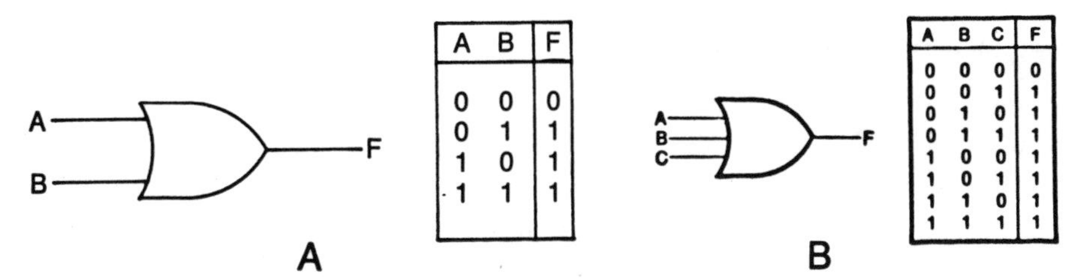


| A | B | F |
|---|---|---|
| 0 | 0 | 0 |
| 0 | 1 | 1 |
| 1 | 0 | 1 |
| 1 | 1 | 1 |

| A | B | C | F |
|---|---|---|---|
| 0 | 0 | 0 | 0 |
| 0 | 0 | 1 | 1 |
| 0 | 1 | 0 | 1 |
| 0 | 1 | 1 | 1 |
| 1 | 0 | 0 | 1 |
| 1 | 0 | 1 | 1 |
| 1 | 1 | 0 | 1 |
| 1 | 1 | 1 | 1 |

Fig. 13-4. The OR gate, shown here in two-input and three-input versions, with truth tables.

No matter how many inputs there are, output F is in the 0 state only when all inputs to the OR gate are at 0. F is a 1 at all other times. For a three-input OR gate, it can thus be written that:

$$F = \overline{AB}C + \bar{A}B\bar{C} + \bar{A}BC + A\overline{BC} + A\bar{B}C + AB\bar{C} + ABC$$

This is determined by stating all factors in the truth table involving the lines where F is 1. In the equation, + is the OR symbol and is *not* a symbol for adding the various factors. F is 1 for the OR circuit if any one of the factors in the expression is true.

Three rules apply to the two input OR gate similar to those used for the AND gate.

1. If A is at one input and 0 is at the second input of a two-input OR gate, the output from the OR gate is whatever A is. Stated in symbols, $A + 0 = A$.
2. If one input to a two-input OR gate is a 1 and the other input is A, the output is 1. Whatever A is, is immaterial. In mathematical terms, $A + 1 = 1$.
3. If both inputs at a two-input OR gate are unknown, but are identical, such as A and A, the output from the OR gate must be the same as the input. Thus $A + A = A$.

There is a variation on the basic OR gate known as the exclusive-OR, or X-OR gate. This gate produces a 1 output when at least one, but not all of its inputs are at 1. This device will be discussed shortly.

## NAND and NOR Gates

Integrated circuits can be composed of several gates on one chip. AND and OR gates are seldom used. More commonly, an IC is comprised of several NAND or NOR gates. A NAND gate is an AND gate followed by a NOT circuit. The AND symbol with a circle at the output is usually used to represent this gate. Similarly, a NOR gate consists of an OR gate followed by a NOT circuit. The OR symbol with a circle at its output represents this gate.

The output of a circuit when the NOT inverters are at the inputs of a gate, differ from the output when the NOT inverter is at the output. The location of the inverter in the circuit cannot be changed, if the circuit is to retain its original characteristics.

### Exclusive OR and Equivalence

Two useful logic arrangements are called exclusive OR and equivalence or exclusive NOR. In exclusive OR, only one item of two (or more) items is true, but both items are not true. It is assumed that A and B are different—never the same. It follows the OR logic with the exception that the output is not true if both A and B are 1. The truth table and symbol for the exclusive OR circuit are in Fig. 13-5. A plus in a circle, $\oplus$, is the symbol indicating the exclusive OR manipulation.

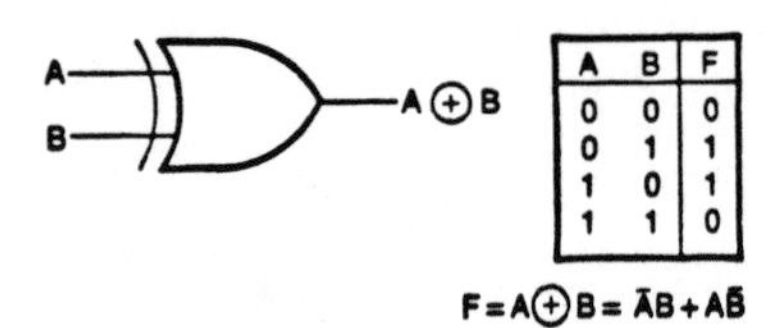


| A | B | F |
|---|---|---|
| 0 | 0 | 0 |
| 0 | 1 | 1 |
| 1 | 0 | 1 |
| 1 | 1 | 0 |

Fig. 13-5. The Exclusive OR gate is a variant of the basic OR gate.

Equivalence can be written as A <=> B. Here, A and B are both 1 if A and B are equivalent. Mathematically, this is written as $A <=> B = AB + \bar{A}\bar{B}$.

Gates may be coupled as shown in Fig. 13-6 to form exclusive OR and equivalence circuits. Gates are coupled to form these and many other types of circuits. Several gate characteristics must be considered when coupling one or more gates to another gate.

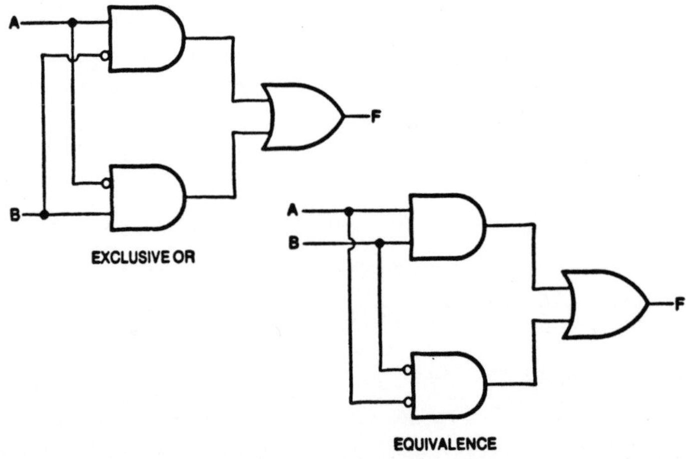


Fig. 13-6. Standard gates can be combined to form Exclusive OR or equivalent circuits.

## Using Boolean Algebra to Manipulate Quantities

In the two-valued (1 and 0) Boolean algebra, factors and quantities derived from logic considerations can be manipulated. These are similar in many instances to the manipulations performed in conventional algebra. Here are the rules.

1. In accordance with the commutative laws, $A + B = B + A$ and $A \bullet B = B \bullet A$. Associative laws are similar for $A(BC) = B(AC) = C(AB)$ and $A + (B + C) = B + (A + C) = C + (A + B)$.
2. Distributive laws are $A(B + C) = (AB) + (AC)$ and $A + (BC) = (A + B)(A + C)$. The second law is the same law as for $A(B + C)$ except a "+" is substituted for a " • " and a " • " is substituted for a "+".
3. Identity elements are $A + 0 = A$ and $A \bullet 1 = A$. The dual theorem states that relationships hold if the OR and AND symbols as well as the 1 and 0 are interchanged. Thus, the dual of $A + 0 = A$ is $A \bullet 1 = A$ as stated above.
4. ORing and ANDing *complements* of quantities are $A + \overline{A} = 1$, and its dual, $A \bullet A = 0$. If the factors fed to the gates are identical and not complements, $A + A = A$ and $A \bullet A = A$.
5. Duals $A \bullet 0 = 0$ and $A + 1 = 1$ were noted above.
6. Using rule 2, it can be shown that $A + AB = A$. First factor the A out of the relationship. This leaves $A(1 + B) = A$. From rule 5, $A + 1 = 1$, so $1 + B = B$. This reduces the factor in the parenthesis to 1 and the relationship becomes $A(1) = A$. Similarly, the dual of the relationship, $A(A + B) = A$, can be determined.
7. Using rule 2, it can be proven that $A + \overline{A}B = A + B$. It should be noted from rule 2 that $(A + \overline{A}B) = (A + A)(A + B)$. As $A + \overline{A} = 1$ from rule 4, the quantity on the left-hand side of the equation reduces to $A + B$.
8. Probably the most useful manipulation procedure in Boolean algebra makes use of *DeMorgan's theorem*. It can be used to take factors out from under a NOT designation. Briefly, the theorem states that $\overline{A + B} = \overline{A \bullet B}$ and that $\overline{A \bullet B} = A + B$. Bars from the NOT symbol are broken up and placed over each individual element of the expression. Should there be one bar over the letter, the bar remains there in the final expression. If two bars are over the letter, they cancel and no bars remain. As for the OR and AND symbols, one bar indicates a change of symbol while two bars indicate no change.

Using all rules, we can prove, for example, that $\overline{AB} = \overline{A}B + A\overline{B} + \overline{AB}$, that $\overline{A} + \overline{B} = \overline{A}B + \overline{A}B + A\overline{B}$, and that $\overline{A + B} = \overline{A} \bullet \overline{B}$. Starting with the first equation:

$$\overline{A}B + A\overline{B} + \overline{AB} = \overline{A}(B + \overline{B}) + A\overline{B} = \overline{A}(1) + A\overline{B} = \overline{A} + A\overline{B} =$$
$$(\overline{A} + A)(\overline{A} + \overline{B}) = 1(\overline{A} + \overline{B}) = \overline{A} + \overline{B} = \overline{\overline{A} \bullet \overline{B}} = \overline{A \bullet B}\text{''}$$

It indicates to us that a NAND gate can be made from an OR gate with the two inputs NOTed.

As for the second equation:

$$\overline{AB} + \overline{A}B + A\overline{B} = \overline{B}(\overline{A} + A) + \overline{A}B = \overline{B} + \overline{A}B = (\overline{B} + \overline{A})$$
$$(B + B) = (B + A)$$

It can also be shown that a NOR gate can be made from an AND gate with the inputs NOTed. Using DeMorgan's theorem.

$$\overline{AB} = \overline{\overline{\overline{A}\,\overline{B}}} = \overline{A + B}.$$

## Gate Characteristics

When working with digital gates, there are several practical characteristics that the designer of digital circuits must consider.

1. *Fan-in* is the number of input terminals at a gate. If there are four input terminals, the fan-in is 4. Gate expanders are frequently connected to an input of a gate. Through use of this device, several inputs can be connected to one of the gate-input terminals. The gate's fan-in figure is thus increased over what it would be without the expander.

2. *Fan-out* depends upon the load presented by the input gate terminals of succeeding gates to the output terminal of a preceding gate. The output from a gate can drive a specific number of subsequent gates without upsetting the 1 to 0 relationship of output and input voltages. The number of gate-input terminals that can effectively be driven by an output from a gate is the fan-out.

3. *Propagation delay* is the time it takes for the rising section and falling section of the signal pulse at the input of a logic circuit to affect its output. The time it takes for a point at the center of the rising portion of the input signal to affect the output is known as the rise propagation delay time, $t_{dpr}$. The fall propagation delay time, $t_{dpf}$, is the time it takes for a point at the center of the falling portion of the input signal to affect the output. Propagation delay time, $t_{dp}$, is usually assumed to be the mean average of the two figures.

4. *Noise immunity.* If a voltage is applied to all inputs of a NAND gate, a NOR gate, or an inverter, a voltage will appear at the output of that gate. This output voltage can be less than, equal to, or more than the input voltage. *Threshold voltage* is an input voltage that is equal to the output voltage. Any voltage less than the threshold voltage can be applied to the input and the output voltage remains unchanged and in saturation. If more than the threshold voltage is applied to the inputs, the output voltage drops. This is true because the output is 1 for all circuits noted above when the inputs are 0 and the output is 0 when the inputs are 1. Noise immunity is equal to the threshold volt-

age. The logic system can tolerate up to this amount of noise and spikes at its inputs, without producing a change in the voltage at its output.

5. Two *noise margin* numbers are associated with combinational logic circuits. One is for the 1 state and the other is for the 0 state. The first number, the high-level noise margin number, $V_{NHI}$, is the difference between the minimum and maximum input voltages that can be applied to a gate that the gate will accept as a 1. The second number, the low-level noise margin, $V_{NLO}$, is the difference between the minimum and maximum input voltages that can be applied to a gate that the gate will accept a 0. Effectively, noise margin is the maximum undesirable voltage that may be at the input of a gate before that input voltage is sufficient to cause that gate to erroneously change its state.

By keeping the above factors in mind, you can successfully couple gates. Two examples are in Fig. 13-6. Circuits are derived from algebraic equations. The simplest circuit using the least number of gates to do the job can be derived by putting the equation in its simplest form using Boolean algebra. *Karnaugh maps* are plots of the original algebraic equations. The plot is in a form that is useful when minimization of circuitry is necessary.

## KARNAUGH MAPS

Truth tables can be charted as maps. Maps are usually limited to situations where there are two, three, or four inputs. Examples of equations for arrangements using two, three, and four inputs, their corresponding truth tables and Karnaugh maps are shown in Fig. 13-7.

The equation, truth table, and Karnaugh map are three different ways of presenting the same information. All terms that must be true, or all factors that must be true, are listed in the expression. Here, each term must be a 1. In the two-input gate in Fig. 13-7, F is 1 when any one of the three terms in the equation is 1. Considering each term individually, $A\overline{B}$ is 1 when A is 1 and B is 0. $\overline{AB}$ is 1 when A is 0 and B is 0, AB is 1 when A is 1 and B is 1. F in the truth table is shown as 1 for all of these conditions. On the Karnaugh map, the columns are headed by the values of A and the rows by the values of B. The box in the map at the upper left refers to the condition at the output of a logic system when A is 0 and B is 0. As F is 1 when this condition exists, a 1 is placed in that box. The next column is headed by an A = 1. As the top row is for B = 0, information described by a listing in the box at the upper right of the map indicates the condition or state of the circuit when A is 1 and B is 0, or $F = A\overline{B}$. As this is in the expression for F, a 1 is placed in this box. The box below this describes the output from the circuit when A is 1 and B is 1 or AB. As this is in the expression, a 1 is also placed in this box. The only remaining box represents the state when input A is 0 and input B is 1, or $\overline{A}B$. This is not in the equation for F. Consequently, a 0 is placed in this box.

The three-input-gate Karnaugh map is only an expansion of the map for the two-input gate. Identical methods and analysis are used here. The columns on the map for

TWO INPUT GATE

EQUATION: $F = A\bar{B} + \bar{A}\bar{B} + AB$

TRUTH TABLE

| A | B | F |
|---|---|---|
| 0 | 0 | 1 |
| 0 | 1 | 0 |
| 1 | 0 | 1 |
| 1 | 1 | 1 |

KARNAUGH MAP

| B \ A | 0 | 1 |
|---|---|---|
| 0 | 1 | 1 |
| 1 | 0 | 1 |

THREE INPUT GATE

EQUATION: $F = AB\bar{C} + \bar{A}\bar{B}C + A\bar{B}C + A\bar{B}\bar{C} + \bar{A}\bar{B}\bar{C}$

TRUTH TABLE

| A | B | C | F |
|---|---|---|---|
| 0 | 0 | 0 | 1 |
| 0 | 0 | 1 | 1 |
| 0 | 1 | 0 | 0 |
| 0 | 1 | 1 | 0 |
| 1 | 0 | 0 | 1 |
| 1 | 0 | 1 | 1 |
| 1 | 1 | 0 | 1 |
| 1 | 1 | 1 | 0 |

| A | 1 | 1 | 0 | 0 |
|---|---|---|---|---|
| C \ B | 0 | 1 | 1 | 0 |
| 0 | 1 | 1 | 0 | 1 |
| 1 | 1 | 0 | 0 | 1 |

FOUR INPUT GATE

EQUATION: $F = ABCD + A\bar{B}C\bar{D} + \bar{A}\bar{B}CD + AB\bar{C}D + A\bar{B}\bar{C}D + \bar{A}\bar{B}\bar{C}D$

TRUTH TABLE

| A | B | C | D | F |
|---|---|---|---|---|
| 0 | 0 | 0 | 0 | 0 |
| 0 | 0 | 0 | 1 | 1 |
| 0 | 0 | 1 | 0 | 0 |
| 0 | 0 | 1 | 1 | 1 |
| 0 | 1 | 0 | 0 | 0 |
| 0 | 1 | 0 | 1 | 0 |
| 0 | 1 | 1 | 0 | 0 |
| 0 | 1 | 1 | 1 | 0 |
| 1 | 0 | 0 | 0 | 0 |
| 1 | 0 | 0 | 1 | 1 |
| 1 | 0 | 1 | 0 | 1 |
| 1 | 0 | 1 | 1 | 0 |
| 1 | 1 | 0 | 0 | 0 |
| 1 | 1 | 0 | 1 | 1 |
| 1 | 1 | 1 | 0 | 0 |
| 1 | 1 | 1 | 1 | 1 |

| | A | 1 | 1 | 0 | 0 |
|---|---|---|---|---|---|
| C | D \ B | 0 | 1 | 1 | 0 |
| 1 | 0 | 1 | 0 | 0 | 0 |
| 1 | 1 | 0 | 1 | 0 | 1 |
| 0 | 1 | 1 | 1 | 0 | 1 |
| 0 | 0 | 0 | 0 | 0 | 0 |

Fig. 13-7. Equations for gates with two, three, and four inputs, along with the truth tables and Karnaugh maps.

the three-input gate represent different coincident inputs for A and B, while the rows represent specific Cs. About the only difference between the Karnaugh maps for the three- and four-input gates is that in the latter drawings the rows represent the different combinations of C and D.

In setting up the Karnaugh maps, each term where F is 1 must include information about all inputs fed to the gating circuit. In the four-input arrangement, for example, there must be A, B, C, and D Boolean variables for each term in the relationship to F, or you do not know in which box on the map to place a 1 and in which box to place a 0. The form relating all factors or terms in each segment of the equation to F is known as the *disjunctive normal form* (dnf).

### Generating the Dnf

Should an equation not be in the dnf, it can be converted simply by ANDing each element by 1. In Boolean algebra rule 4, it was noted that A + A = 1. Using this rule, any equation can be converted to the dnf. Let's try some examples.

Example 1: $F = AB + \overline{B}C$

Here, a C is missing from the first term and an A is missing from the second term. So AND the first term with a 1 = C + C and AND the second term with a 1 = A + A. The dnf is then determined as follows:

$$
\begin{aligned}
F &= AB(C + \overline{C}) + (A + \overline{A})\overline{B}C \\
&= ABC + AB\overline{C} + A\overline{B}C + \overline{A}\overline{B}C
\end{aligned}
$$

Example 2:

$$
\begin{aligned}
F &= A + \overline{B}C \\
&= A(B + \overline{B})(C + \overline{C}) + (A + \overline{A})\overline{B}C \\
&= (AB + A\overline{B})(C + \overline{C}) + A\overline{B}C + \overline{A}\overline{B}C \\
&= ABC + AB\overline{C} + A\overline{B}C + A\overline{B}\overline{C} + A\overline{B}C + \overline{A}\overline{B}C
\end{aligned}
$$

Example 3:

$$
\begin{aligned}
F &= (A + \overline{B})(C + \overline{D}) \\
&= AC + A\overline{D} + \overline{B}C + \overline{B}\overline{D} \\
&= AC(B + \overline{B})(D + \overline{D}) + A\overline{D}(B + \overline{B})(C + \overline{C}) + \\
&\quad \overline{B}C(A + \overline{A})(D + \overline{D}) + \overline{B}\overline{D}(A + \overline{A})(C + \overline{C}) \\
&= (ABC + A\overline{B}C)(D + \overline{D}) + (AB\overline{D} + A\overline{B}\overline{D}) \\
&\quad (C + \overline{C}) \\
&\quad + (A\overline{B}C + \overline{A}\overline{B}C)(D + \overline{D}) + (AB\overline{D} + \overline{A}\overline{B}\overline{D}) \\
&\quad (C + \overline{C}) \\
&= (ABCD + ABC\overline{D} + A\overline{B}CD + A\overline{B}C\overline{D}) \\
&\quad + (ABC\overline{D} + AB\overline{C}\overline{D} + A\overline{B}C\overline{D} + A\overline{B}\overline{C}\overline{D}) \\
&\quad + (A\overline{B}CD + A\overline{B}C\overline{D} + \overline{A}\overline{B}CD + \overline{A}\overline{B}C\overline{D}) \\
&\quad + (A\overline{B}C\overline{D} + A\overline{B}\overline{C}\overline{D} + \overline{A}\overline{B}C\overline{D} + \overline{A}\overline{B}\overline{C}\overline{D})
\end{aligned}
$$

Because there are several identical groups repeated in the final expression, it can be reduced. Any repeated group can be dropped. The following equation for F remains after this manipulation.

$$
F = ABCD + ABC\overline{D} + A\overline{B}CD + A\overline{B}C\overline{D} + AB\overline{C}\overline{D} + \overline{A}\overline{B}\overline{C}\overline{D} + \overline{A}\overline{B}CD + \overline{A}\overline{B}C\overline{D} + \overline{A}\overline{B}\overline{C}\overline{D}
$$

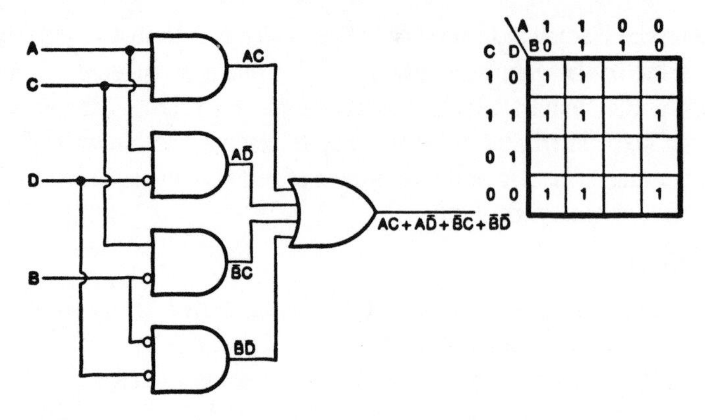


Fig. 13-8. A Karnaugh map for Example 3 and combinational logic-gating circuit.

This relationship is plotted on the Karnaugh map in Fig. 13-8. All boxes without 1s are assumed to be 0s.

When an equation is in the dnf, all Boolean variables are included in each term. Each term is referred to as a *minterm* using the symbol m. A subscript is placed next to the m indicating the binary value of that term. In the last expression for F, minterm ABCD = $m_{15}$, minterm ABCD = $m_2$, minterm ABCD = $m_{11}$, and so on.

## Simplifying Equations Through Use of the Karnaugh Map

Observing the Karnaugh map in Fig. 13-8, two of the terms in the equation for F are $A\overline{B}C\overline{D}$ and $ABC\overline{D}$. If you factor out $AC\overline{D}$, the equation becomes $F = AC\overline{D}(B + \overline{B})$. As B + B = 1, the $A\overline{B}C\overline{D}$ and $ABC\overline{D}$ terms become $AC\overline{D}$. The terms, and the Boolean variables in these terms, can be dropped from the final equation when there are adjacent 1s on the map. Similarly, the top two terms in the first column are F = $A\overline{B}C\overline{D}$ and $A\overline{B}CD$. Here, $A\overline{B}C$ can be factored and $F = A\overline{B}C$. When all four entries in the upper left-hand corner of the map are considered, the equation can be simplified to F = AC. Bs and Ds differ and thus they drop out from the final statement of the equation.

Pursuing this procedure, the two 1s at the bottom of the first two columns are taken into account by dropping the B variable. The resulting equation, is $F = A\overline{C}\overline{D}$. However, this can be simplified further. If you consider the bottom row on the map as being adjacent to the top row, the Cs also differ. Variable C is 0 for the bottom row and 1 for the top row. Considering the 1s in the bottom and top rows of the two left-hand columns, both the B and C variables can be dropped and the final statement for F due to the 1s in these four boxes on the map is $F = A\overline{D}$. Note that the two 1s in the top row of the first two columns were used twice to simply the final equation. All 1s must be considered at least one time, but may be used as often as necessary to help simplify the final expression.

In a similar manner, the top two 1s in the column at the extreme right may be used in conjunction with the top two 1s in the column at the extreme left to establish the term BC. The 1 in the bottom row of the column at the extreme right may be used with the 1s at the remaining three corners of the map, to provide us with the term BD. The final equation for F becomes:

$$F = AC + A\overline{D} + BD + \overline{B}\overline{D}$$

This is exactly the second statement of the expression in Example 3. It is not a simplification, as no further simplification is possible. A drawing of gates to satisfy this relationship (combinational logic circuit) is also shown in Fig. 13-8.

Different combinations of terms can be formed from information found on the map. The simplest and best combination can be found by trial-and-error methods. When choosing a combination, try to eliminate the maximum number of variables you can from each term. Expressions can be further simplified if there are complete rows or columns of 1s on the map. If the top row had all 1s in Fig. 13-8, a term in the equation for F would have been $C\overline{D}$, for variables A and B would both have been eliminated. If the top two rows were all 1s, also the D variable would have been eliminated and the portion of the equation representing these two rows would have been simply F = C. Appropriate variables are eliminated when there are adjacent 1s on the map.

Another way of simplifying equations and circuits is through the use of *don't cares*. In some designs, one or more sets of terms in the final equation are unimportant. In this case, it does not matter if a box on the map has a 1 or a 0. If a don't care term exists, an X is placed in the box representing that term. When selecting combinations to form the simplest equation, the X is used either as a 0 or as a 1, as desired. The choice depends entirely upon which bit goes furthest in simplifying the final equation.

## FLIP-FLOPS

Up to this point, the 1s or 0s at the input of the circuits were the sole determining factors as to what the output from the circuit will be. Beginning with this section, we add a memory to the circuit. In these *sequential* circuits, the output not only depends upon the inputs being applied to the logic circuit at the present moment, but also depends upon previously applied inputs and the previous state of the circuit. The flip-flop or bistable multivibrator is the basic element of a sequential circuit. Two flip-flops—one using two NAND gates and the other using two NOR gates, along with their truth tables, are shown in Fig. 13-9.

### R-S Flip-Flop

Flip-flops can be made from two inverters, two NAND gates, or two NOR gates. Regardless of the type of gates or inverters chosen, information from their outputs must

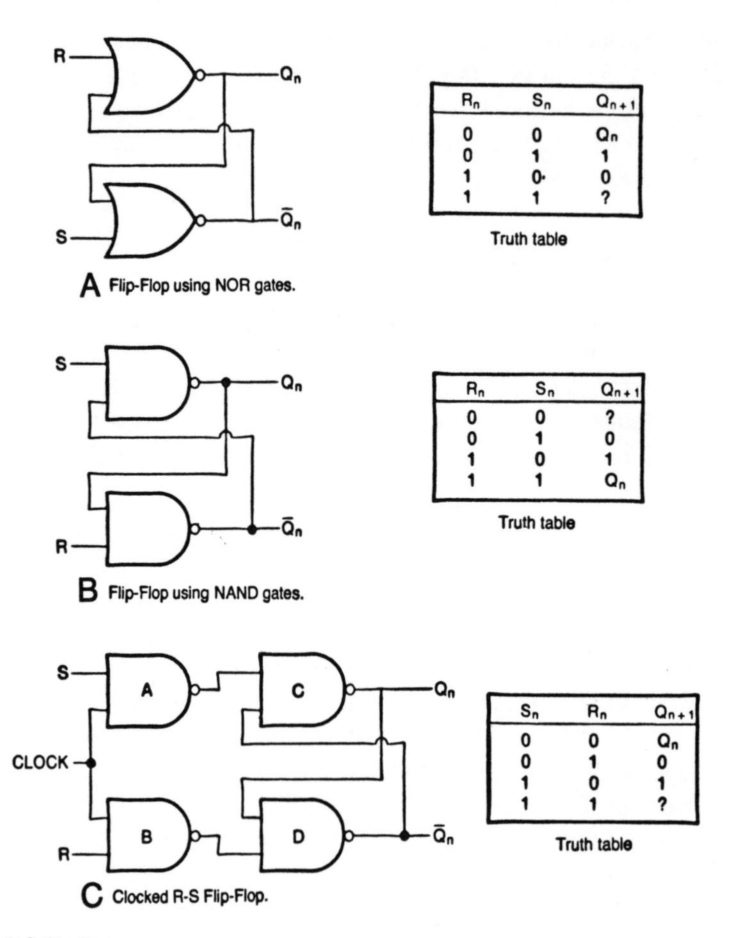


| $R_n$ | $S_n$ | $Q_{n+1}$ |
|---|---|---|
| 0 | 0 | $Q_n$ |
| 0 | 1 | 1 |
| 1 | 0· | 0 |
| 1 | 1 | ? |

Truth table

| $R_n$ | $S_n$ | $Q_{n+1}$ |
|---|---|---|
| 0 | 0 | ? |
| 0 | 1 | 0 |
| 1 | 0 | 1 |
| 1 | 1 | $Q_n$ |

Truth table

| $S_n$ | $R_n$ | $Q_{n+1}$ |
|---|---|---|
| 0 | 0 | $Q_n$ |
| 0 | 1 | 0 |
| 1 | 0 | 1 |
| 1 | 1 | ? |

Truth table

Fig. 13-9. R-S flip-flops.

be fed back to their inputs. Starting with the NOR gate flip-flop, assume it is initially in the condition where Q is 1 and $\overline{Q}$ is 0. By applying a 1 to the R input and leaving input S at 0, the upper gate is turned on. Q becomes 0. Q is now a 1 because both inputs at the lower gate are 0. Should a 1 be applied to the S input and a 0 to R, the lower gate is turned on. $\overline{Q}$ reverts to the 0 it originally was assumed to be, and Q turns back into a 1, because now both inputs to the upper gate are 0. Should R and S both be made 0s, Q and $\overline{Q}$ do not change from their original 1 and 0 states, respectively, because the 1 originally at Q feeds the lower gate so that Q remains at 0 and the 0 at $\overline{Q}$ feeds the upper gate so that Q remains at 1. If R and S inputs are both 1s, the output is undefined for either output can be at 1 or at 0. 1s should never be fed to both inputs simultaneously when the flip-flop is composed of NOR gates. The data just described is summarized in the truth table. The subscript n at each letter indicates the input applied to, or the output from the flip-flop, before it is triggered. Subscript n + 1 at the Q indicates the Q output from the flip-flop after it has been triggered.

A similar logical sequence can be stated for a flip-flop composed of NAND gates. Both R and S may be 1 when NAND gates are used, but the output is undefined when 0s are applied to both inputs. This is opposite to the situation that exist when NOR gates are used.

NAND logic circuits are most commonly used because they lend themselves to *clocking* more readily than do flip-flops made of NOR logic gates. (By clocking we refer to a circuit arrangement in which an extraneous pulse is required at some predetermined input before the Q and $\overline{Q}$ output will switch.) NAND gate circuits are used throughout the discussion of sequential circuits with the assumption that R must be 0 if the output $Q_{n+1}$ is to be 0. It can then be said that by making R into a 0, $Q_{n+1}$ is *reset* to 0. Similarly, S must be 0 if the output $Q_{n+1}$ is to be 1, so that S must be 0 if $Q_{n+1}$ is to be *set* to 1. If both inputs are 1s, the state of the flip-flop remains unchanged and $Q_{n+1} = Q_n$. Should both inputs be 0s, the output $Q_{n+1}$ is undefined.

A clock circuit may be added to the R-S flip-flop as shown in Fig. 13-9C. Here, two NAND gates precede those already in the flip-flop circuit. Information at the S and R inputs never reach the flip-flop composed of NAND gates C and D, until a square-wave type of pulse or a 1 is applied to the clock inputs of gates A and B.

### Toggle (T) and Gate Latch (D) Flip-Flops

A feedback circuit with a clocking input variation on the circuit in Fig. 13-9C is shown in Fig. 13-10A. Here Q and $\overline{Q}$ interchange states each time T goes negative. Diodes are off until a negative pulse from T turns them on. There is no change of state until the pulse is supplied, for otherwise signals from Q and $\overline{Q}$ cannot reach the S and R inputs. This change from 0 to 1 and from 1 to 0 continues as long as pulses are applied to clock input T.

The toggle flip-flop circuit does not in itself require the R and S inputs. If R and S inputs are available they take priority over other inputs to the circuit, even if they are not used in the actual operation of the flip-flop. If R is made 0, the flip-flop resets to 0. Stated otherwise, if R = 0, Q = 0. Similarly, S = 0, Q is set to 1, or Q = 1.

In the D or gated latch flip-flop, an extra terminal is added to the circuit. Information is retained at that D terminal. This information affects the output only after a clock pulse has been applied to the flip-flop. After the clock pulse, Q becomes a 1 if the D input is a 1 and a 0 if that happens to be the input at D.

### Ripple Counter

Q and $\overline{Q}$ outputs change with each clock pulse fed to a flip-flop when that flip-flop is arranged in a T circuit. Should the output from Q be fed to the T input of succeeding

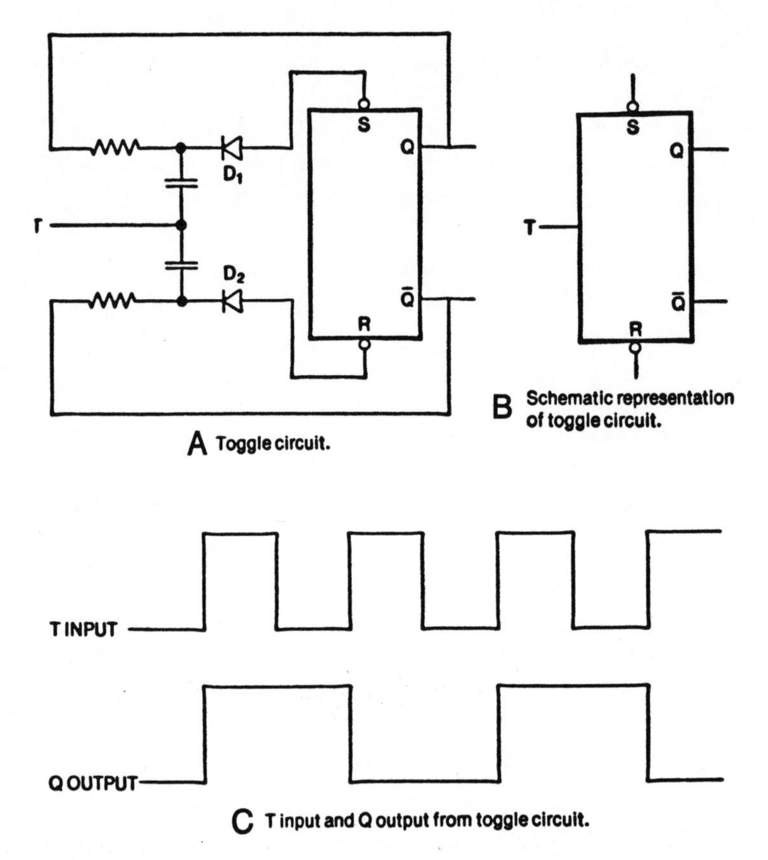


Fig. 13-10. A toggle circuit.

flip-flop, its state will change each time the Q output of flip-flop #1 changes state. It should be noted that the output from the first flip-flop goes from a 1 to a 0 or from a 0 to a 1 half as often as does its input, because it only switches when a 1 is applied to its input. This also holds true for the succeeding flip-flop. Graphically, the characteristics can be seen in Fig. 13-10C, where the clock input T and the Q output from the T circuit is shown. The state is shown as changing on the rising portion of each clock pulse. Instead of this, the circuit can be designed for a change of state to occur on the falling portion of each pulse.

Basic ripple counters using the T circuit are capable of counting a quantity of binary numbers equal to 2 to an exponent. (Bits at the Q outputs of all flip-flops are taken in sequence to form these numbers.) The exponent is equal to the number of flip-flops in the circuit. If one flip-flop is used, the power of 2 is 1, so the amount of numbers that can be counted using this circuit is $2^1 = 2$. Its outputs are only 0 and 1. Should two flip-flops be in the circuit with the Q output of the first flip-flop connected to the T input of the next flip-flop, the power of 2 is 2 or $2^2 = 4$. Four binary numbers can then be counted by using bits at the Q outputs from the two flip-flops—00, 01, 10, and 11 (0, 1, 2, and 3 in the decimal system). After counting to 11, the counter reverts back to 00 and repeats the counting cycle. Because the two flip-flops go through four states before returning to the original 00 state, it is referred to as a modulo-4 counter. The circuit using one flip-flop is a modulo-2 counter. Should three flip-flops be in a circuit, the total count that can be made is $2^3 = 8$. The binary numbers counted are 000, 001, 010, 011, 100, 101, 110, and 111. It then reverts back to 000 to begin the repeat cycle of the sequence. This, then, is a modulo-8 counter. Bits are determined from the 1s and 0s at the Q terminals of the various flip-flops. The flip-flop to which the clock pulse is fed provides us with the least significant bit at its Q output. Input pulses do not have to be evenly spaced. Random pulses can be fed to the system and counted.

A three-stage ripple counter is shown in Fig. 13-11A. Here, the clock pulse triggers $FF_3$. Pulses at Q of $FF_3$ trigger $FF_2$. Pulses at Q of $FF_2$ trigger $FF_1$. Each succeeding flip-flop has half the number of pulses of the preceding one. The flip-flop with the smallest number of pulses, $FF_1$, delivers the MSB at its Q output, output 1. Output 3 from $FF_3$ presents the maximum number of pulse changes, and hence supplies the LSB. Eight states from 000 to 111 (0 to 7) are completed through this circuit. Thereafter, the sequence is repeated. Outputs at the Qs of the various FFs are shown in Fig. 13-11B.

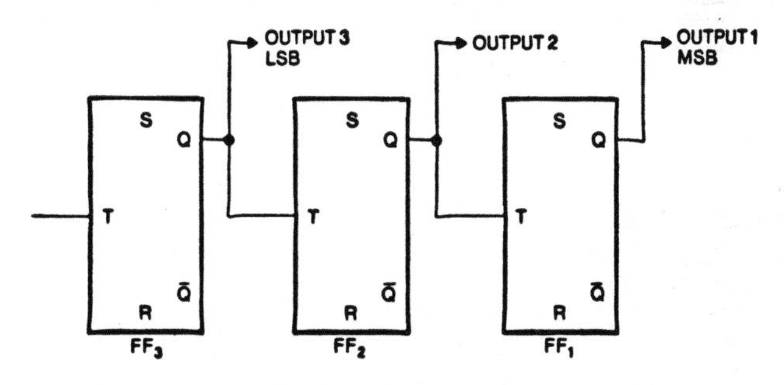


A Three stage ripple counter.

| OUTPUT 1 AT Q OF $FF_1$ | OUTPUT 2 AT Q OF $FF_2$ | OUTPUT 3 AT Q OF $FF_3$ | DECIMAL EQUIVALENT OF STATES |
|---|---|---|---|
| 0 | 0 | 0 | 0 |
| 0 | 0 | 1 | 1 |
| 0 | 1 | 0 | 2 |
| 0 | 1 | 1 | 3 |
| 1 | 0 | 0 | 4 |
| 1 | 0 | 1 | 5 |
| 1 | 1 | 0 | 6 |
| 1 | 1 | 1 | 7 |
| 0 | 0 | 0 | 0 |
| 0 | 0 | 1 | 1 |
| 0 | 1 | 0 | 2 |
| 0 | 1 | 1 | 3 |
| | | | etc. |

B Table of states for three-stage ripple counter.

Fig. 13-11. Flip-flops can be combined to form a ripple counter.

Circuits can be added to reduce the number of states in the sequence. From zero to eight ($2^3$) states are possible when three flip-flops are used. Should only two flip-flops be in the circuit, up to $2^2$ or 4 states are possible. Hence a modulo-5 through modulo-8 counter requires at least three flip-flops. Should a modulo-10 counter be required, more than three flip-flops are needed. As $2^4 = 16$, four flip-flops will suffice. The modulo numbers must be equal to or less than $2^n$ when n is the number of R-S flip-flops in the circuit.

As an example of a counter with a sequence of reduced length, let us design a modulo-6 ripple counter. Three R-S flip-flops are required for 6 is between $2^2 = 4$ and $2^3 = 8$. Use the drawings and tables in Fig. 13-12. Start by indicating the various states

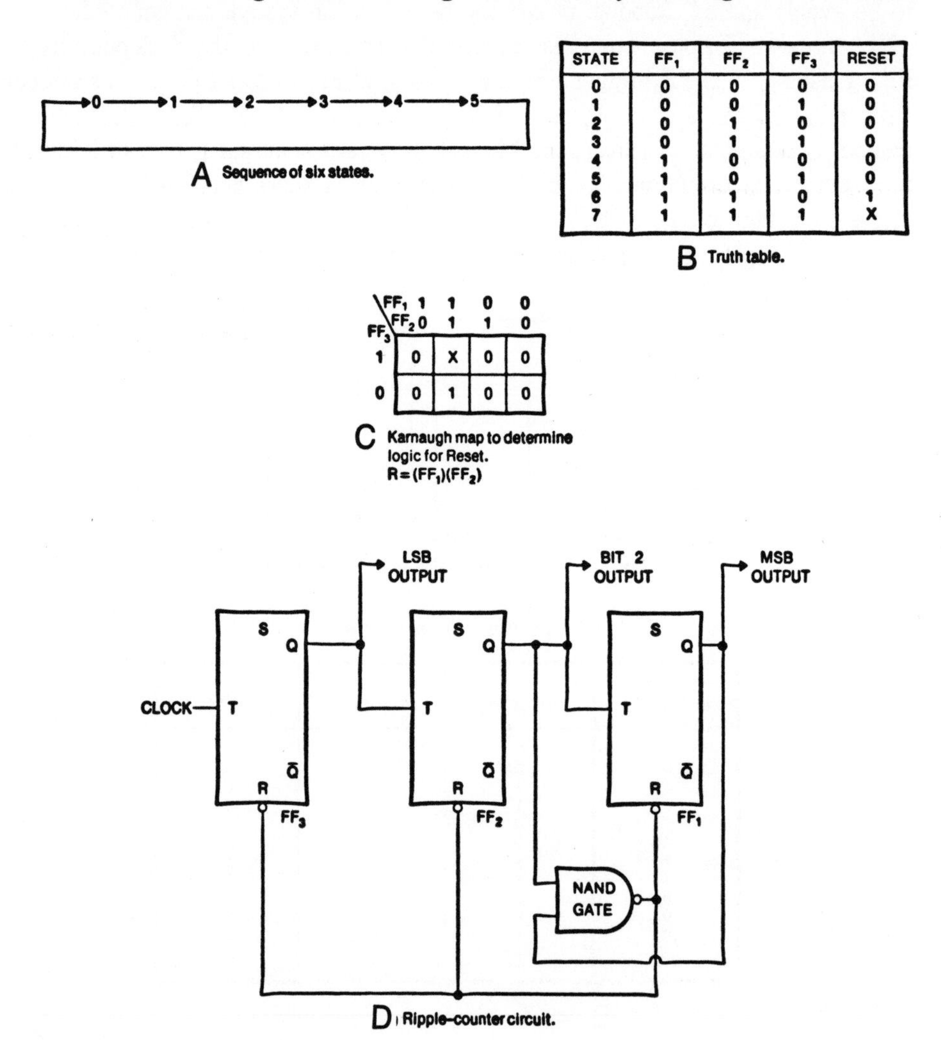


| STATE | $FF_1$ | $FF_2$ | $FF_3$ | RESET |
|---|---|---|---|---|
| 0 | 0 | 0 | 0 | 0 |
| 1 | 0 | 0 | 1 | 0 |
| 2 | 0 | 1 | 0 | 0 |
| 3 | 0 | 1 | 1 | 0 |
| 4 | 1 | 0 | 0 | 0 |
| 5 | 1 | 0 | 1 | 0 |
| 6 | 1 | 1 | 0 | 1 |
| 7 | 1 | 1 | 1 | X |

B Truth table.

| $FF_3$ \ $FF_1$ $FF_2$ | 1 0 | 1 1 | 0 1 | 0 0 |
|---|---|---|---|---|
| 1 | 0 | X | 0 | 0 |
| 0 | 0 | 1 | 0 | 0 |



Fig. 13-12. A Modulo-6 ripple counter.

that the counter must pass through. There are six states starting at 0 and proceeding through 5. The count must then return to 0 to repeat the sequence. Next, write a complete truth table showing all eight possible states that a three-flip-flop circuit may go through.

After going through six states and #5 (101) has been counted, pulses must be applied to all reset (R) inputs to reset the Q outputs of all flip-flops to 0. State 7 (or #6-110) should not be counted. The instant the flip-flops provide a 110 at the Q outputs (indicating the undesirable number 6), the flip-flop outputs should immediately be switched to 000 to recycle the 0 to 5 sequence of the six states. To indicate this, place a 1 in the "reset" column of the line showing binary number 110. This indicates that there must be a logic circuit to sense 110 at the various Q outputs and deliver this information to the reset inputs of the flip-flops. An X is placed in the 111 row of state 7. As this is an unused state, it does not matter what might have been fed to the R inputs of the flip-flops from this state. The X indicates a "don't care" condition. All other states are shown with a 0. If a 1 would be in the reset column for any other state, the logic circuit would also reset the flip-flops to a 000 output when it flips into that state.

Next plot the information from the truth table on a Karnaugh map. If you consider the X as a 1 on the map, $FF_3$ drops out from the final equation as the X and 1 are in adjacent boxes. The equation then becomes $R = (FF_1)(FF_2)$. A plot of this logic gate along with the balance of the flip-flop circuit is shown. Q outputs from $FF_1$ and $FF_2$ are fed to this NAND gate. The output from the NAND gate is fed to the R inputs of all flip-flops so that Qs at all outputs are reset to 0. Note that instead of using an AND gate determined by the equation $R = (FF_1)(FF_2)$, a NAND gate is used. This is because R must be 0 if the flip-flops are to be reset. After applying the 0s to all R inputs, all required 0s appear at the Q outputs.

In order to simplify further discussions, we will refer to the condition or state of Q as the output from a flip-flop. Thus, if Q is 0, the flip-flop is referred to as being a 0. Putting a 0 at R resets the flip-flop to 0. Obviously, if Q is 0, $\overline{Q}$ is 1 and vice versa. Also, the flip-flop providing the MSB will be referred to simply as A. All flip-flops delivering less significant bits will be labeled B, C, D, etc., in order of bit significance, with the earlier letters being the more significant bits.

We can now proceed to design a ripple counter where states in the midst of a sequence are skipped. Assume we wish to go through states 5-6-7-9-10-11-5-6-7, etc. If we use three flip-flops, we can have only $2^3 = 8$ states. This is insufficient as we must be able to read up to 11 at the Q outputs. Hence we must use four flip-flops where a maximum of $2^4 = 16$ states are possible.

Start by plotting the sequence of states along with a table covering all 16 states. As two sets of logic are necessary to reset the counters—to 9 after 7 and to 5 after 11—two reset columns are on the truth table. In the column marked "reset 8/9," as 1 is on the line with 1000 to indicate that the instant the flip-flop outputs supply 1000,

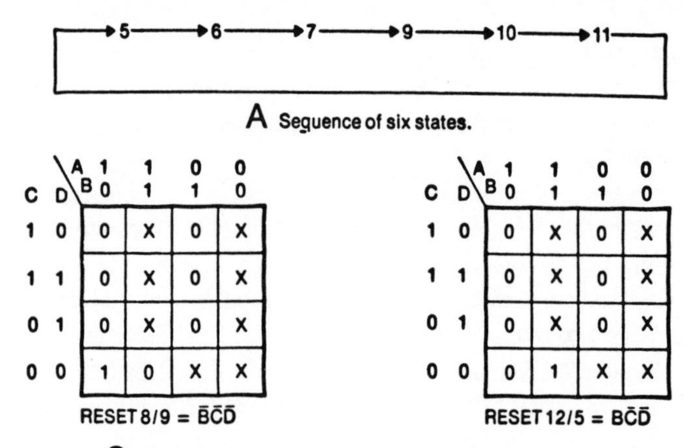


C Karnaugh maps to determine logic circuits for reset.

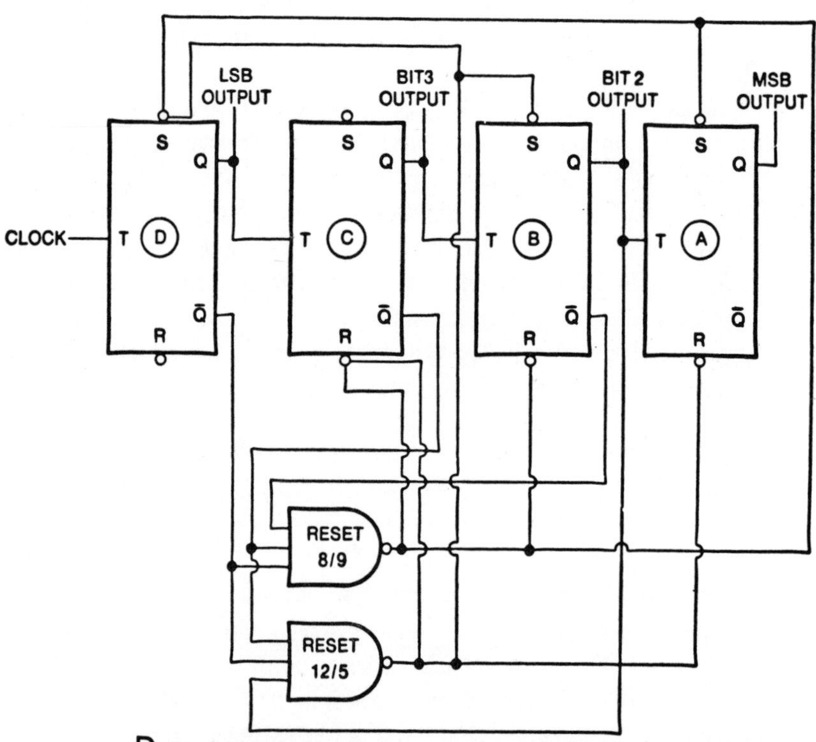


D Flip-Flop using basic logic circuitry. Fallacy in circuit is that outputs from both NAND gates are connected to each other at R input of flip-flop C and S input of flip-flop D.

Fig. 13-13. A ripple counter that skips states.

| DECIMAL NUMBER | A | B | C | D | RESET 8/9 | RESET 12/5 |
|---|---|---|---|---|---|---|
| 0 | 0 | 0 | 0 | 0 | X | X |
| 1 | 0 | 0 | 0 | 1 | X | X |
| 2 | 0 | 0 | 1 | 0 | X | X |
| 3 | 0 | 0 | 1 | 1 | X | X |
| 4 | 0 | 1 | 0 | 0 | X | X |
| 5 | 0 | 1 | 0 | 1 | 0 | 0 |
| 6 | 0 | 1 | 1 | 0 | 0 | 0 |
| 7 | 0 | 1 | 1 | 1 | 0 | 0 |
| 8 | 1 | 0 | 0 | 0 | 1 | 0 |
| 9 | 1 | 0 | 0 | 1 | 0 | 0 |
| 10 | 1 | 0 | 1 | 0 | 0 | 0 |
| 11 | 1 | 0 | 1 | 1 | 0 | 0 |
| 12 | 1 | 1 | 0 | 0 | 0 | 1 |
| 13 | 1 | 1 | 0 | 1 | X | X |
| 14 | 1 | 1 | 1 | 0 | X | X |
| 15 | 1 | 1 | 1 | 1 | X | X |

B Truth table.

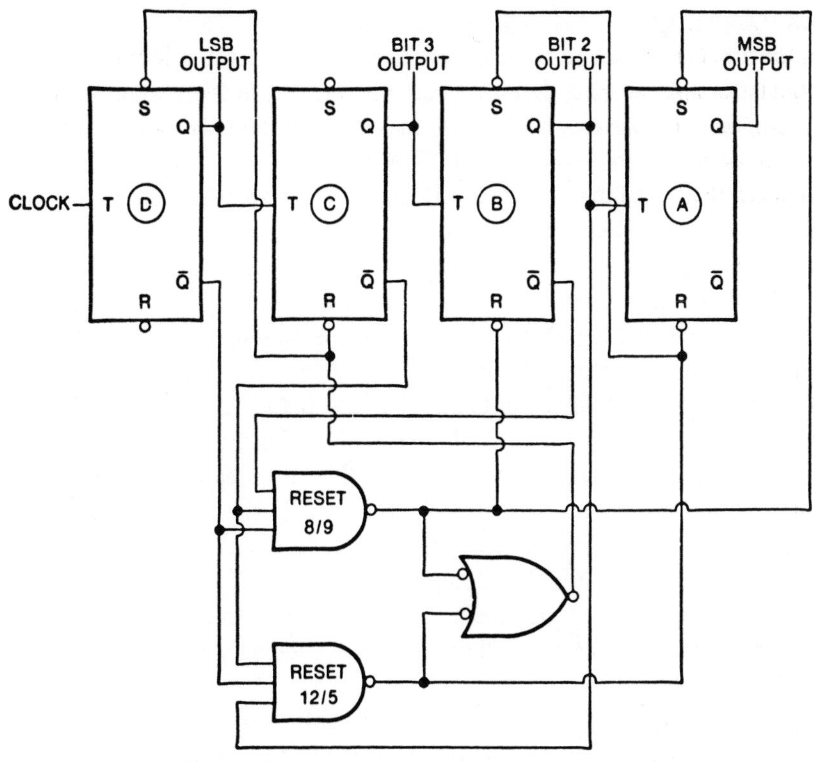


E Corrected circuit adding NOR gate to arrangement in Fig. 13-13D.

the output must be shifted instantly from 1000 (state 8) to 1001 (state 9). Listings in the "reset 8/9" column must be 0 on the lines that have numbers shown in the sequence diagram (Fig. 13-13A), so that the logic circuit meant only to shift 1000 to 1001 will not be activated when these numbers are at the Q outputs.

In the column "reset 12/5," place a 1 in row 1100 (state 12) to indicate that there must be an instantaneous shift to 0101 (state 5) when 1100 is at the Q outputs. In all other active states there must be a 0 in this column. Also add a 0 in the column for state 8 so that there will be no shift to 0101 when this number, 1000, appears at the Q outputs. Enter a 0 on the 1100 line in the "reset 8/9" column, so that the logic circuit will not cause the output to shift to 1001 when it enters the 1100 state, and will only shift from the 1100 state to 0101 due to the 1 in the "reset 12/5" column. As none of the other states have any effect on the sequence because they are skipped, enter Xs in both reset columns on all of these remaining lines.

Using the information in the truth table, draw two Karnaugh maps—one for each function that must be fulfilled. In the "reset 8/9" map, the 1 may be coupled with the X in the 0000 box. Output A drops out to produce the requirement that reset 8/9 = $\overline{B}\,\overline{C}\,\overline{D}$. Similarly, the 1 in the "reset 12/5" map may be coupled with the X in the 1101 or 0100 box. As the latter box has many factors in common with those determined from the "reset 8/9" map, and because this coincidence may simplify the final logic circuit, we will use the X in the 0100 box. Now reset 12/5 = $\overline{B}\,\overline{C}\,\overline{D}$.

From this we learn that when the outputs are $\overline{B}\,\overline{C}\,\overline{D}$, the logic must reset the flip-flop outputs to state 9 or 1001. Hence a 0 must be fed from the logic circuit to the S input of the first and last flip-flops A and D, to produce 1s at their Q outputs. 0s must be fed to the R inputs of flip-flops B and C to produce 1s at their Q outputs. Similarly, when the outputs are $\overline{B}\,\overline{C}\,\overline{D}$, the logic circuit must reset the flip-flop output to state 5 or to 0101. Hence, a 0 must be fed to the R inputs of flip-flops A and C to produce 0s at their outputs and a 0 to the S inputs of flip-flops B and D to produce 1s at their outputs. A drawing of the flip-flops with the logic circuits is in Fig.13-13D.

Outputs from both NAND gates are connected to the S input of flip-flop D and the R input of flip-flop C. This puts the outputs from the two NAND gates in parallel. In order to properly apply the outputs from the logic circuits to the R and S inputs on the flip-flops, the outputs from the two NAND gates must be isolated. This can be done by inverting a portion of these outputs and feeding them both to independent inputs of a NOR gate. The output from the NOR gate is then fed to the R and S inputs in question. This is shown in Fig. 13-13E.

The counting sequence of circuits discussed thus far starts at a low number and proceeds to the highest number before the circuit is reset to once again provide the lowest number and repeat the sequence. These are known as *up* counters. There are also *down* counters, where the counting sequence is from the largest number in a group down to the smallest number. A counter using circuits similar to those shown above will count from the highest to the lowest number, if one or all of the following conditions have been met. If only two of these criteria are satisfied, you will once again have an *up* counter.

1. The output signal that is to be fed to the clock input of a succeeding stage must be taken from the $\overline{Q}$ terminal of the preceding flip-flop rather than from the Q terminal.

2. The output level for each bit must be taken from the $\overline{Q}$ terminal rather than from the Q terminal.

3. The flip-flop should switch states on the zero to maximum transition (leading edge) of the clock pulse rather than on the maximum to zero-level transition.

Complications arise in all designs when the flip-flop that may be available does not have a *reset* input, but is supplied by the manufacturer only with a *set* terminal. In this case, the *set* becomes a *reset* terminal if the Q and $\overline{Q}$ terminals on the IC are interchanged. You can relabel the *set* input with the designation *reset* if you relabel the Q and $\overline{Q}$ terminals to show that they have been interchanged. Then proceed as before. Similarly, if you have an IC in which the clock flips the state on the wrong edge of the pulse (such as in the zero-to-maximum output transition rather than in maximum-to-zero output transition), connect the output from the $\overline{Q}$ terminals of the flip-flop preceding the one in question to the clocking input of that reverse-performing flip-flop.

## J-K Flip-Flop

Three types of flip-flops have been discussed. These were the R-S, T-type or Toggle, and the D-type or gated latch. A fourth type is the J-K flip-flop. Its truth table is shown in Fig.13-14A. Note that any of the four states of the J-K flip-flop may be used, for now, when J = K = 1, the Q output has a determinable form changing states with each clock pulse.

The J-K flip-flop symbol, is in Fig. 13-14. Two terminals not noted as such on other flip-flops, are shown here—the Clear (Cr) and the Preset (Pr). The initial state of the flip-flop can be established using these inputs. By setting Cr at 0 and Pr at 1, Q becomes

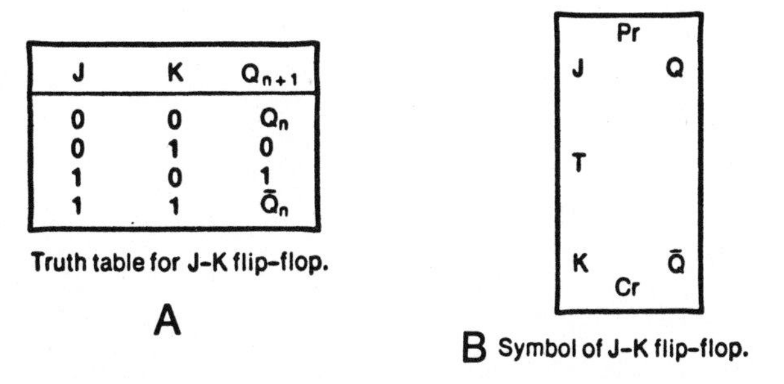

| J | K | $Q_{n+1}$ |
|---|---|---|
| 0 | 0 | $Q_n$ |
| 0 | 1 | 0 |
| 1 | 0 | 1 |
| 1 | 1 | $\overline{Q}_n$ |

Truth table for J-K flip-flop.

A



B Symbol of J-K flip-flop.

Fig. 13-14. The J-K flip-flop.

a 0. When Pr = 0 and Cr = 1, Q is 1. Information applied to these terminals takes preference over information applied to any of the other inputs. The flip-flop will not change states if both Cr = 0 and Pr = 0. Present and clear functions are performed when the clock pulse is at zero. After having been cleared or preset, Pr and Cr must both be set to a 1 level before the clock pulse becomes a 1 if the flip-flop is not to be influenced by these inputs. The Pr and Cr terminals may or may not be part of a particular IC J-K flip-flop supplied by a manufacturer.

Problems may arise if the duration of the clock pulse exceeds that of the 1s and 0s applied to the J-K inputs. When this happens, the flip-flop may change states during the time the clock pulse is present. This is known as a *race-around* condition. It can be alleviated by using a master-slave flip-flop arrangement such as the one shown in Fig. 13-15. The first flip-flop in the circuit is the master or *first rank* of a dual-rank flip-flop and the second one is the slave or *second rank*. The clock pulse turns on the master while the slave is off because the clock pulse is fed through an inverting NOT gate to the slave. This gate inverts the phase of the clock pulse, so while a 1 is fed to the master, the slave sees a 0. After the master has flipped to its desired state, the clock pulse reverts to 0. The master is off and the slave is turned on. Only then will the slave flip. Whatever pulse the slave feeds back to the input of the master is immaterial at the moment, as the master's clock pulse is 0 and the flip-flop cannot change states.

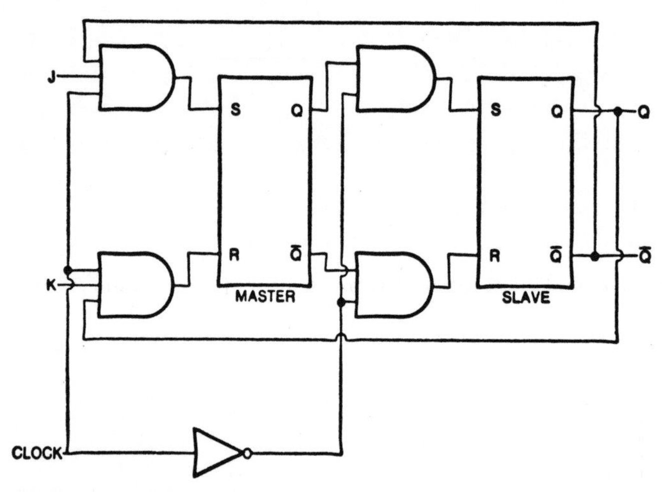


Fig. 13-15. A master-slave flip-flop.

## J-K, T-Type, and D-Type Flip-Flops

The T-type flip-flop changes states each time a clock pulse is applied. The J-K flip-flop serves as a T-type flip-flop. The J and K inputs are connected to each other and a 1 is applied to the combination. Each time a clock pulse hits, the states flip.

Output from a D-type flip-flop assumes the state of one of the inputs. Noting the truth table in Fig. 13-14A, $Q_n$ is in the same state as the J input if the K input is in the

Table 13-2. The Transition from One State to the Next Stage Using Various Versions of the Flip-flop.

| Type-T | | | | Type-D | | | | J-K flip-flop | | | |
|---|---|---|---|---|---|---|---|---|---|---|---|
| $Q_n$ | $J$ | $K$ | $Q_{n+1}$ | $Q_n$ | $J$ | $K$ | $Q_{n+1}$ | $Q_n$ | $J$ | $K$ | $Q_{n+1}$ |
| 0 | 1 | 1 | 1 | 0 | 0 | 1 | 0 | 0 | 0 | x | 0 |
| 1 | 1 | 1 | 0 | 1 | 0 | 1 | 0 | 0 | 1 | x | 1 |
| 0 | 0 | 0 | 0 | 0 | 1 | 0 | 1 | 1 | x | 1 | 0 |
| 1 | 0 | 0 | 1 | 1 | 1 | 0 | 1 | 1 | x | 0 | 1 |

opposite state. So to turn the J-K flip-flop into a D-type circuit, decide what output you want at Q. Feed that signal to the J input of the flip-flop and the opposite signal to the K input. To get a signal of the opposite states at the inputs, connect a NOT circuit between these inputs. Should you want the output to differ from the input each time the clock flips, just feed the undesired 1 or 0 to the K input, place a NOT circuit between the K and J inputs, and observe the output at Q.

Information concerning the various versions of J-K flip-flops, is summarized in Table 13-2. The J-K flip-flop table is not only a combination of rows from the original J-K truth table, but can be derived by combining the two specialized T- and D-types of J-K flip-flops.

## Parallel Counters

Ripple counters have the disadvantage that transition from one state to another is not always clean. To overcome this, the clock terminals on all flip-flops in the circuit can be connected in parallel and logic used to dictate the switching. T-type, D-type, and J-K flip-flops can be used for this purpose. In this type of design, the Q or Q outputs are connected to the J and K inputs of the appropriate flip-flops through logic circuits. All flip-flops simultaneously flip to their next states at the instant the clock pulse is applied. Using this type of arrangement, logic circuits can be designed around these parallel J-K flip-flop circuits for counting in any sequence—up, down, skipping numbers—or in combinations of these sequences. The logic circuits need only be arranged to feed the proper 1 or 0 bit to the appropriate flip-flop at the right time.

Let's start with a simple example. Use a T-type flip-flop to count the sequence from 0 to 7. This requires three flip-flops. The sequence of states is shown in the diagram in Fig. 13-16A and in the table in Fig. 13-16B. Three columns are necessary to determine the logic circuits required between the outputs of the various flip-flops and their various inputs. Hence, one column for the logic circuit is required for each flip-flop. $A_J$ is the J input of the A flip-flop while $A_K$ is that flip-flop's K input. They are shown as one item as they are both connected together in a T configuration. Similarly, the logic circuit must be found to feed the J and K inputs of the B flip-flop ($B_J/B_K$), and to feed the J and K inputs of the C flip-flop ($C_J/C_K$). A represents the flip-flop that delivers the MSB.

The T-type arrangement flips at the application of each clock pulse only when both the J and K inputs are 1. It retains its previous state if the inputs are both 0s. Noting

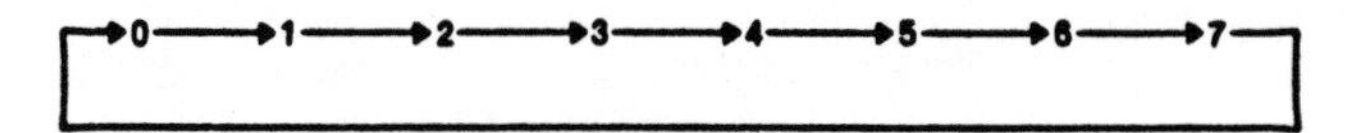


A Sequence of eight states.

| STATE | A | B | C | $A_J/A_K$ | $B_J/B_K$ | $C_J/C_K$ |
|---|---|---|---|---|---|---|
| 0 | 0 | 0 | 0 | 0 | 0 | 1 |
| 1 | 0 | 0 | 1 | 0 | 1 | 1 |
| 2 | 0 | 1 | 0 | 0 | 0 | 1 |
| 3 | 0 | 1 | 1 | 1 | 1 | 1 |
| 4 | 1 | 0 | 0 | 0 | 0 | 1 |
| 5 | 1 | 0 | 1 | 0 | 1 | 1 |
| 6 | 1 | 1 | 0 | 0 | 0 | 1 |
| 7 | 1 | 1 | 1 | 1 | 1 | 1 |

B Truth table.

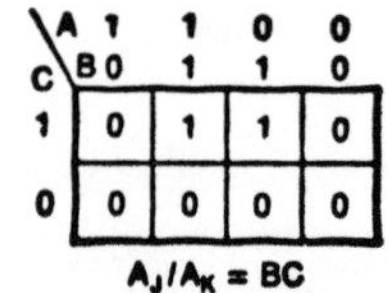


C Karnaugh map to determine J and K inputs to flip-flop A. Because C and $B_J/B_K$ columns in truth table are identical, it is obvious that inputs to flip-flop B is Q output from flip-flop C. Also input to flip-flop C is 1 at all times, because it changes from 1 to 0 and from 0 to 1 with each change of state.

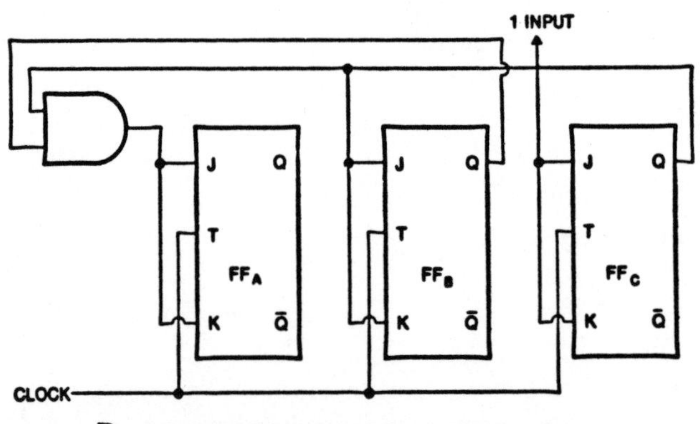


D Counting circuit. Outputs for bits are taken at Q of each flip-flop.

Fig. 13-16. A T-type counter using a parallel circuit.

flip-flop A, no flipping is required from the 0 to the 1 state, from the 1 to the 2 state, nor from the 2 to the 3 state. Outputs are at 0 in these four states. Hence a 0 should be at the J and K inputs of flip-flop A in these four states, so that the flip-flop will not flip when the next clock pulse arrives. Place 0s in the $A_J/A_K$ column in the 0, 1 and 2 rows.

A 1 is required at the output of flip-flop A in state 4. Therefore, as the output of flip-flop A is a 0 when the circuit is in state 3, a 1 input must be presented to the J and K terminals of flip-flop A in order for it to shift to a 1 in state 4. Place a 1 in the $A_J/A_K$ column on row 3 so that a 1 will be at the J and K inputs of flip-flop A when the next clock pulse hits, to put it into state 4.

As flip-flop A must remain at 1 in states 4, 5, 6, and 7, place a 0 in the $A_J/A_K$ column in the rows for states 4, 5, and 6. A 1 is used in row 7 as the output from flip-flop A must change from a 1 to 0 when the states change from 7 to 0 to repeat the cycle.

Columns $B_J/B_K$ and $C_J/C_K$ can be filled in using similar reasoning for the required 1 and 0 inputs at flip-flops B and C, as was used above for flip-flop A. The Q output of flip-flop B changes from 0 to 1 when going from states 1 to 2 and 5 to 6. Thus a 1 is necessary in the $B_J/B_K$ column in rows 1 and 5. This flip-flop switches from 1 to 0 when going from states 3 to 4 and from states 7 to 0. Consequently, a 1 is necessary in rows 3 and 7. As there is no switching of states from 0 to 1, 2 to 3, 4 to 5, and 6 to 7, place 0s in rows 0, 2, 4, and 6 to complete all requirements for the J and K inputs of flip-flop B.

Flip-flop C changes from a 1 to a 0 and from a 0 to a 1, when going from one state to the next. The J and K inputs of this flip-flop must be 1 in each case. Place 1s in each row in the $C_J/C_K$ column. With the table completed, we can now determine the logic circuits required.

Every entry in the $C_J/C_K$ column in Fig. 13-16B is a 1. Thus the J and K inputs of flip-flop C must be 1s at all times. These inputs should therefore be connected to a voltage that will keep them at 1. The $B_J/B_K$ column is identical to the C column. Therefore, the J and K inputs of flip-flop B should be connected to the Q output terminal of flip-flop C.

Only the $A_J/A_K$ column does not seem to have an easy answer as to the information to be fed to the J and K inputs of flip-flop A. We can determine just what should be fed to this input through use of a Karnaugh map, as drawn in Fig. 13-16C. It shows that these inputs must be connected through an AND gate to the Q terminals of flip-flops B and C. A complete circuit diagram of the counter is shown in Fig. 13-16D.

Now let's design a T-type circuit with a more complex output sequence. Let's count in a sequence 1, 2, 3, 5, 4, 7. The sequence diagram is in Fig. 13-17A and the truth table is in Fig. 13-17B. Only three flip-flops are necessary as no number is larger than 7. Numbers 0 and 6 are not in the sequence. Place an X in the $A_J/A_K$, $B_J/B_K$ and $C_J/C_K$ columns (for the moment disregard the bits shown in the parentheses next to the Xs on the table) to indicate that states 0 and 6 are not involved and that their outputs do not (or should not) affect the counting sequence. Outputs from these states are consequently don't cares.

Next supply the 1s and 0s in the columns. As there is no change in output from $FF_A$ (column A) when the states go from 1 to 2 and from 2 to 3, place 0s in rows 1

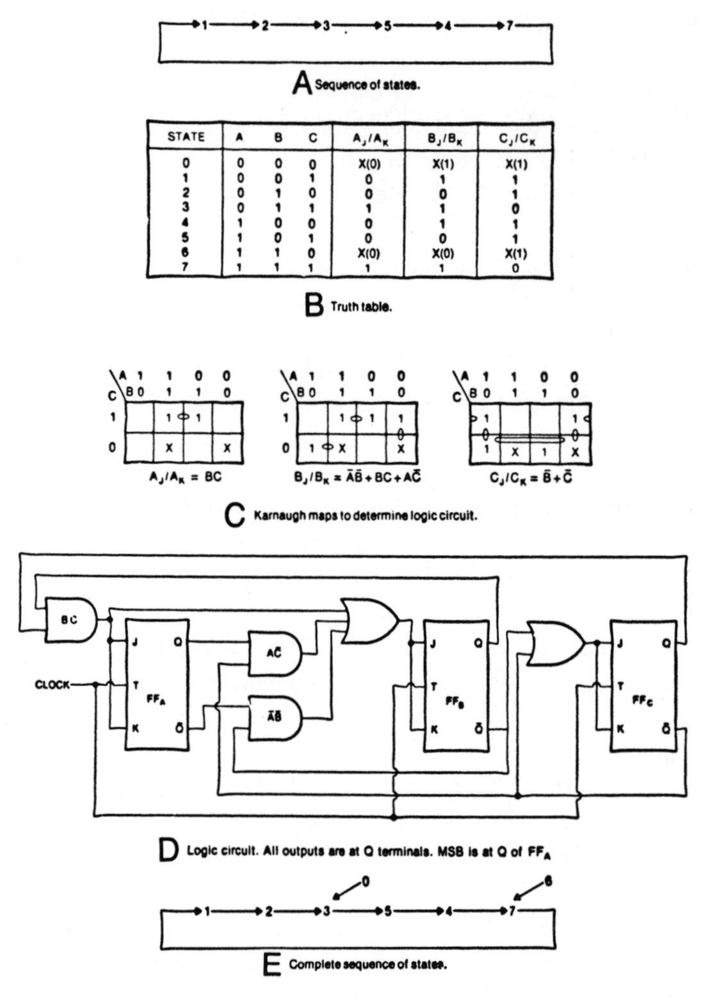


| STATE | A | B | C | $A_J/A_K$ | $B_J/B_K$ | $C_J/C_K$ |
|---|---|---|---|---|---|---|
| 0 | 0 | 0 | 0 | X(0) | X(1) | X(1) |
| 1 | 0 | 0 | 1 | 0 | 1 | 1 |
| 2 | 0 | 1 | 0 | 0 | 0 | 1 |
| 3 | 0 | 1 | 1 | 1 | 1 | 0 |
| 4 | 1 | 0 | 0 | 0 | 1 | 1 |
| 5 | 1 | 0 | 1 | 0 | 0 | 1 |
| 6 | 1 | 1 | 0 | X(0) | X(0) | X(1) |
| 7 | 1 | 1 | 1 | 1 | 1 | 0 |

Fig. 13-17. This T-type design uses J-K flip-flops.

and 2 in the $A_J/A_K$ column. When going from state 3 to state 5, the output from $FF_A$ changes from a 0 to a 1, so place a 1 in the $A_J/A_K$ column in row 3. A remains at 1 when the states shift from 5 to 4 and from 4 to 7, so place 0s in the $A_J/A_K$ column on rows 5 and 4, As going from state 7 to state 1 requires a change of output from $FF_A$, place a 1 in the $A_J/A_K$ column in row 7. Use similar logic for placing 1s and 0s in the $B_J/B_K$ and $C_J/C_K$ columns.

Now plot Karnaugh maps using the data in each of the columns. These are shown in Fig. 13-17C.

You may find that the flip-flops start in an unused state such as in 0 or 6 in our example. It may stay in one of the unused states or it may flip from one of the unused states to another. If the circuit does this without entering an active state, the circuit will never progress through the required cycle. We must check each state to see if it will lead to any one of the active states. If it does, then the cycle commences and keeps providing only the required sequence as described in Fig. 13-17A. In this case, the circuit is referred to as *bush*. If a check of the unused states indicates that they do not lead to one of the states in the active cycle and the sequence cannot be initiated, the circuit is referred to as *bushless*. To correct a bushless circuit, a 1 must be substituted for one or more of the Xs in the truth table and the analysis is tested by putting this 1 in place of the appropriate X on the Karnaugh map.

To check if the circuit to be used is bush, determine whether all Xs in the table are 1s or 0s. To do this, substitute the 1s and 0s on the relevant lines in the table into the relationships derived from the Karnaugh maps. The numbers calculated from these substitutions are shown in parentheses next to the Xs in Fig. 13-17B. From the first Karnaugh map, we know that $A_J/A_K = BC$. In state 0, $B = 0$ and $C = 0$. Hence $A_J/A_K = 0$. The X in state 0 in the $A_J/A_K$ column is consequently 0. In the $B_J/B_K$ equation derived from the second Karnaugh map, if any term in the equation is 1, $B_J/B_K = 1$. In state 0, $A = 0$ and $B = 0$, so $A = 1$ and $B = 1$. As $A \bullet B$ is a term in the $B_J/B_K$ equation and it is equal to 1, $B_J/B_K = 1$. So place a 1 next to the X in the $B_J/B_K$ column on the state 0 line. Use this procedure to determine if all the remaining Xs are 0s or 1s.

With this information, we can see that the A flip-flop will not change states when starting in state 0, because there is a 0 in the $A_J/A_K$ column on that row. Because of the 1s next to the Xs in the $B_J/B_K$ and $C_J/C_K$ columns on the state 0 row, the B and C flip-flops will change states when starting in state 0. The next state after 000 will consequently be 011 because the A flip-flop remains at 0 and the B and C flip-flops change from their initial 0s to 1s. This is state 3. Because state 3 is in the counting sequence, the numbers will then be able to shift properly from state 3 through all the rest of the active states.

As for state 6, only the C flip-flop will change states because there is a 1 only in the $C_J/C_K$ column in this row. As the states of flip-flops A and B remain unchanged (and at 1) because of the 0s in columns $A_J/A_K$ and $B_J/B_K$, the succeeding state will be 111. As state 6 leads into state 7, and 7 is a number in the active sequence, there is no problem. The circuit is bush.

The example just detailed is relatively simple. Here the unused numbers went directly into the counting sequence. This is not a requirement. One unused number may lead to one or more other unused numbers. The last number in the chain must lead to any one of the numbers in the active sequence or the circuit is not bush.

Let us proceed to design a D-type flip-flop using the same number sequence shown in Fig. 13-17A. The various drawings and charts are shown in Fig. 13-18. To start, note that the output Q, of a D-type flip-flop is the same as the input at J. In the truth table in Fig. 13-18B, consideration is given to the fact that the input level waiting at the J terminal in one state generates the output for the succeeding state. These outputs are shown in columns A, B, and C for the appropriate flip-flop. For example, A is 0 in states

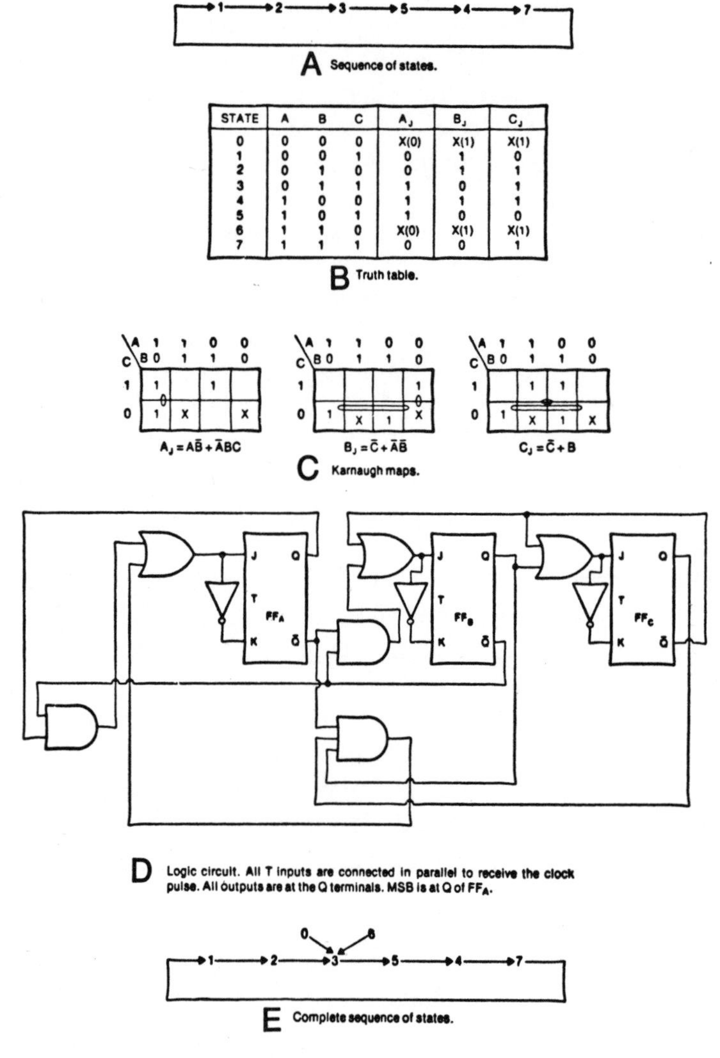


| STATE | A | B | C | $A_J$ | $B_J$ | $C_J$ |
|---|---|---|---|---|---|---|
| 0 | 0 | 0 | 0 | X(0) | X(1) | X(1) |
| 1 | 0 | 0 | 1 | 0 | 1 | 0 |
| 2 | 0 | 1 | 0 | 0 | 1 | 1 |
| 3 | 0 | 1 | 1 | 1 | 0 | 1 |
| 4 | 1 | 0 | 0 | 1 | 1 | 1 |
| 5 | 1 | 0 | 1 | 1 | 0 | 0 |
| 6 | 1 | 1 | 0 | X(0) | X(1) | X(1) |
| 7 | 1 | 1 | 1 | 0 | 0 | 1 |

**Fig. 13-18. A D-type design using J-K flip-flops.**

1, 2, and 3. Therefore the notation in the $A_J$ column for the states preceding these must be a 0. So place a 0 in the $A_J$ column on line 7 because the next state is 1, on line 1 because the next state is 2, and on line 2 because the next state is 3. As A is 1 in states 5, 4, and 7, place a 1 in the $A_J$ column on line 3 because the next state is 5, on line 5 because the next state is 4, and on line 4 because the next state is 7. Place Xs in the $A_J$, $B_J$ columns for rows 0 and 6, as they are not in the sequence. Using similar reasoning, complete the truth table for columns $B_J$ and $C_J$. Note that $A_K$, $B_K$ and $C_K$ inputs are now shown in the table. They are understood to be 180° output of phase with the J inputs. A NOT circuit must be placed between the J and K inputs at each flip-flop to accomplish this situation.

Draw Karnaugh maps and sequential logic circuits, as before. Check if the circuit is bush. Here, it is bush. As a final example, use basic J-K flip-flops. We will apply information noted in Table 13-2.

The various steps in the sequence to determine and check the circuit are shown in Fig. 13-19. Inputs fed to the individual J and K terminals of the various flip-flops are determined from the Karnaugh maps and are not related to each other. The resulting circuit involving the flip-flops and logic is much simpler than the ones derived for the T-type and D-type flip-flops.

A careful check must be made to see if the circuit is bush in states 0 and 6, because of the large number of Xs in the table and on the map. First determine the J and K inputs for the flip-flops. If the J and K inputs of any one flip-flop are identical, and they are both 1s or both 0s, the circuit behaves as a T-type flip-flop. Ensuing states will change with each pulse if the J and K of that flip-flop are both 1s and will stay in the original state if both J and K are 0s. Should J be a 1 and K a 0, or should K be a 1 and J a 0, the circuit behaves as a D-type flip-flop. The output will assume the state of the J input terminal.

In *state 0* (or 000), $A_J$ and $A_K$ are both 0, so in the next state the output from $FF_A$ remains at the 0 of state 0. $B_J$ is 1 and $B_K$ is 0, so the $FF_B$ output shifts from 0 in the 0 state to 1 in the next state. $C_J$ and $C_K$ are both 1s so the output at $FF_C$ shifts from the 0 in the state 0 to a 1 in the ensuing state. Due to the shifts and lack of shifts of the states of $FF_A$, $FF_B$, and $FF_C$ from the 000 state, the next state becomes 011, or 3. This state is in the active number sequence.

In *state 6* (or 110), $A_J$ is 0 and $A_K$ is 1, so in the next state, the output from $FF_A$ shifts from a 1 to a 0. $B_J$ is 0 and $B_K$ is 0, so the $FF_B$ output remains in the 1 state and does not change when the overall state shifts from 6 to the next number. As $C_J$ is 1 and $C_K$ is 0 in state 6, the output from $F_{CC}$ is 1. Due to the shifts and lack of shifts of the states of $FF_A$, $FF_B$, and $FF_C$ from the 110 state, the next state becomes 011 or 3. As the states following states 0 and 6 both lead to 3, and 3 is in the active number sequence, the circuit is bush.

### Shift Register

If a number of flip-flops are in a circuit, an arrangement can be made where a bit that is fed to the first flip-flop is shifted one flip-flop at a time to other flip-flops in the circuit. The shift may be to a flip-flop at the right or left of the first one. The shift takes place with each clock pulse. This operation can be accomplished easily using a D-type group of flip-flops, as shown in Fig. 13-20.

Assume a 1 is fed to the J input of $FF_A$. After the next clock pulse, the Q output of this flip-flop and the J input of $FF_B$ will be a 1. Should the next pulse at the J input

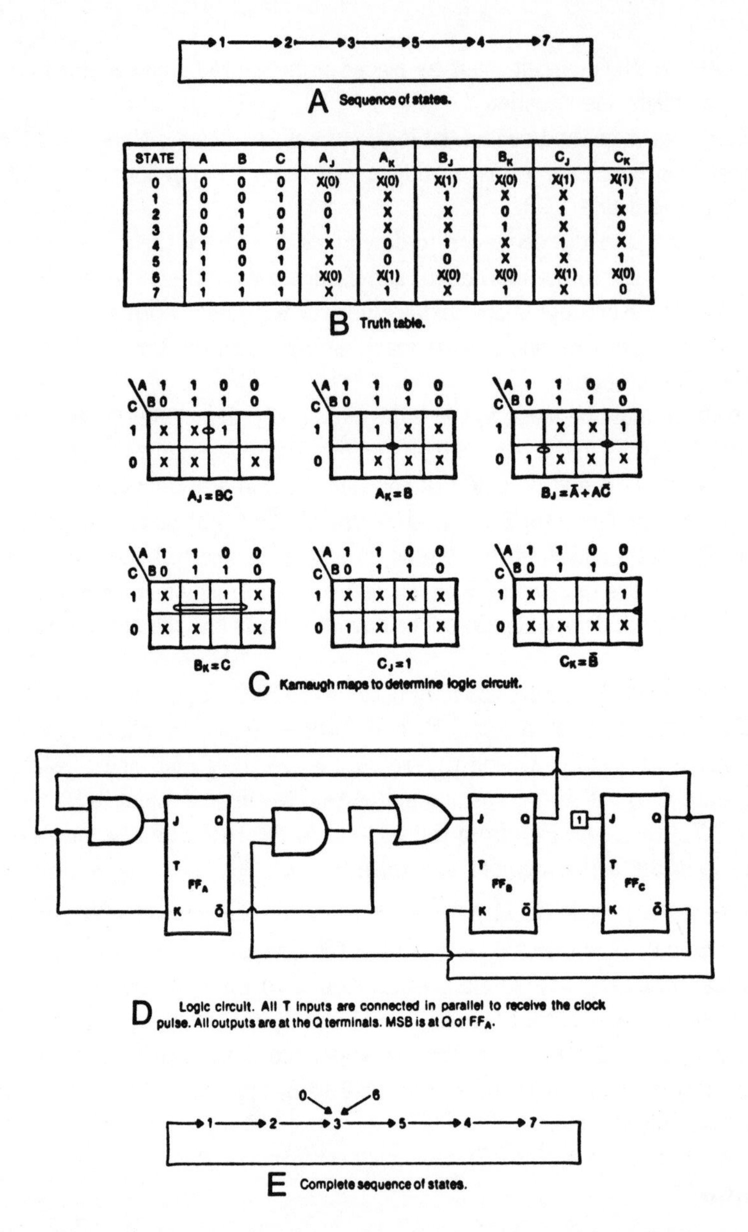


A Sequence of states.

| STATE | A | B | C | $A_J$ | $A_K$ | $B_J$ | $B_K$ | $C_J$ | $C_K$ |
|---|---|---|---|---|---|---|---|---|---|
| 0 | 0 | 0 | 0 | X(0) | X(0) | X(1) | X(0) | X(1) | X(1) |
| 1 | 0 | 0 | 1 | 0 | X | 1 | X | X | 1 |
| 2 | 0 | 1 | 0 | 0 | X | X | 0 | 1 | X |
| 3 | 0 | 1 | 1 | 1 | X | X | 1 | X | 0 |
| 4 | 1 | 0 | 0 | X | 0 | 1 | X | 1 | X |
| 5 | 1 | 0 | 1 | X | 0 | 0 | X | X | 1 |
| 6 | 1 | 1 | 0 | X(0) | X(1) | X(0) | X(0) | X(1) | X(0) |
| 7 | 1 | 1 | 1 | X | 1 | X | 1 | X | 0 |

B Truth table.



C Karnaugh maps to determine logic circuit.



D Logic circuit. All T inputs are connected in parallel to receive the clock pulse. All outputs are at the Q terminals. MSB is at Q of $FF_A$.

E Complete sequence of states.

Fig. 13-19. A circuit for generating a number sequence using clocked J-K flip-flops.

of $FF_A$ be a 0, Q at its output will be 0 and Q at the output of $FF_B$ will be 1. The 1 has shifted from the output of $FF_A$ to the output of $FF_B$. This continues through the entire line of flip-flops. All pulses fed to J at $FF_A$ will appear in sequence at the outputs of the succeeding flip-flops. If pulses are fed, for example, in a sequence 100 to J at

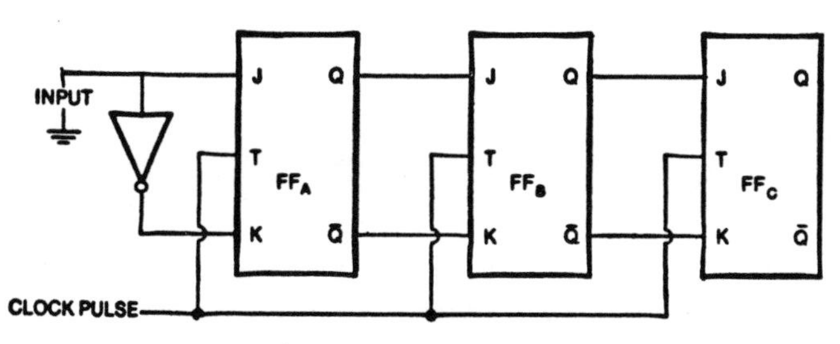


Fig. 13-20. This shift register has outputs at terminal Q of the flip-flops.

$FF_A$, after the first clock pulse a 1 will be at the output of $FF_A$; after the second clock pulse a 1 will be at the output of $FF_B$ and a 0 at the output of $FF_A$; after the third clock pulse a 1 will be at the output of $FF_C$ and 0s will be at the outputs of the other flip-flops. The bit at the output of $FF_C$ is the MSB. The Q outputs of the flip-flops can be assumed to be the original number fed bit-by-bit to the input of $FF_A$.

*Ring counters* are flip-flops into which only a single 1 pulse is fed. This 1 pulse is shifted in step from $FF_A$ through the entire sequence of flip-flops in the circuit, right to the last flip-flop. Instead of falling out of the sequence, a circuit ties the output from the last flip-flop to the input of $FF_A$. If there are N flip-flops in the circuit, each circuit had its turn at being a 1 after N clock pulses have passed. The time between succeeding 1s, at any one flip-flop, is 1/N. Ring counters can thereby be used as dividers (divide by N) in timing circuits.

## ARITHMETIC OPERATIONS

All basic operations—addition, subtraction, multiplication, and division—can be implemented by using addition. The basic circuit is the *half adder*. Here, two binary numbers are totaled. There is always a sum. If two binary numbers like 1 and 1 are added, the sum is 10. The 0 is considered the sum while the 1 is a carry to be added to the next bit at the left. Adding a 0 and 1, the total is a 1 with no carry. Similarly, adding two 0s, the total is a 0 and there is no carry to the next bit. The symbol for a half adder and its truth table are in Figs. 13-21A and B, respectively. Note that in the table the bit at the right in the word is the sum and the one at the left is the carry.

When analyzing the truth table, it can be seen that $S = \overline{A}B + \overline{A}B$, the equation for an exclusive OR gate. Also, $C = AB$, the equation for an ordinary AND gate. Using this information, we can implement a half adder from logic gates, as shown in Fig. 13-21C.

In order to do an arithmetic operation, not only must the circuit be capable of adding two bits, but it must also be able to add the carry from a less significant bit to the sum of the two bits. This capability of adding three bits is required of all but the LSB in the operation, as all arithmetic starts here and hence no carry is available to it. To accommodate all three items, two half adders and some supplementary logic must be used in a circuit. The final circuit is referred to as a *full adder* (F.A.). The symbol, truth

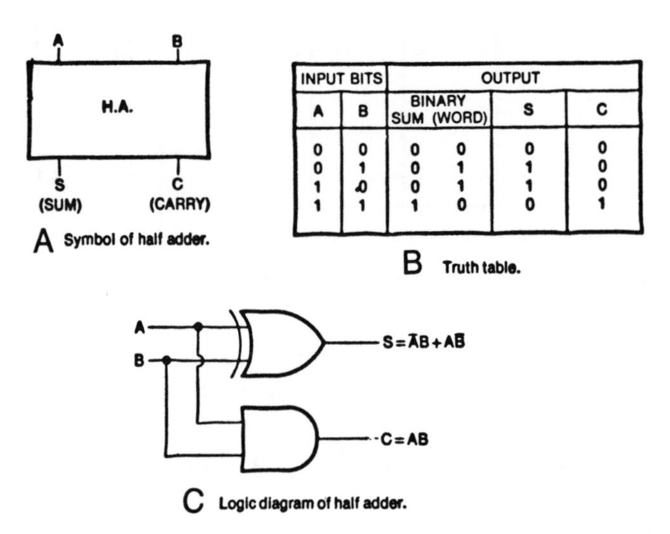


| INPUT BITS | | OUTPUT | | | |
|---|---|---|---|---|---|
| A | B | BINARY SUM (WORD) | | S | C |
| 0 | 0 | 0 | 0 | 0 | 0 |
| 0 | 1 | 0 | 1 | 1 | 0 |
| 1 | 0 | 0 | 1 | 1 | 0 |
| 1 | 1 | 1 | 0 | 0 | 1 |

Fig. 13-21. Half-adders.

table, and implementation are shown in Fig. 13-22. A circuit using adders for binary number with more than just one bit is also shown. Here the subscript 0 refers to the least significant bit of the binary number, while higher-order bits are referred to by numerical subscripts increasing progressively in numerical size. Numbers $A_3A_2A_1A_0$ and $B_3B_2B_1B_0$ are added together in the circuit shown. There is evidently a carry left over after all arithmetic operations have been completed. This carry is used as the fifth and most significant bit in the sum. The sum of the numbers with a carry is $C_3S_3S_2S_1S_0$.

To subtract two binary numbers, the 2's complement of the negative number or subtrahend is usually used. The new subtrahend and the original minuend are then added together using circuits similar to those just described or their IC equivalents. The 2's complement of a number is obtained by writing the original number in binary form, changing all 0s to 1s and all 1s to 0s in that number, and then adding a 1.

Multiplication of binary numbers may be accomplished by adding one number as many times as indicated by the second number. Thus if 18 is to be multiplied by 23, 18 can be added 23 times to itself to produce the product. Long division processes used for decimal system numbers may also be employed when dividing one binary number by another.

## DIGITAL-TO-ANALOG (D/A) CONVERTERS

Many applications require changing a voltage or current to a binary word and a binary word to a voltage or current of the real world. These are referred to as conversions of analog to digital (A/D) and digital to analog (D/A), respectively.

In a D/A conversion application, a binary number is applied to the circuit. The voltage or current at the output from the circuit is equal to or proportional to the size of

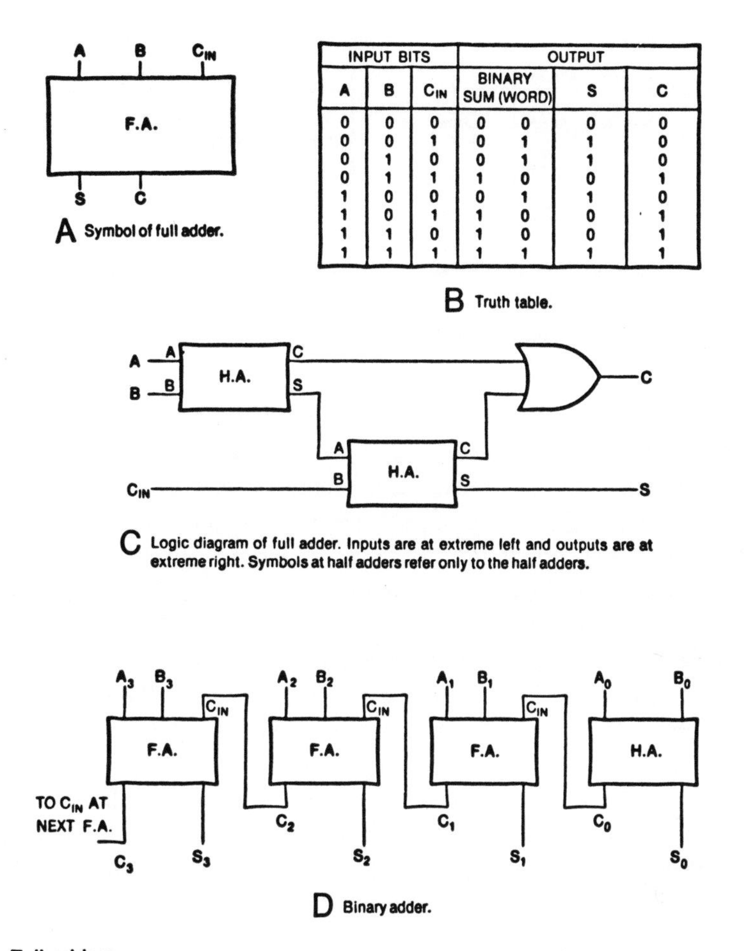


A Symbol of full adder.

| INPUT BITS | | | OUTPUT | | |
|---|---|---|---|---|---|
| A | B | $C_{IN}$ | BINARY SUM (WORD) | S | C |
| 0 | 0 | 0 | 0 0 | 0 | 0 |
| 0 | 0 | 1 | 0 1 | 1 | 0 |
| 0 | 1 | 0 | 0 1 | 1 | 0 |
| 0 | 1 | 1 | 1 0 | 0 | 1 |
| 1 | 0 | 0 | 0 1 | 1 | 0 |
| 1 | 0 | 1 | 1 0 | 0 | 1 |
| 1 | 1 | 0 | 1 0 | 0 | 1 |
| 1 | 1 | 1 | 1 1 | 1 | 1 |

B Truth table.

C Logic diagram of full adder. Inputs are at extreme left and outputs are at extreme right. Symbols at half adders refer only to the half adders.

D Binary adder.

Fig. 13-22. Full adders.

the number applied to its input. The circuit could be programmed to deliver a voltage equal to a fraction of the binary number at its input. This can be accomplished in the digital or analog portion of the circuit.

A D/A converter can be made simply through use of the summing network in Fig. 13-23A. Assume the voltage at all Q outputs of a register are fixed at a regulated voltage, V. Apply this voltage from the various flip-flop outputs of the register individually to inputs A, B, C, and D. When the output of $FF_A$ is 1, input A of the D/A converter is turned on. Current because of V at the Q output of that flip-flop flows through the meter. This current is V/R. Similarly, if there is a 1 at the Q output of $FF_B$, input B of the D/A converter is turned on. Current now flows through resistor 2R and the meter of the D/A converter because of the voltage, V, present at the Q output of $FF_B$. This current is V/2R. When input C is turned on by a 1 at the Q output of $FF_C$, the current flowing through the meter is V/4R. Similarly, V/8R flows through the meter when input

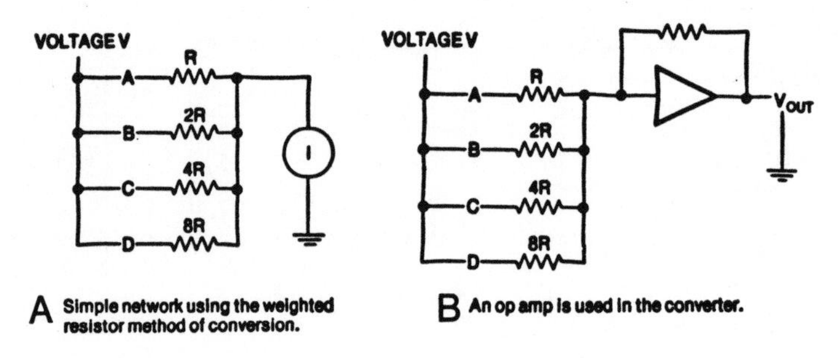


Fig. 13-23. A simple D/A converter.

D is turned on because of a 1 at the output of $FF_D$. The size of the resistors determine the portion of the meter current derived at each output from the flip-flops and at each corresponding input of the D/A converter. Currents are weighted through use of different size resistors, so that the overall current reading is proportional to the significance of the bit that initiates the current flow. All currents pass through the meter and it indicates the total current represented by the binary number or states at the input. The voltage developed across the meter is also proportional to the applied binary input, so the D/A output can be considered either as a voltage or a current.

A more sophisticated D/A converter uses the op amp in the summing circuit discussed in Chapter 7. In the D/A converter drawn in Fig. 13-23B, the Q outputs from the register are connected to the appropriate resistors. The gain of the operational amplifier with feedback is equal to the feedback resistor divided by the input resistor and is equal to $V_{out}/V$. If V is applied to more than one of the resistors at the same time, it indicates that more than one of the Qs at the flip-flops of the counter or shift register are at 1. Output voltage, $V_{out}$, is equal to the sum of the output voltages from all sources.

Basically, D/A converters operate as described. In practical situations, two major variations exist in an IC comprising converters. For one, the converter is usually not driven directly from the Q outputs of a counter or shift register. These outputs are used only to drive some types of electronic switches that do all the work. When a switch is used, a well-regulated voltage may be applied to the converter circuit so that there will be a high degree of conversion accuracy. Accuracy is further enhanced by increasing the number of bits in the counter, so that each bit represents a smaller fraction of the total voltage that is applied to the converter.

The performance of a D/A converter can be judged by a number of factors.

1. *Resolution* is the smallest change in voltage at the output of the D/A converter caused by a change from 0 to 1 or from 1 to 0 in the LSB at the input. For example, an eight-bit converter has $2^8 = 256$ steps from minimum to maximum. No significant digit at the output can be less than 1/256 of the total voltage that this binary input

represents. The percent resolution of the converter is therefore (1/256) × 100 = 0.39% of full scale. Thus a change of one bit in the LSB column will change the output by 0.39 percent.

2. *Accuracy* is the percentage by which the actual output voltage deviates from the ideal output voltage. It is usually stated as a percentage of the full-scale output.

3. *Linearity*, or to be more exact in our choice of words—nonlinearity, is a measure of how much the actual output from the D/A converter differs from the theoretical output. As an example, assume the output is permitted to change under the ideal situation by 1 volt when there is a change of 1 bit at the input in the LSB column. If the change were 2 volts instead of the ideal 1-volt, the percentage differential linearity is [(2 – 1)/1] × 100 = 100%. The differential linearity percentage error is usually much smaller, usually in the neighborhood of 1% or 2%.

4. *Gain error* or *scale factor error* is related to the linearity of the conversion. This error can be corrected easily by adjusting the converter feedback circuit.

5. *Offset* error of the op amp represents a deviation from 0 volt at the output when 0 volt is applied to the inputs.

6. *Speed* is the time it takes for a complete change in the output voltage from a D/A converter for a shift of one bit in the word at the input.

7. *Settling time* is the time it takes for a voltage to reach its final value after the input has changed from one level to the other.

8. *Slew rate* is the voltage per microsecond change possible with the D/A converter in question.

9. *Glitches* are spikes or transients that are at the output of a D/A converter when the magnitude of the output voltage changes. The glitch is approximated by the area of the spike. It is the product of the peak output voltage and the time that the pulse exists. D/A converters should be designed for a minimum transient glitch, as measured in volt-seconds.

## ANALOG-TO-DIGITAL (A/D) CONVERTERS

It is frequently desirable to apply information about voltage, temperature, mechanical strain, liquid level, and numerous other factors to a computer, or to note these factors on digital readouts. To accomplish this goal, the A/D converter is used. A converter frequently involves a ramp generator and a voltage comparator in its circuitry. Examples of each of these are shown in Fig. 13-24.

The open loop op amp serves as a *voltage comparator*. Because of its extremely high gain, the output voltage from the op amp is equal to $-V_{CC}$, the negative supply voltage, when V1 is greater than V2. The output voltage is equal to the positive supply voltage, $+V_{CC}$, when V2 is greater than V1. A fixed or regulated voltage may be applied to either the V1 or V2 terminal as a reference for comparison purposes. The $+V_{CC}$ or $-V_{CC}$ output may be used as a 0 or as a 1 in the logic circuit.

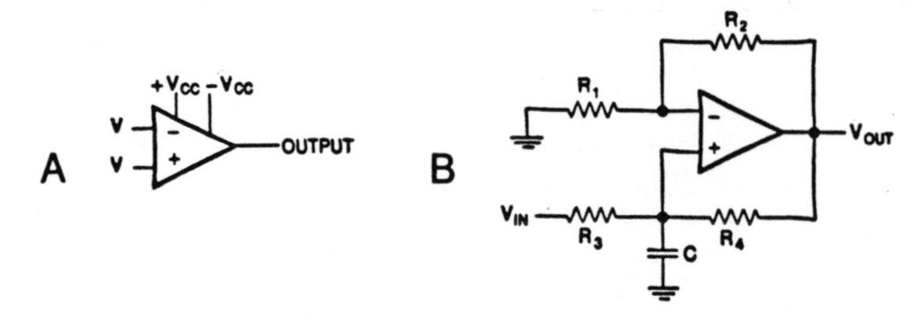


Fig. 13-24. A voltage comparator and ramp generator. (A) $V_1$ and $V_2$ are compared. $+V_{cc}$ and $-V_{cc}$ are the supply voltages. (B) The ramp generator's output voltage varies linearly with time.

An ideal *ramp generator* provides a voltage that increases linearly with time. It consists of an input voltage that charges a capacitor through a device with an infinite resistance. If the resistance is not very large, the ramp is not linear. To produce a linear ramp, one of the current sources described in an earlier chapter can be used, because a current source presents a high impedance to its output. The op amp circuit in Fig. 13-24B is ideal in this application as it behaves as an extremely high impedance for charging capacitor C. The ramp at output $V_{OUT}$ is given by the equation:

$$\frac{V_{OUT}}{t} = \frac{(R_1 + R_2)V_{IN}}{R_1R_3C} \tag{13-1}$$

where,

C is the capacity in farads,

R is the resistance in ohms.

$V_{IN}$ and $V_{OUT}$ are input and output voltages,

t is time in seconds.

A comparator and ramp generator are used in the basic A/D converter in Fig. 13-25A. After the counter has been reset to zero, the ramp generator applies its output to the comparator. The input voltage to be measured is also applied to the comparator. The output from the comparator is high until the voltage from the ramp exceeds $V_{IN}$, the input voltage. During the time that the output is high, a 1 is applied to the AND gate. The AND gate passes accurately spaced clock pulses to the counter.

Getting back to the ramp generator, once the voltage at its output exceeds $V_{IN}$ the comparator outputs a 0 and turns off the AND gate so that clock pulses can no longer pass through to the counter. Pulses were passed on to the counter only during the specific period when the output from the comparator was at 1. The output from the counter is now applied to a circuit for displaying a number proportional to the count. The circuit is frequently designed for the displayed number to be equal to $V_{IN}$.

The counter in Fig. 13-25A is limited in accuracy by the precision or lack of precision of the clock pulses and by the linearity of the ramp generator. The dual ramp or dual

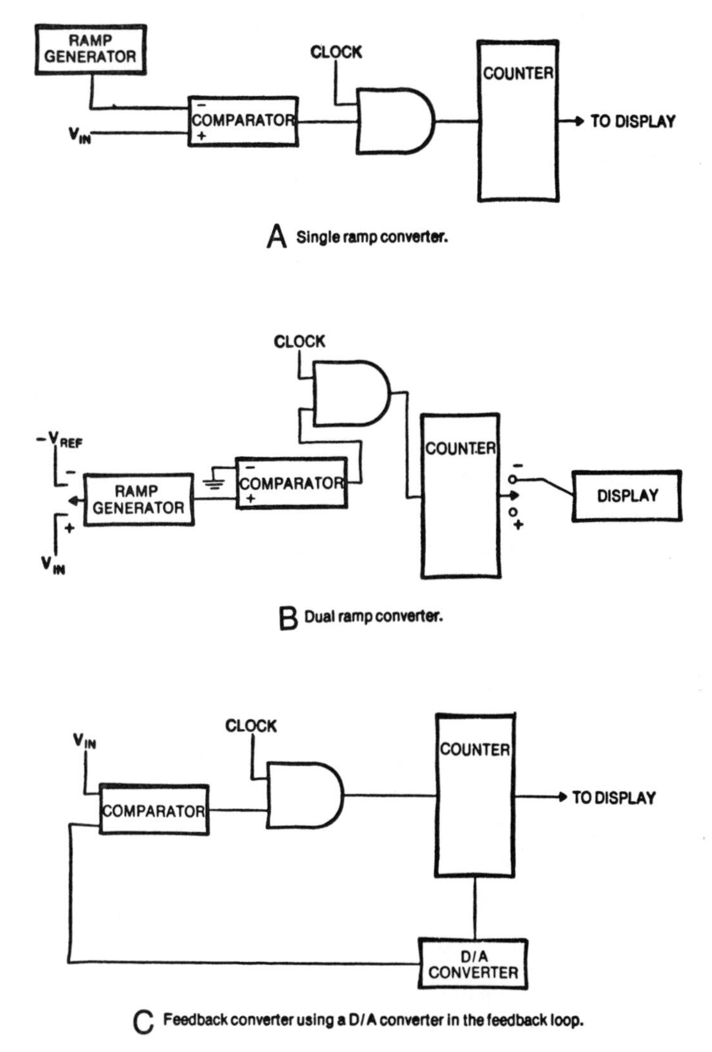


Fig. 13-25. Several approaches to A/D conversion.

slope generator converter is more accurate. A representative drawing of this type of circuit is in Fig. 13-25B. When the switch is in the + setting, $V_{IN}$ is applied to the ramp generator. The output is positive and keeps increasing with time. The comparator delivers a 1 to the AND gate. Clock pulses pass through the AND gate and are counted. The count starts at 0 and continues until the MSB at the output of the counter becomes a 1. These pulses are not displayed on the readout as the switch between the counter and display is open in the + setting. When the MSB from the counter becomes a 1, the switch is automatically set to the − position. The output from the ramp now begins to decrease at a constant rate with a slope proportional to the magnitude of $-V_{REF}$.

Clock pulses pass through the AND gate until the ramp voltage drops to 0 and the AND gate is turned off. These are the clock pulses that are displayed.

When $V_{IN}$ is applied to the ramp generator, the output from the ramp generator keeps increasing. As noted, output increases until the MSB from the counter is a 1. The time it takes for a 1 to be the MSB is solely dependent upon the number of clock pulses fed to the counter and has nothing to do with the magnitude of the voltage, $V_{IN}$, being measured. What does depend upon $V_{IN}$ is the size of the maximum output voltage from the ramp generator. In the period of time required for the MSB to become 1 (as determined by the clock pulses), the output from the ramp generator increases and reaches a maximum level. This level is determined solely by $V_{IN}$. The time it takes for the ramp output voltage to drop from its peak or maximum level to 0 volt is proportional to $V_{IN}$. As clock pulses pass through the counter and are connected to the display during the second period when the ramp voltage drops, the digital readout is an indication of $V_{IN}$. The slope of the rolloff or negative ramp is constant and set by the size of $-V_{REF}$. Because the slope is constant, it affects the time during which the counter is on and its output is being displayed. However, output is still related to $V_{IN}$, as the time it takes for the negative slope to reach 0 volt is a function only of the peak voltage reached, so long as the rolloff slope is constant regardless of the magnitude of the voltage at its starting point.

Switches are not normally moved from the + to the − position and back, as may be inferred from the drawing. The 1 pulses at the MSB from the counter triggers a logic circuit and electronic switches to do the job. Note that at the instant the MSB becomes a 1, all lesser significant bits are 0s. It is only these lesser bits that are displayed at the instant the switch is set to −. The magnitude of the numbers in the readout increases until the output indicates the number of clock pulses that pass through the circuit during the period of time when the negative slope exists.

A feedback type of A/D converter is shown in Fig. 13-25C. When the output of the D/A converter is 0, the output from the comparator is a 1 so that clock pulses pass through to the counter. These are displayed as well as applied to the D/A converter. Voltage at the output of the D/A converter increases because of the pulses applied to it from the counter. Voltage increases until it just exceeds $V_{IN}$. At this instant, the output from the comparator becomes a 0. It turns off the AND gate and clock pulses can no longer reach the counter. As the counter continues until the D/A output just exceeds $V_{IN}$, this voltage is the determining factor of the time it takes the clock pulses to pass on to the counter. The readout is consequently related to $V_{IN}$.

All the counters described above must be reset. If an up-down counter can be substituted for the up counter used up to this point, the counter continues repetitively, without the necessity of resetting the counter to zero after each cycle.

When the output from the D/A is higher than $V_{IN}$, the output is applied in such a form that the counter counts down. Should the reverse relationship be true, the counter counts down. Should the reverse relationship be true, the counter counts up. As the count is uninterrupted, the reset function is no longer required.

Of all A/D converters, the fastest one uses the successive approximation method. Here a programmer is used rather than a counter. A 1 is fed to the MSB in the register. The D/A output from this is compared with $V_{in}$. If the resulting digital number is smaller than $V_{in}$, the 1 is left in the MSB; if not, it is removed from the MSB and fed to the next bit in order of significance in the register. A 1 is left in the first most significant bit where the resulting digital number will be less than $V_{in}$. If the 1 is left in any bit, other 1s are tried in succeeding bits until the total number presented by the D/A converter is almost equal to $V_{in}$. This method uses complex logic circuitry. It can easily be applied to fast A/D circuits through the use of an IC capable of performing the converting function. After the proper information is at the output of the D/A converter, it is fed to a converter similar to that shown in Fig. 13-25C to get the digital readout.

The following factors are used to judge the performance of an A/D converter.

1. *Accuracy* is frequently plagued by a ±1 LSB uncertainty. Other causes of inaccuracy are *scale factor error* or *gain error*, where the full-scale reading differs from what it should be because the reference voltage is not precise; *offset error*, where the readout does not indicate zero for a 0-volt input—an error easily corrected by a circuit adjustment; *hysteresis error*, where the final reading depends upon whether the code is changing from a larger number to a smaller number or vice versa; and *quantizing error*, where the digital reading does not change for every infinitesimal change of the analog input voltage. This last error is due to the digital steps present after the conversion.
2. The *linearity error* is the difference between the actual and ideal voltage required to change the digital output reading by one bit.
3. *Precision* is a measure of how accurately the readings repeat themselves for the same analog input signal.
4. *Resolution* is a measure of a converter's ability to change readings for successive analog inputs. For example, a four-bit A/D converter can be designed for a change of 1 LSB for each change of 1 volt at the input, but can the A/D converter circuit actually respond to this 1 volt input change? Its ability to respond is its resolution characteristic.
5. *Drift* is a circuit characteristic where its reading changes with time.
6. *Speed* is the time needed for the A/D converter to do its job after the analog input has been applied.

## MISCELLANEOUS CONVERTERS

Besides the D/A and A/D converters, other digitally related converters with various characteristics are available as integrated circuits. Among these are the *voltage-to-frequency (V/F) converter* and a device that acts in the opposite direction, the F/V converter. Chips are available to perform both functions. These are referred to as VFV devices.

I-f-to-digital converters are devices made to read an i-f frequency from a digital display.

*Voltage-to-pulse-width converters* are designed to feed information to a digital or analog computer. *Pulse-width-to-voltage converters* are useful as items in a pulse-duration transmitter.

*Synchro-to-digital* and *digital-to-synchro converters* are used to indicate angles and speeds of synchro systems employing rotating devices.

## Multiplexers

A *multiplexer* selects data or a train of pulses from one of several inputs and routes it to one output. Different numbers of inputs and outputs compose the various multiplexer models. A four-input-to-one-output multiplexer is shown in Fig. 13-26. *Data inputs* A, B, C, and D may be the Q outputs from four registers in independent circuits. Each

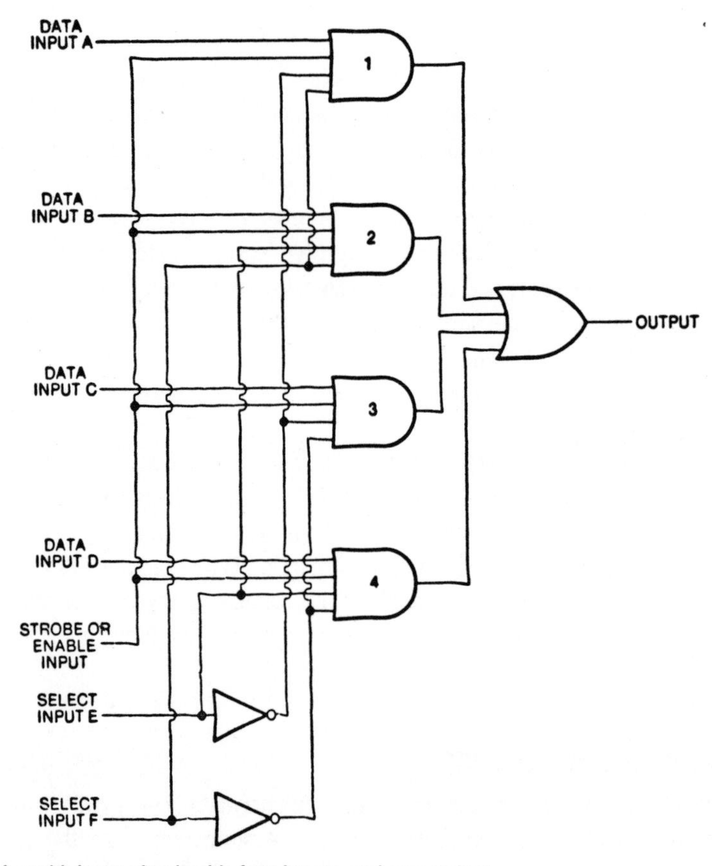


Fig. 13-26. A multiplexer circuit with four inputs and one output.

input is applied to a different AND gate of the multiplexer. 1s and 0s are also applied to the AND gates through *select inputs* E and F. Note that two of the AND gates receive one of their select inputs directly from E and the remaining two AND gates receive their select inputs from the inverted output of the E line. Inputs to the latter two gates are $\overline{E}$. F and $\overline{F}$ are applied in a similar fashion to the four AND gates. All gates receive inputs from different pairs of selected inputs. Gate 1 receives signals from $\overline{E}$ and F, gate 2 from E and F, gate 3 from $\overline{E}$ and $\overline{F}$, and gate 4 from E and $\overline{F}$. These input pairs are known as the *address*. The information derived from the E and F inputs to all AND gates provides 1s at two of the inputs of only one AND gate. A third 1 may be available from the data input terminal. The fourth 1 must come from the *strobe or enable input*. When four 1s are presented at all four inputs of any one of the gates, that gate is turned on.

The only input not discussed as yet is the strobe input. The function of the strobe line is to give all inputs a chance to settle to their final states before the appropriate AND gate is turned on. Once all inputs are in their final states, the strobe supplies a 1 to all gates. Now, one gate has 1s at all its select inputs and at its strobe line input. If the data at the input to that AND gate is at 1, that gate is turned on, because now there are 1s at all its inputs. The 1 that is at the output of that AND gate because it was turned on is fed through the OR gate to the output terminal. If the data at the input to that AND gate is at 0, the AND gate remains off and a 0 if fed through the OR gate to the output terminal. Because information as to which gate was activated by the select input has been noted, the information at the output is attributed to the appropriate data input signal.

We indicated that the input from one gate is selected by the multiplexer so that information available at that gate can appear at the output. This is only part of the story. A train of pulses can be fed to the E and F inputs to select a series of gates to be activated during a specific period of time. Gates may be selected in a random fashion or in a specific pre-determined sequence. Data supplied from the gates appear at the output. This data can be recorded on one track of magnetic tape. Information from E and F can be recorded simultaneously on other tracks. This information can then be applied along with the recorded data to a demultiplexer for retrieval of the original data and information as to which source is involved with which group of data.

### Demultiplexers

Opposite in action to that of a multiplexer, a *demultiplexer* takes a binary signal from one source and directs it to one of several different outputs. An example of a device that takes one input and directs it to four outputs is shown in Fig. 13-27. The train of input pulses is applied to all input data terminals. The AND gate to be activated is determined by the address or states of select inputs E and F as applied to the gates. Gates E and F may be activated from independent sources or from information recorded on

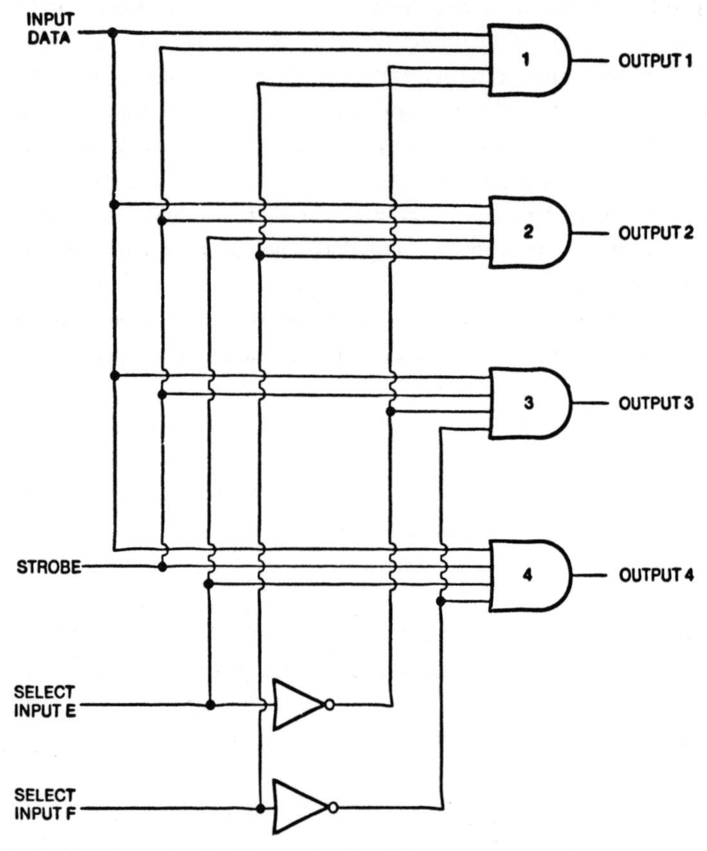


Fig. 13-27. A demultiplexer circuit with one input and four outputs.

tape. The AND gate with 1s at its select inputs delivers its input data to its output terminal after the strobe input is in the 1 state.

## Decoder

A *decoder* circuit accepts a binary number at its input and delivers a decimal number at its outputs. The output (or outputs) used depends upon the form the decimal number must take. A practical example of a decoder is a device that changes BCD numbers to the decimal digits represented by these numbers.

BCD numbers and their decimal equivalents are shown in Table 13-3. Because there are ten decimal digits, there must be ten gates at the output of the decoder. The output from each gate is used to indicate the decimal digit that is at the input of the decoder in BCD form.

AND gates can be used in the decoder circuit. Bits are applied to all gates, but in different 1 and 0 combinations. When the combination is proper to supply 1s to the inputs

Table 13-3. A Comparison of BCD Numbers and Their Decimal Equivalents.

| BCD code | | | | Decimal |
|---|---|---|---|---|
| *A* | *B* | *C* | *D* | |
| 0 | 0 | 0 | 0 | 0 |
| 0 | 0 | 0 | 1 | 1 |
| 0 | 0 | 1 | 0 | 2 |
| 0 | 0 | 1 | 1 | 3 |
| 0 | 1 | 0 | 0 | 4 |
| 0 | 1 | 0 | 1 | 5 |
| 0 | 1 | 1 | 0 | 6 |
| 0 | 1 | 1 | 1 | 7 |
| 1 | 0 | 0 | 0 | 8 |
| 1 | 0 | 0 | 1 | 9 |

of one of the AND gates, that gate is turned on. Some device, for example a bulb, may be illuminated to indicate the gate that is on. An example of a circuit used to convert BCD to decimal digits is in Fig. 13-28. Let us use several BCD numbers to see how they pass through this circuit.

When 0000 is fed to the circuit, the inputs are $\overline{A}\ \overline{B}\ \overline{C}\ \overline{D}$. The 1s are only at the $\overline{A}$, $\overline{B}$, $\overline{C}$, and $\overline{D}$ outputs from the inverters at the input circuit. Four of the eight terminals from the input circuit are connected to inputs on each AND gate. Different combinations from the input are connected to each gate, as shown at the gates in the drawing. A particular AND gate is turned on only when the input circuit supplies 1s to all inputs of that gate. Because 1s are only at the A, B, C, and D output terminals of the input circuit, the AND gate with a 0 output is connected to all four of these terminals, so only this AND gate is turned on.

Similarly, if the input is 0101, the only terminals of the input circuit that are 1s are $\overline{A}$, $\overline{B}$, $\overline{C}$, and D. The only AND gate connected to all of these terminals has a 5 output. Consequently, this is the only gate turned on when 0101 is fed to the input circuit as 1s are at all the inputs of only this AND gate.

### Encoder

An *encoder* takes information present at a number of its inputs and converts it into a binary number. As an example, assume there are 26 letters at 26 inputs to a circuit. Each letter is connected to a voltage through an individual switch. By closing the switch at a particular letter, a 1 is applied to an input of the circuit. The circuit can be designed so that the 1 turns on specific gates to establish a series of 1s and 0s atthe output of the encoder. Different groups of 1s and 0s are generated when different switches are closed. Each group of 1s and 0s can be used to represent a specific letter.

This method can be used to change decimal numbers into numbers in the BCD system. An example of this type of circuit is in Fig. 13-29. A, B, C, and D are 1s only

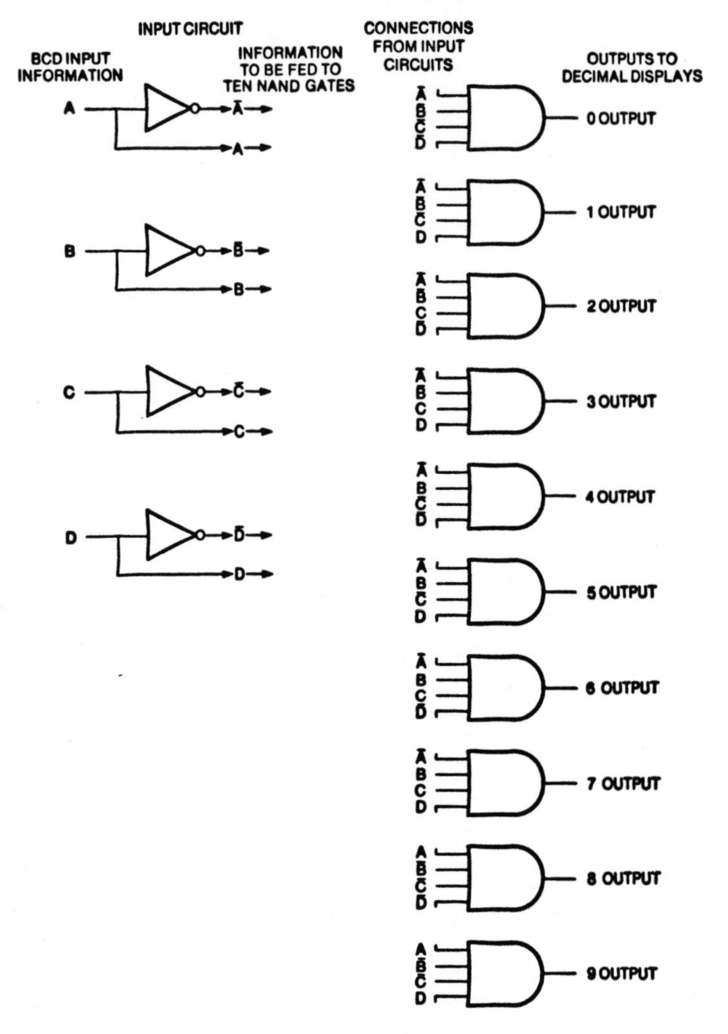


Fig. 13-28. This circuit converts BCD numbers to decimal numbers.

when the appropriate switch is closed turning on specific diodes in the circuit. Otherwise, the outputs are 0s.

BCD is only one of many codes that can be generated. Some of the more useful four digit codes that are used in various applications are shown in Table 13-4. Here, bits may have different weights or significance than they do in the BCD code. The weighted value of each bit in a particular code category is shown at the head of each column.

The last two columns differ from the others. Numbers in the "Excess 3" column are 0011 (3 in the decimal system) units larger than are numbers in the standard BCD (8421) column. In the "Gray" column, numbers are chosen almost haphazardly. Here,

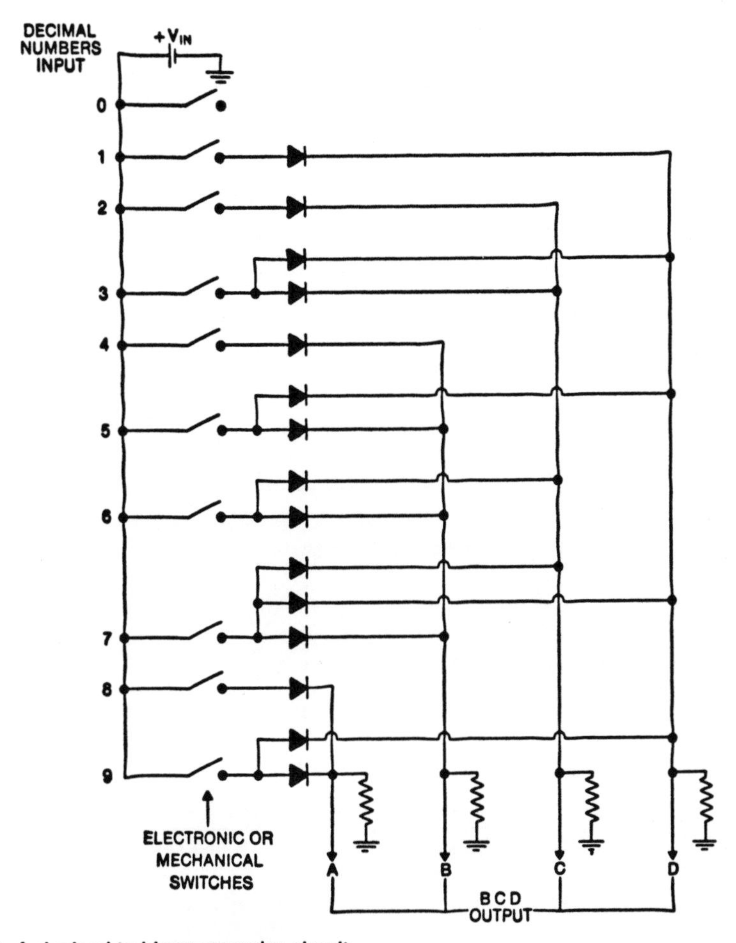


Fig. 13-29. A decimal-to-binary encoder circuit.

Table 13-4. Some Useful Four-digit Codes.

| | 8 4 2 1 | 4 2 2 1 | 5 2 1 1 | 6 3 2 1 | Excess 3 | Gray |
|---|---|---|---|---|---|---|
| 0 | 0 0 0 0 | 0 0 0 0 | 0 0 0 0 | 0 0 0 0 | 0 0 1 1 | 0 0 0 0 |
| 1 | 0 0 0 1 | 0 0 0 1 | 0 0 0 1 | 0 0 0 1 | 0 1 0 0 | 0 0 0 1 |
| 2 | 0 0 1 0 | (0 0 1 0) | (0 1 0 0) | 0 0 1 0 | 0 1 0 1 | 0 0 1 1 |
| 3 | 0 0 1 1 | (0 0 1 1) | (0 1 0 1) | (0 1 0 0) | 0 1 1 0 | 0 0 1 0 |
| 4 | 0 1 0 0 | (1 0 0 0) | 0 1 1 1 | 0 1 0 1 | 0 1 1 1 | 0 1 1 0 |
| 5 | 0 1 0 1 | (1 0 0 1) | 1 0 0 0 | 0 1 1 0 | 1 0 0 0 | 0 1 1 1 |
| 6 | 0 1 1 0 | (1 1 0 0) | (1 0 0 1) | (1 0 0 0) | 1 0 0 1 | 0 1 0 1 |
| 7 | 0 1 1 1 | (1 1 0 1) | (1 1 0 0) | (1 0 0 1) | 1 0 1 0 | 0 1 0 0 |
| 8 | 1 0 0 0 | (1 1 1 0) | (1 1 0 1) | (1 0 1 0) | 1 0 1 1 | 1 1 0 0 |
| 9 | 1 0 0 1 | 1 1 1 1 | 1 1 1 1 | (1 1 0 0) | 1 1 0 0 | 1 1 0 1 |

Numbers in parentheses are not the only ones that can fill the slots. For example, in the 4221 column, 2 can be 0100.

| INPUT BCD CODE | | | | DECIMAL NUMBER | OUTPUT 6321 CODE | | | |
|---|---|---|---|---|---|---|---|---|
| E | F | G | H | | A | B | C | D |
| 0 | 0 | 0 | 0 | 0 | 0 | 0 | 0 | 0 |
| 0 | 0 | 0 | 1 | 1 | 0 | 0 | 0 | 1 |
| 0 | 0 | 1 | 0 | 2 | 0 | 0 | 1 | 0 |
| 0 | 0 | 1 | 1 | 3 | 0 | 1 | 0 | 0 |
| 0 | 1 | 0 | 0 | 4 | 0 | 1 | 0 | 1 |
| 0 | 1 | 0 | 1 | 5 | 0 | 1 | 1 | 0 |
| 0 | 1 | 1 | 0 | 6 | 1 | 0 | 0 | 0 |
| 0 | 1 | 1 | 1 | 7 | 1 | 0 | 0 | 1 |
| 1 | 0 | 0 | 0 | 8 | 1 | 0 | 1 | 0 |
| 1 | 0 | 0 | 1 | 9 | 1 | 1 | 0 | 0 |
| 1 | 0 | 1 | 0 | 10 | X | X | X | X |
| 1 | 0 | 1 | 1 | 11 | X | X | X | X |
| 1 | 1 | 0 | 0 | 12 | X | X | X | X |
| 1 | 1 | 0 | 1 | 13 | X | X | X | X |
| 1 | 1 | 1 | 0 | 14 | X | X | X | X |
| 1 | 1 | 1 | 1 | 15 | X | X | X | X |

A Table relating BCD inputs to 6321 code outputs. Note the don't cares in states 10 through 15.

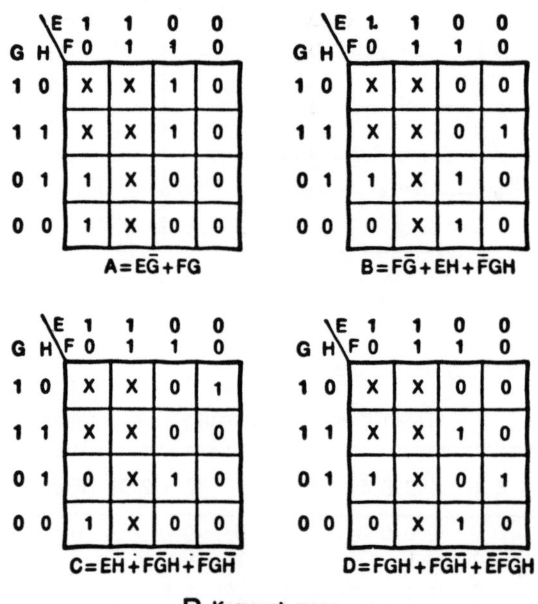


B Karnaugh maps.

Fig. 13-30. The procedures for designing a logic circuit to convert from BCD to 6321 code.

only one bit changes in the transition from one number to the next number. Only one gray code defined by this single bit change rule is shown here. Other gray codes with these characteristics are possible. Code converters are used for converting from one binary code to the other.

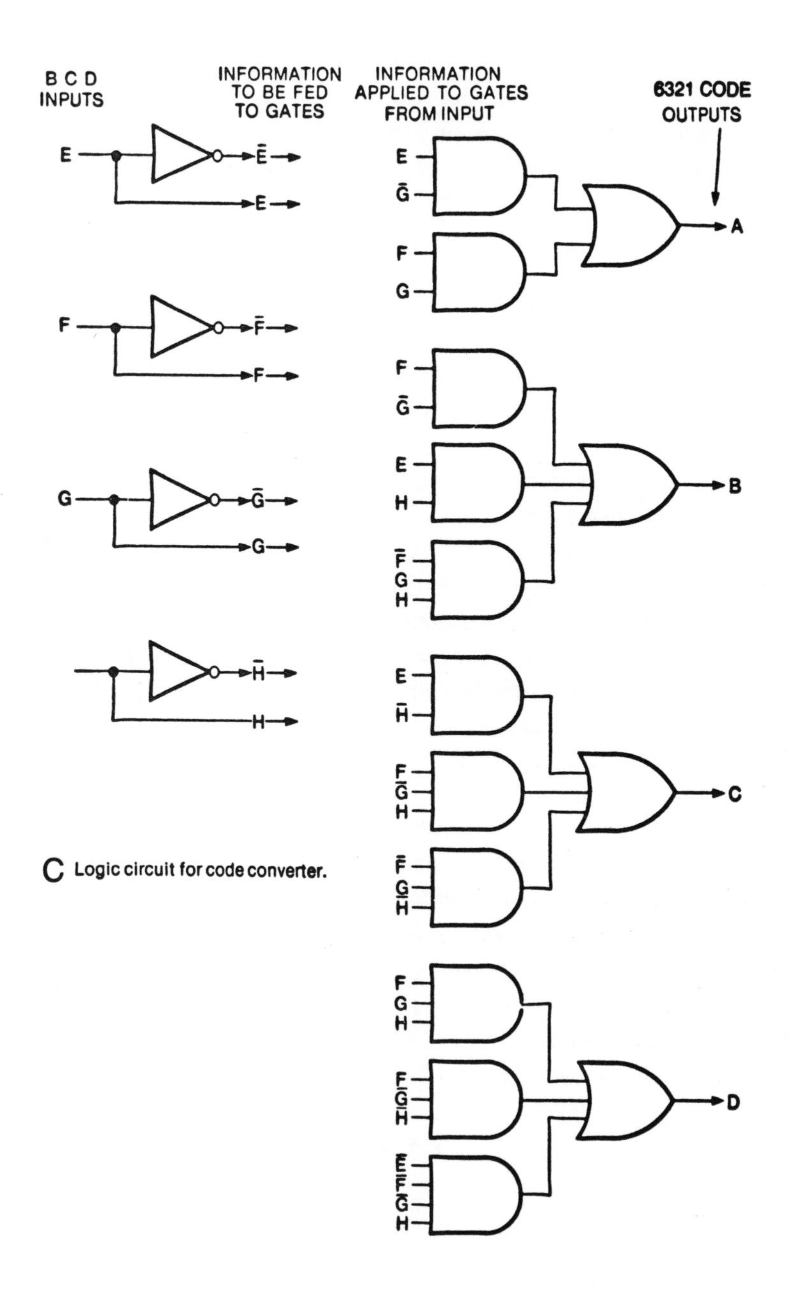


C Logic circuit for code converter.

## CODE CONVERTERS

Logic circuits can be used for converting from one code to another. A procedure for determining the circuitry required for converting from a BCD number to the 6321 code is shown in Fig. 13-30. Steps in the procedure are as follows:

1. Set up a table showing the relationships between the two codes. Assume the outputs from the logic circuit with the new code are ABCD and the inputs to the logic circuits with information about the old code are EFGH.

2. Draw a Karnaugh map for each output bit. Determine the relationships of bits A, B, C and D to each BCD word, from the maps.

3. Using information derived from the Karnaugh maps, draw the logic circuit.

A Read Only Memory (ROM) can be used to convert from one binary code to another. The relationship between the output and input is stored permanently in the memory. A ROM consists essentially of two sections—a decoder to change the input code into individual words and an encoder to use these words to deliver the desired information to the output. Encoders and decoders are similar to those described earlier, while the words may be equivalent to decimal digits or larger decimal numbers.

ROMs are programmed by the manufacturer of the memory to perform the job the customer wants it to do. PROMs are Programmable ROMs. These are not programmed, but have fuses in series with the various circuits. These fuses can be opened by the user so that the PROM can be programmed to do a specific job. In either case, the job the memory performs cannot be changed. The Random Access Memory (RAM) can be programmed repeatedly. Information can be written into it and altered as needed. The readout from the RAM can also be altered as needed to satisfy a particular need.

The EEROM or Electrically Erasable ROM stores information permanently as does the ROM, but the program can be altered by changing voltages at the various terminals.

Getting back to the basic code converter, IC circuits have been designed to convert from large binary numbers to BCD numbers. These are used in decoders to establish the decimal readouts, but modern readouts do not use the numbers as described earlier, instead they use seven-segment digits. Different segments are illuminated to form different numbers. A seven-segment array with its breakdown into the ten basic decimal digits is shown in Fig. 13-31. Each segment is lit only when required as a portion of a specific number. For example, segment F must be on when numbers 0, 4, 5, 6, 8, and 9 are needed. This information can be written in digital form as follows:

$$F = 0 + 4 + 5 + 6 + 8 + 9$$

because F must be at 1 when any one of these digits must be displayed. This type of information can be derived for all segments through analysis of the drawings in Fig. 11-31B. Complex ICs have been designed to perform the task of providing for this type of decimal readout from BCD inputs.

Elements of seven-segment digits are commonly composed of LEDs or LCDs (liquid crystal displays). LCDs are required where low power drain is an important consideration. Both types of digital displays are available as individual digits or in a group

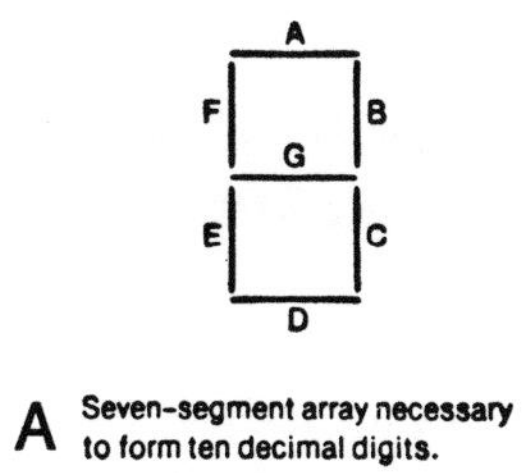


A Seven-segment array necessary to form ten decimal digits.

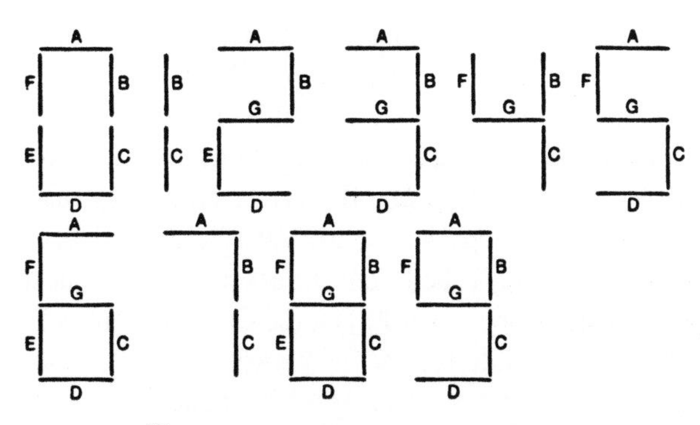

B Decimal digits formed from seven segment array.

Fig. 13-31. Decimal numbers are formed by seven segment displays.

of several digits. While LED displays use a steady current to light the segment, pulses are required to illuminate LCD segments. The primary disadvantage of the LCD is its sensitivity to extremes of temperature.

## MICROPROCESSORS

A microprocessor is basically a central processing unit. By itself, it can do nothing. It does not add or subtract and does not do calculations. It takes information and instructions from outside sources and feeds it to (or directs it to) address lines where the information can do its job.

Basically, there are three sets of terminals on a microprocessor. There are *data lines*, which are used to feed information or data to the microprocessor. Information can also be fed from inside the microprocessor to the data lines. A ROM may be connected to these data lines to supply information for processing.

A second set of terminals is supplied to *control* the action of the microprocessor. Among these terminals, you are likely to find a read/write line. Coded information supplied here directs the microprocessor to either read the information from the data line or to write information (put out information) through the data line. Directions may also be provided at these terminals for the microprocessor to *interrupt* what it is doing and change over to some other more immediately vital activity, to *load* the microprocessor with information originating in a memory, to *halt* or stop carrying out the instructions it has received, to *jump* over or switch over to another source for its instructions, and so on.

The third set of terminals is composed of *address lines*. The microprocessor takes the data and control instructions and directs the processed information to a set of binary-coded outputs. The information handled by the microprocessor is supplied through these address-line outputs (in a binary code) to a memory unit such as a RAM, to a print-out machine for supplying data on printed papers, or to other electronic or mechanical peripheral devices.

Microprocessors are useful in many applications. They can be found as the central processing unit in a computer. Some of these chips take data from sensors and feed information back to circuits to control the temperature of fluids or to limit the percentage of impurities in a fluid. The microprocessor is also used in the "miles-to-empty" circuit found in some automobiles, to advise the driver how many miles he can drive before his gas tank gets empty. Applications are virtually without bounds.

Microprocessors are frequently called "computers on a chip," but this is generally not quite accurate. A microprocessor is usually just the CPU, or Central Processing Unit. This is the section of the computer that actually executes the commands. Practical computers require several additional stages. At a bare minimum, a computer is made up of four stages, as shown in Fig. 13-32. These sections are:

CPU (or microprocessor)
memory
input port
output port

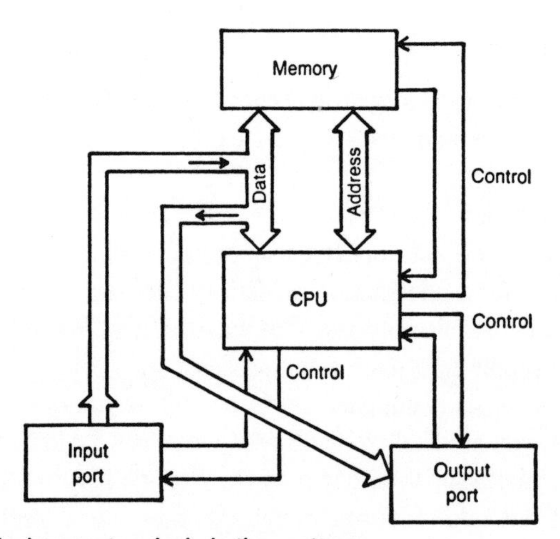


Fig. 13-32. Most practical computers include these stages.

The CPU, of course, does the actual computing. The memory stores the program commands and the data used in executing the program. The input and output ports allow the computer to communicate with the outside world (anything that is not an integral part of the computer itself). The input port permits data from some external device (keyboard, modem, floppy disc, etc.) to be fed into the computer. Similarly, the output port lets the computer feed its results out to some external device (display screen, printer, modem, floppy disc, etc.) Some peripheral devices (such as the modem and the disc drive) utilize both the input and output ports.

A functional computer system also requires a clock signal source (a precise square-wave oscillator) to synchronize the various sections of the system, so that data can be reliably interchanged from section to section.

Communication between the various internal sections of the computer is accomplished via buses. A bus is nothing more than a digital signal line that can carry coded binary data. The binary data can represent anything at all—numeric values, alphanumeric characters, or machine-language commands. The only difference is how the CPU is instructed to interpret the binary information.

The data bus connects the processor section to everything else. Data can flow from the input to the CPU, from the CPU to the output port, or between the CPU and the memory in either direction. The CPU determines which piece of data goes where.

There is also an address bus between the CPU and the memory. The data on this bus determines which portion of the memory the processor is using. The concept is simple enough if you think the CPU needs to know the address or location of whatever it is looking for, just as you need to know the address to find a friend's home.

The third and final bus is used for system control and synchronization. The signals on this bus keep the various sections of the computer functioning simultaneously. For example, the input port uses this bus to let the CPU know there is some incoming data available from an external input device.

The CPU itself is a complicated device made up of a number of sub-circuits, usually within a single IC chip.

The internal structure of the popular Z80 CPU is illustrated in Fig. 13-33. This unit is fairly typical in the sections it contains.

The ALU is the Arithmetic Logic Unit. This circuitry performs arithmetic (addition, subtraction, multiplication, etc.) operations and logic (AND, OR, NOT, etc.) operations. Dedicated ALU ICs are available. To a limited extent they can be considered very crude, stripped-down CPUs.

The registers are temporary memory locations. They are very limited in size. Most hold just a single eight-bit byte. Some special purpose registers hold sixteen bits. These registers are utilized during various CPU operations. Some form of external memory is required for practical applications.

Since the Z80 handles data in eight-bit chunks, it is called an eight-bit processor.

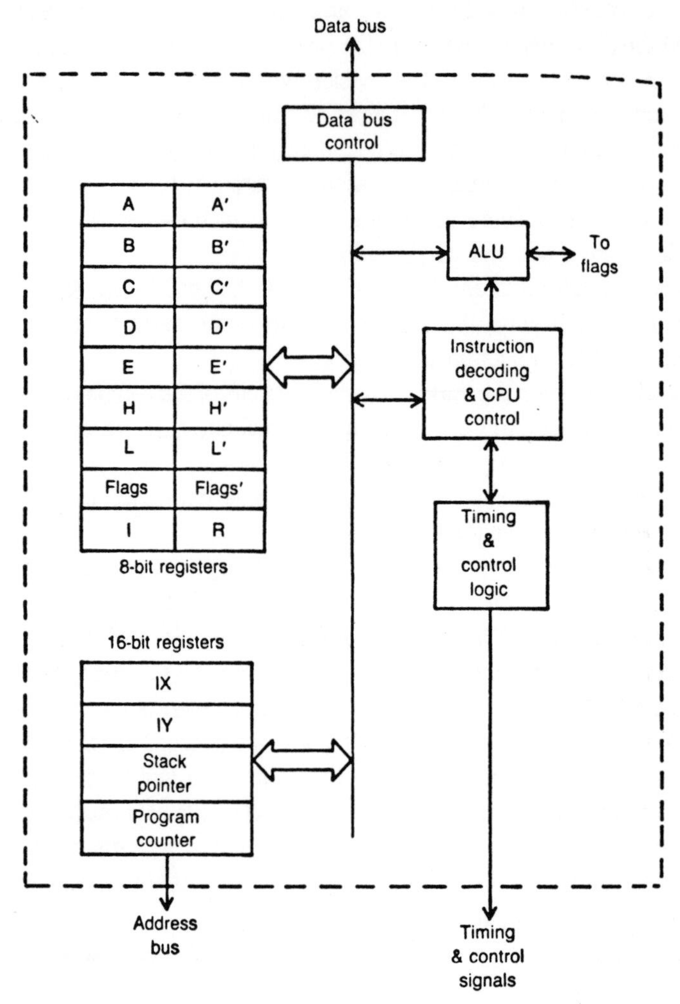


Fig. 13-33. The Z80 is a typical CPU IC.

More recently developed CPUs are sixteen-bit devices. They work on sixteen bits at a time, offering faster operation and greater precision in certain applications.

The other sections of the CPU are pretty much self-explanatory:

- Data bus control.
- Instruction decoding and CPU control.
- Timing and control logic.

The CPU is a very complex device, and it is impossible to treat it in depth in this volume. A number of books on CPUs are available, including *30 Customized Microprocessor Projects* by Delton T. Horn (TAB BOOKS #2705).

## MEMORY

The simplest digital memory device is the flip-flop, discussed earlier in this chapter. A flip-flop "remembers" its state (high or low) indefinitely, as long as power is not interrupted. A single flip-flop can store one bit. The bit being stored can be changed to its opposite state at any time, simply by triggering the flip-flop.

Obviously, a one bit memory is of limited usefulness. By combining multiple flip-flops, a larger memory unit can be created. Several flip-flops can be combined to form a shift register, essentially a memory circuit for one binary word.

While conceptually simple, flip-flops and shift registers are inadequate for serious memory applications, such as in a computer system. Specialized digital circuits called memories have been developed for such applications. Many different types and sizes of memory are widely available in IC form today.

Generally speaking, solid-state memory devices can be divided into two broad classes:

- RAM
- ROM

RAM (*Random Access Memory*) can be read from or written to. The CPU can examine or change the data stored at any specific location.

ROM (*Read-Only Memory*) on the other hand, can only be read from, it cannot be written to. The CPU can examine any data stored anywhere within a ROM, but it cannot alter it in any way.

Within each of these broad categories, there are several variations.

### RAM Devices

Any specific location in the memory can be contacted without stepping through any other memory locations. This permits very fast exchange of data between memory and the CPU.

A shift register on the other hand, is an example of a sequential memory. The various memory locations must be accessed in a specific, non-variable order.

In a random access memory, each individual memory location is assigned a unique address. This type of memory can be considered analogous to a post office box system. Any individual box (or memory location) can be uniquely defined by an address identifying its column and row. This concept is illustrated in Fig. 13-34.

In RAM, we can either look at the value at a specific location without changing it (read), or we can replace the old value with a new value (write). For this reason, RAM is occasionally called *Read/Write Memory*, or RWM. Many technicians argue that this is a better name, since ROM can also be randomly accessed.

There are two basic types of RAM in common use, depending on the circuitry used:

- static RAM
- dynamic RAM

| | A | B | C | D |
|---|---|---|---|---|
| 1 | A1 | B1 | C1 | D1 |
| 2 | A2 | B2 | C2 | D2 |
| 3 | A3 | B3 | C3 | D3 |
| 4 | A4 | B4 | C4 | D4 |

Fig. 13-34. Memory addressing can be thought of as a post office box system.

A static RAM is essentially made up of a series of addressable flip-flop-like units. Data can be stored in a static RAM virtually indefinitely, unless the stored values are erased or changed, or the power supply is interrupted. A static RAM cannot store data without continuously applied power.

The major disadvantage of static RAM is the relatively bulky circuitry required for each memory location. For a given memory size, static RAMs tend to be considerably larger and more expensive than a comparable dynamic RAM.

In a dynamic RAM, each bit is stored as a charge on a capacitor (or, more commonly in practical units, the etched equivalent within an IC). A charged capacitor represents a logic 1, while a discharged capacitor represents a logic 0.

A dynamic RAM is much simpler than a static RAM, so it tends to be smaller and cheaper.

Dynamic RAM has its disadvantages. The most important limitation of this type of RAM is that no practical capacitor can hold a charge indefinitely. Eventually the charge will tend to leak off, erasing the stored data.

Electronically reading the value stored in a dynamic memory cell tends to recharge partially charged capacitors, refreshing the memory. All practical dynamic memory systems require special refreshing circuits that automatically read all of the memory locations at regular intervals, to prevent the charged capacitors from leaking off too much voltage. Many modern CPUs include on-board refresher circuitry. If the power supply is interrupted for any reason all data stored in either a dynamic or static memory will be irretrievably lost.

### ROM Devices

ROM is used for permanent data storage. Stored data cannot be changed or erased, even if power is interrupted. A computer's operating system (basic system programming) is almost always stored in ROM, so the computer will be operational immediately upon power-up.

ROM memory cells are considerably simple than either type of RAM memory cells. Like RAM, ROM is usually supplied in IC form. All of the data in a ROM chip is permanently determined by the manufacturer when the device is made. Obviously, this means that ROMs are only practical for applications where many identical units are required. Otherwise, the manufacturing costs would be too high per unit. In mass production ROM chips are relatively cheap.

In some applications, especially for circuit designers and hobbyists, only a few copies are needed. True ROM devices are not practical.

Fortunately, a compromise is available. For those applications where permanent (non-erasable) storage is needed, but only a few copies (or perhaps just a single prototype) will be made, a user-programmable ROM-like device has been developed. This type of memory is called PROM, or *Programmable Read-Only Memory*.

Each memory cell in a PROM is similar to those in a ROM. The absence or presence of a diode unit, as illustrated in Fig. 13-35, determines whether the cell contains a 1 or a 0. In a PROM the diodes are a special fused type, and the chip is supplied with a diode in every cell. All address locations contain 1's. The user can program the chip by selectively blowing out the fuses on the unwanted diodes (where 0's are to be stored).

Once programmed, a PROM behaves exactly like a ROM. Data can be read from it, but cannot be erased or changed during operation. Of course, additional fuses can be blown (additional circuitry is required to program a PROM) to change more 1's to 0's. There is no way to ever replace a blown diode (change a 0 back to a 1).

If a mistake is made during programming the PROM, or even a small change is needed, the entire chip must be discarded, and a new one must be programmed from scratch. Once it is correctly programmed, the data stored in a PROM will theoretically last forever.

There is a special type of PROM that allows the chips to be erased and reused,

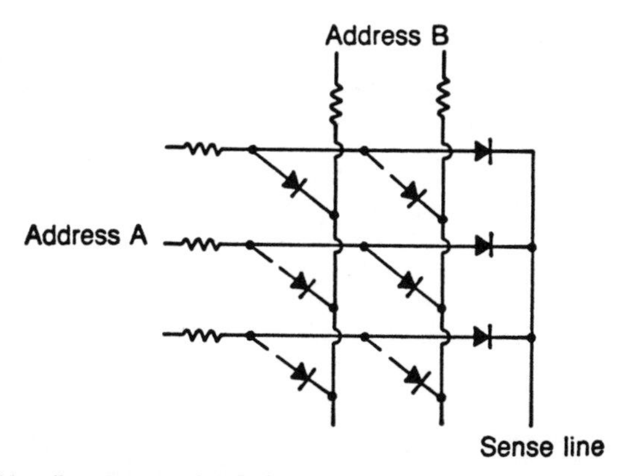


Fig. 13-35. ROM cells are very simple in concept.

known as EPROM, or *Erasable Programmable Read-Only Memory*. The stored data can be erased by exposing the chip to a strong ultraviolet light. All data is erased at once. Individual memory locations cannot be selectively changed.

The IC housing of an EPROM contains a small window. Ordinarily, it is covered with a light-proof lid, removed only to erase the chip.

### Memory Size

Since binary numbers are used to define the memory location addresses, the number of cells in a memory system is almost invariably a power of two, like 256 ($2^8$), 1024 ($2^{10}$), or 4096 ($2^{12}$).

In large computer systems, memory size is usually given in K. The letter K (kilo) is normally used to indicate a factor of one thousand. However, 1000 is not a power of two. The nearest power of two of 1024, so in memory systems, K actually represents a factor of 1024. That is:

1 K = 1024
4 K = 4096
16 K = 16384
64 K = 65536
128 K = 131072
256 K = 262144
1000 K = 1 Meg = 1024000

You do have to be careful about what is being counted. For many memory ICs, 1 K indicates a storage capacity of 1024 *bits*, while in computer systems, the quantity in question is the number of *bytes* that can be stored.

### TRI-STATE LOGIC

In ordinary digital or logic systems, signals can only take on one of two possible states, 1 or 0. However, some devices also offer a third high-impedance state, this is known as tri-state logic, and is very useful in certain applications. In the high impedance state, the signal is neither a 1 nor a 0. In fact, it appears to other gates as the absence of any signal at all. This could be called the *IGNORED* state.

# 14

# Semiconductor Materials and Logic Families

Look through almost any electronics distributor catalog and you'll see that semiconductor devices are made from a variety of materials. In addition, similar devices manufactured in different ways will perform in similar, but subtly different ways. This chapter will briefly cover some commonly used semiconductor materials, and how the material used affects performance.

Later in this chapter we will examine the different types of IC construction techniques used in digital devices, where compatability can be a special problem to the uninitiated.

## CONDUCTORS, INSULATORS, AND SEMICONDUCTORS

An electric current can be forced through any material. Electric current depends on the movement of electrons. Some materials have a significant number of easily movable electrons. Such materials are called conductors. Copper is an example of a good conductor. It offers a very low resistance to an electric current.

Other materials have their electrons tightly held in place, resistant to movement (electric current), such materials are called insulators. Rubber and slate are two good insulators.

There is a third important class of substances with properties somewhere between conductors and insulators. These substances are known as semiconductors.

It is important to remember that *all* substances, whether they are conductors, insulators, or in between, will offer some resistance to current flow. Conductors present a very small resistance, while insulators present a very large resistance. As might be expected, a semiconductor offers a moderate amount of resistance to the flow of electrons through it.

For example, a cubic centimeter of copper has a resistance of about $1.7 \times 10^{-6}$ (0.0000017) ohm. This is clearly an almost negligible amount of resistance.

On the other hand, a cubic centimeter of slate has a resistance in the neighborhood of 100 megohms (100,000,000 ohms). Compared to copper, virtually no current can flow through slate.

Now, compare both of these substances to germanium. A cubic centimeter of this material has a resistance of approximately 60 ohms. Germanium is a semiconductor.

## DOPING

In order to be useful in electronic components, a semiconductor material must be *doped* with a suitable impurity. To understand how this works, you have to know just a little bit of how the atom is constructed.

A simple model of an atom is shown in Fig. 14-1. At the center is the nucleus consisting of protons (positive charge) and neutrons (no electric charge). The nucleus

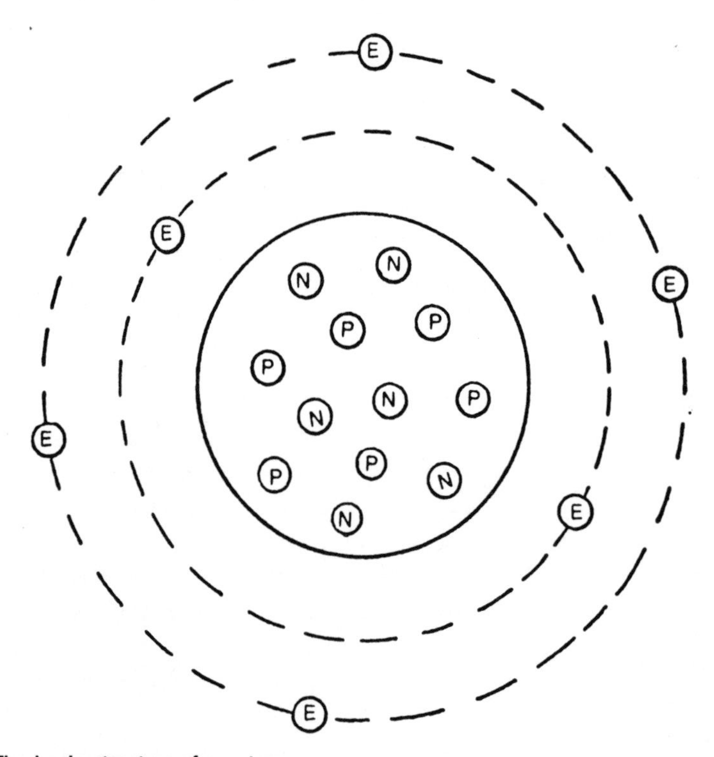


Fig. 14-1. The basic structure of an atom.

contains a number of other types of particles, but they aren't important for our purposes here.

The nucleus is surrounded by electrons that carry a negative charge. The electrons are arranged in orbits called shells. Each shell has a specific number of electrons that it can hold. All inner shells hold their maximum number of electrons. The outermost shell may or may not be full.

The electrons in the outermost shell are called *valence electrons.* Some, or all, of the valence electrons may be shared with adjoining atoms to create molecules.

Electric current is the movement of valence electrons.

A germanium atom has four valence electrons. In pure germanium, the valence electrons pair up with the valence electrons of other germanium atoms in a crystalline structure, as illustrated in Fig. 14-2. The atoms within the crystal are held together by a force called the covalent bond. As this term suggests, adjacent atoms share their valence electrons.

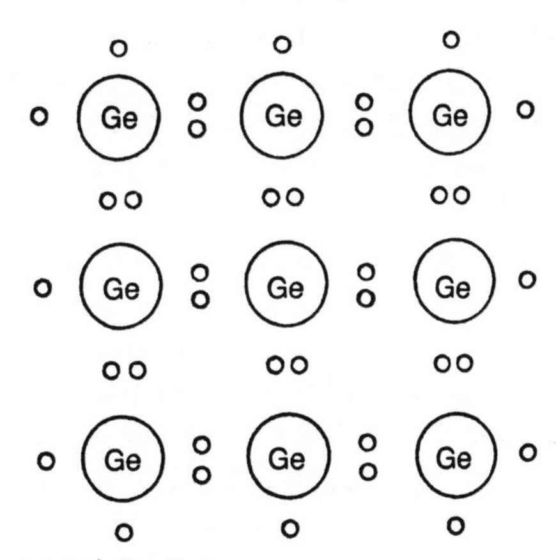


Fig. 14-2. The germanium crystal structure.

Fig. 14-3 shows what happens when we dope a pure germanium crystal with a small amount of arsenic. The arsenic atom will try to act like a germanium atom, but since arsenic has five valence electrons instead of four, there will be an extra electron left over for each arsenic atom.

This extra electron can drift freely from atom to atom throughout the crystal. The crystal as a whole is still electrically neutral, because the total protons equal the total electrons.

If a voltage is applied across the crystal, the extra drift electrons will be drawn to the positive terminal of the voltage supply and removed from the crystal. Now, the crystal as a whole has fewer electrons than protons. It has a positive electric charge, and draws new electrons out of the negative terminal of the power source. These electrons will move through the crystal and out to the positive terminal. This is an n-type semiconductor.

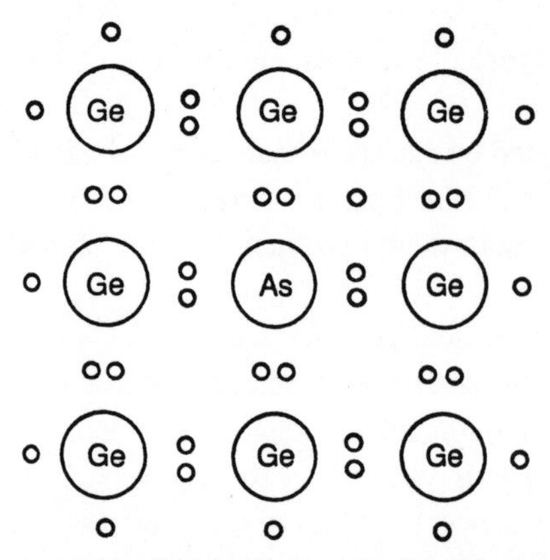

Fig. 14-3. When germanium crystal is doped with arsenic, there is an extra electron left over.

Indium is also used as an impurity element for doping germanium crystals. Indium has just three valence electrons, so the result is as shown in Fig. 14-4. For each indium atom added to the germanium, the crystal will be one electron short. There is an extra "hole" or space for an electron. Nearby electrons will try to fill this hole, leaving holes in their previous positions. In effect, the extra hole drifts through the crystal. The net charge of the crystal will remain zero (electrically neutral). This is called a p-type semiconductor.

If we connect a voltage source across this material, electrons will be drawn out of the negative terminal to fill the crystal's drifting holes. The crystal has more electrons than protrons, so it has a negative charge. Excess electrons are drawn off to the positive terminal of the voltage source, leaving more holes behind.

Neither the n-type nor the p-type semiconductors are particularly exciting or interesting in themselves. But when they are put together to form a pn junction they exhibit the unique characteristics that make diodes, transistors, and ICs possible.

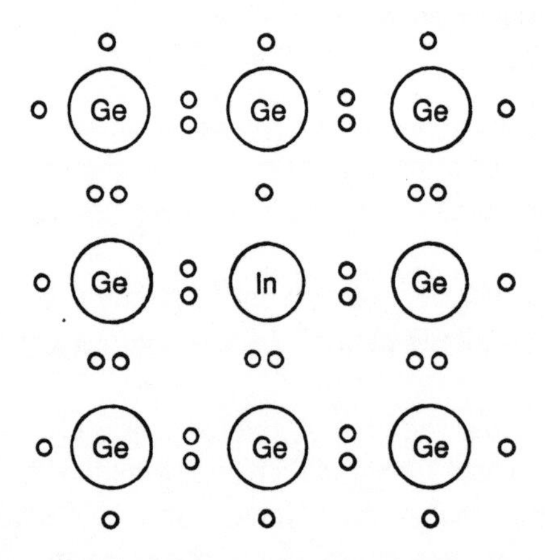

Fig. 14-4. When germanium crystal is doped with indium, there is an extra hole left over.

To form an n-type semiconductor, an impurity with extra valence electrons is added to the semiconductor crystal. This is called a donor impurity.

A p-type semiconductor is created by adding an impurity element with too few valence electrons, creating extra holes. This is called an acceptor impurity.

## SEMICONDUCTOR MATERIALS

A number of materials can be used as semiconductors. The most widely used are germanium and silicon.

Early semiconductor components were usually made from germanium, but germanium is more heat sensitive than silicon, and such devices could often be damaged during soldering. Germanium is still used in some special purpose semiconductor components. The forward drop across a germanium pn junction is typically about 0.3 volt.

Silicon, being more durable, is used in the vast majority of modern semiconductor devices. Another reason for the widespread use of silicon is that it is quite a common element in the earth's crust, although it is usually found combined with oxygen or other elements and must be purified for use in semiconductors.The forward drop across a silicon pn junction is typically about 0.7 volt.

Other elements that exhibit semiconductor properties include boron, carbon, sulfur, and tellurium. In many newer semiconductor devices, special compounds are used, including gallium arsenide, indium antimony and a number of metal oxides.

Gallium arsenide has been receiving a lot of attention lately because of its use in semiconductor components for microwave and laser applications.

## LOGIC FAMILIES

As discussed in Chapter 13, digital circuits are made up of units known as gates. These gates can be made with a variety of different circuits. The type of circuitry used within a digital IC determines the logic family the chip belongs to. While they perform similar functions, gates from different logic families are not 100 percent compatible.

### RTL Gates

Most of the earliest digital ICs belonged to the RTL family. RTL is short for *Resistor-Transistor Logic.* As the name suggests, the basic circuit is made up of a resistor and a transistor. Figure 14-5 shows a RTL inverter. The base resistor is used to limit input current. Most RTL circuits were designed for a power supply of 3.6 volts.

Since RTL circuits are fairly easy to implement in chip form, they were popular in the early days of digital integrated circuits. Faster and more powerful technologies have since been developed at competitive costs, so today RTL is pretty much an obsolete logic family. You can still find some RTL ICs available from surplus dealers.

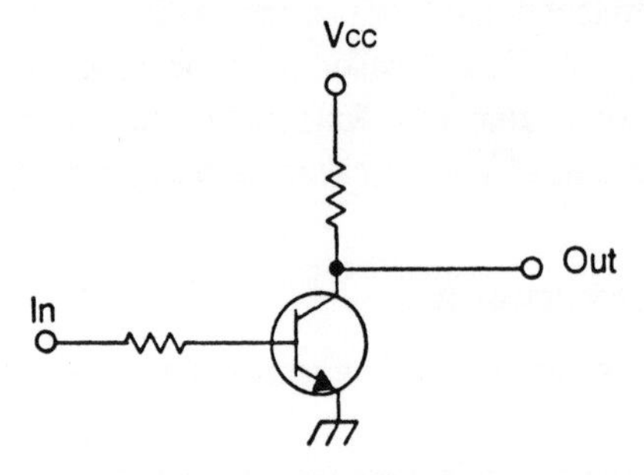


Fig. 14-5. A RTL inverter is made up of an input resistor and a transistor.

## DTL Gates

Another popular early logic family was DTL, or *Diode-Transistor Logic.* Basically, DTL gates are quite similar to RTL circuits with the addition of input diodes. A DTL NOR gate is shown in Fig. 14-6. Due to its relatively slow switching speed, DTL, like RTL, is more or less an obsolete technology.

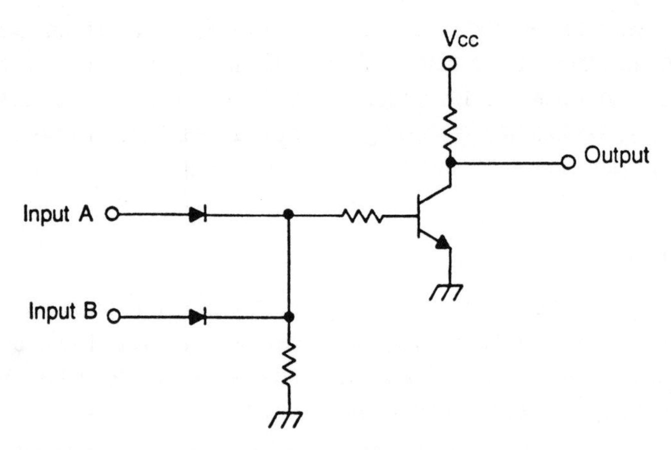


Fig. 14-6. A DTL NOR gate uses input diodes instead of input resistors.

## TTL Gates

Digital electronics didn't really begin to take off in a big way until the development of the TTL logic family. TTL stands for *Transistor-Transistor Logic.* This logic family is based upon the ability to form transistors with multiple emitters on an IC chip. Such a multiple emitter transistor is shown in Fig. 14-7.

TTL gates offer relatively high speed operation, low power consumption, and a reasonably small susceptibility to noise. Until recently TTL was the unquestioned king of logic families. While it is no longer the most popular logic family, it is still in widespread use.

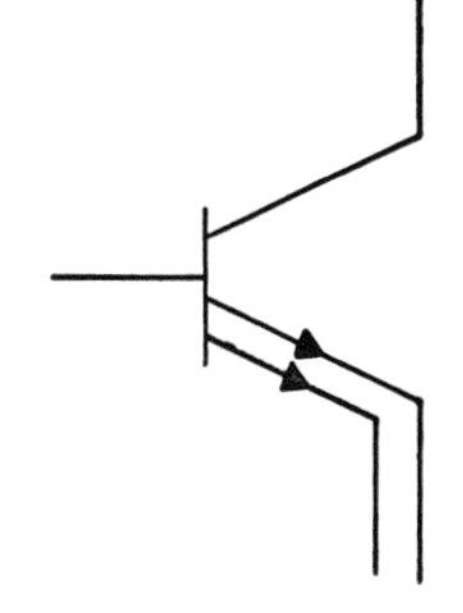

Fig. 14-7. TTL circuits are built around multi-emitter transistors.

Figure 14-8 shows a basic TTL AND gate circuit. The multiple emitters essentially take the place of the diodes in comparable DTL circuits.

The power supply voltage for TTL devices is rather critical. These ICs require a tightly regulated +5 volt power supply. As a rule, power supply deviations greater than ±0.5 volt cannot be tolerated. An incorrect supply voltage will almost certainly result in improper or unreliable operation, and is quite likely to actually damage the ICs. This is especially true of over-voltages.

An input voltage below 0.2 volt is interpreted by a TTL gate as a logic 0. A logic 1 is defined as anything above 2.4 volt. An input voltage between 0.2 volt and 2.4 volt is undefined in TTL logic. This wide gap provides a comfortable noise margin to avoid false triggering.

A logic 1 output from a TTL gate will not be much more than about half the supply voltage (nominally 2.5 volt). A logic 0 output will require the input device to sink a fairly substantial amount of current. 16 mA is a typical figure.

If an input to a TTL device is left unconnected (or "floating"), it will generally behave like a logic 1 input. A floating input "pulls itself up" to the logic 1 voltage level.

A TTL logic 1 output voltage can be raised with a simple pull-up resistor, as shown in Fig. 14-9. This trick is frequently used by digital circuit designers.

A large number of TTL devices are readily available in IC form. Generally they are numbered 74xx, with the xx being a two or three digit number identifying the specific device.

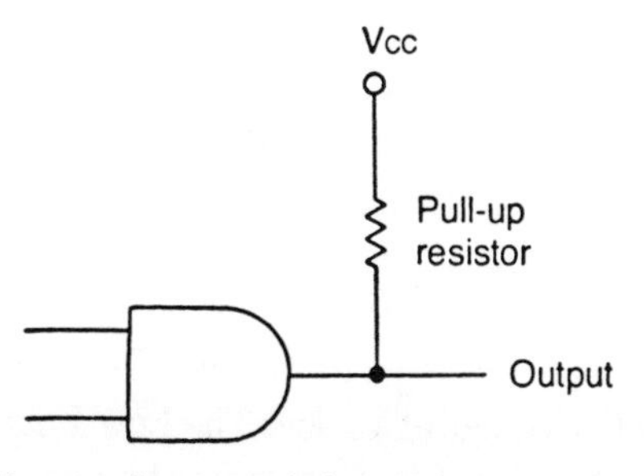


Fig. 14-8. A pull-up resistor is sometimes needed at the output of TTL gates.

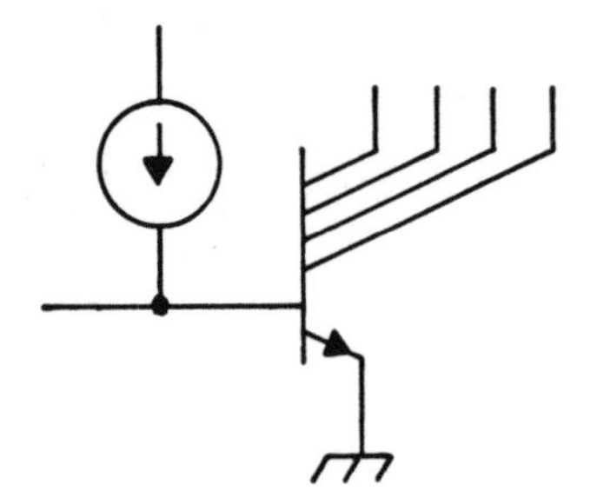

Fig. 14-9. $I^2L$ gates use multi-collector transistors.

You might also encounter TTL ICs numbered 54xx. Such a device is essentially identical to a similarly numbered 74xx unit, except it is designed to meet stricter military specifications (primarily a wider range of operating temperatures). For the vast majority of applications, there is no difference between the 54xx and 74xx lines.

There are a number of important advantages offered by TTL circuitry. For one thing, its very popularity has ensured the wide-spread availability of literally hundreds of SSI (*Small-Scale Integration*) and MSI (*Medium-Scale Integration*) devices at reasonably low prices.

TTL gates are relatively immune to noise, and consume fairly small amounts of power during steady state conditions (when the output is not changing states).

Perhaps the most important advantage of TTL technology is the circuitry's capability for operating at high switching speeds. Most TTL devices can handle 20 MHz signals without problems. A few specialized units can even accept signals ranging up to 125 MHz.

No technology is perfect. TTL gates have their disadvantages too. The most obvious is their strict power supply requirements.

While TTL is fine for SSI and MSI ICs, this technology is not very well suited to LSI (Large-Scale Integration).

While TTL gates consume very little power during steady state conditions, when the output switches from one state to the other, a large current spike is pulled from the power supply for a fraction of a second. This can be insignificant in a small circuit with just a handful of gates, but in a large system with dozens to hundreds of gates, the accumulated power drain could be quite severe.

This current spike can even be dangerous to the delicate circuitry within the chip itself. For this reason a bypass capacitor should always be placed across the power lines on each and every TTL IC in every circuit. A 1 pF to 10 pF capacitor connected between the ground and $V_{cc}$ pins should be sufficient in most cases. This capacitor should be mounted as close as is physically possible to the body of the IC.

Actually, TTL is not just a single logic family. A number of subfamilies have been developed, each with its own special advantages and disadvantages.

Low-Power TTL gates are designed to consume less power during switching. This reduced current consumption comes at a price. The Low-Power devices cannot operate at speeds as high as regular TTL gates. Generally, the Low-Power units use about 0.1

Table 14-1. Comparing Operating Speed and Power Consumption of TTL and Its Subfamilies Illustrates the Speed / Power Trade-off.

| | Standard TTL | Low-power TTL | High-speed TTL | Schottky TTL | Low-power Schottky TTL |
|---|---|---|---|---|---|
| speed | 1 | 0.1 | 2 | 3.5 | 1 |
| power | 1 | 0.1 | 2 | 2 | 0.2 |

of the current, but their maximum speed is only about 10 to 25 percent of a comparable TTL device. Low-Power TTL is numbered 74Lxx.

The opposite trade-off is offered by High-Speed TTL. These devices can handle frequencies up to 50 MHz, but consume more current. Fewer High-Speed gates can be connected directly to a gate output. Standard TTL gates have a fan-in of 1 and a fan-out of 10. (One gate output can drive up to ten gate inputs.) The fan-in for High-Speed TTL gates is typically 1.3. High-Speed TTL devices are numbered in the 74Hxx format.

A more advanced form of TTL uses Schottky diodes for a better speed/power trade-off. These devices are called Schottky TTL, and are numbered in the 74Sxx format. A Schottky TTL gate uses about the same amount of power as a comparable standard TTL gate, but can offer switching speeds up to 3.5 times faster.

There is also a Low-Power Schottky TTL subfamily that runs at the same speed as comparable standard TTL devices, but consumes only about 20 percent of the power. Low-Power Schottky TTL devices are numbered in the 74LSxx format.

The various TTL subfamilies are compared in Table 14-1.

## CMOS Gates

While TTL continues to hold an important place in digital electronics, in the past few years it has become overshadowed by the CMOS logic family. CMOS stands for *Complementary Metal-Oxide Semiconductor.* ICs in this logic family are constructed in a manner similar to IGFETs, that were discussed back in Chapter 2. Occasionally, this logic family is called COS/MOS.

CMOS gates feature low power consumption and reliable operation over a wide range of supply voltages (typically about 3 to 15 volts) at a relatively low cost.

Part of the price, however, is a trade-off in operating speed. The higher the power supply voltage (assuming it doesn't exceed the maximum level), the faster CMOS gates can operate. But even at their fastest, CMOS circuits are still slower than standard TTL.

CMOS circuitry is well suited for LSI devices, as well as for SSI and MSI devices. Most CMOS ICs offer a guaranteed noise immunity of ±3 volts, and typical units often exceed this specification. Lower supply voltages lead to narrower noise margins.

*All* inputs to a CMOS gate *must* be connected to something. If they are not being fed with a digital signal source they should be tied to either $V_{cc}$ or ground. A floating input can make a CMOS gate quite unstable, and the output state can become unpredictable.

Many CMOS devices use tri-state logic. This was discussed in Chapter 13.

Several numbering schemes are used for CMOS ICs. The two most common are CD4xxx (the CD is sometimes omitted) and 74Cxx. Devices numbered 74Cxx are equivalent in function to similarly numbered TTL chips.

One potential problem with CMOS ICs is that they tend to be sensitive to static electricity. Newer units have better built-in protection, but it is still smart to take precautions.

Unused CMOS chips should be stored with their pins shorted together to prevent accidental static discharge, that can damage the delicate on-chip components. This shorting can be accomplished by inserting the pins in a special conductive foam, or by storing the ICs in anti-static containers. Ordinary aluminum foil also works fine.

Be careful not to touch the pins of a CMOS IC, unless your body is securely grounded. We've all had a spark jump from our fingertips to a metal doorknob, or something similar. Imagine what such a burst of static could do to the thin metal-oxide film in a CMOS chip!

If you solder the pins of a CMOS IC directly (rather than using a socket), the soldering iron should also be properly grounded.

Static discharge was more of a problem with early CMOS chips. Newer devices are more durable, but should still be handled with care, just to be on the safe side.

## A TTL-CMOS Hybrid

A fairly recent development has been a new hybrid logic family known as high-speed CMOS. The 74HCxx numbering system is usually followed for these devices.

The high-speed CMOS family offers many of the advantages of both the ordinary CMOS and TTL families. These devices require a slightly higher supply current than their regular CMOS counterparts, but the power drain is still lower than for Schottky TTL ICs at low frequencies.

In terms of speed, the high-speed CMOS family offers approximately a ten-fold increase in performance over standard CMOS devices. High-speed CMOSs can operate at frequencies comparable to low-power Schottky TTLs.

High-speed CMOS ICs are designed for best performance with a power supply in the range of 3 to 6 volts, but they can be operated with anything from −0.5 to 7.0 volts.

## ECL Gates

Another important logic family is ECL, or *Emitter-Coupled Logic.* ECL is not as widespread as TTL or CMOS, but they are invaluable in applications where very high operating speeds are called for. Some ECL devices can handle frequencies up to 200 MHz.

ECL ICs put an almost constant drain on the power supply. There are no steep current spikes when the output changes state, unlike the other major logic families. ECL devices are usually numbered 10xxx.

### I²L Gates

One final logic family is I²L, or *Integrated Injection Logic.* This type of gate is very compact and has a minimum power drain. It is good for low to moderate speed applications.

You should recall that TTL gates use transistors with multiple emitters. In I²L gates we find transistors with multiple collectors, as shown in Fig. 14-9.

### Combining Logic Families

It is often necessary to mix devices from different logic families within a single circuit. Certain unusual devices might be available in only one logic family. Often different portions of a large system have different requirements. For example, in a circuit using a great many ICs, keeping the power consumption down is likely to be an important design consideration. However, a few stages within the system might need to be operated at high speeds. It would probably be wasteful to use high-speed ICs throughout the entire circuit if the high frequencies will be encountered by only a few of the devices. The solution is to mix logic families.

Different logic families usually cannot be directly wired to one another. Different power-supply voltages are used, and the logic levels are often defined at different levels. Input and output currents can also be mismatched, creating potential fan-in / fan-out problems.

By using a few basic tricks, the various logic families and subfamilies can be successfully used together within a single circuit.

The addition of a simple pull-up resistor, as shown in Fig. 14-10, allows a TTL gate to drive a CMOS device without much trouble. The supply voltage should be suitable for the TTL IC (nominally +5 volts). CMOS devices will work with this lower voltage,

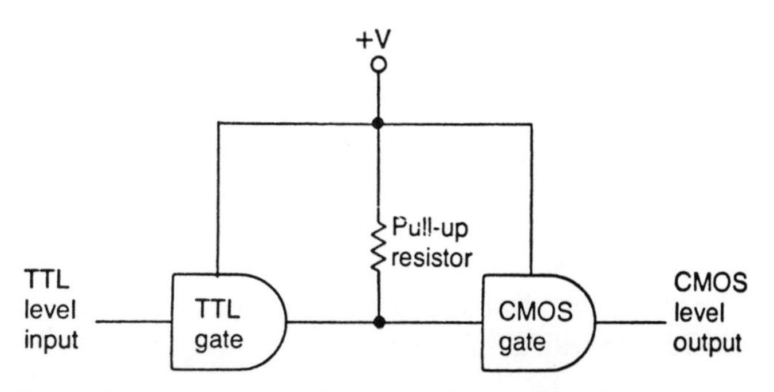


Fig. 14-10. A pull-up resistor can allow a TTL gate to drive a CMOS device.

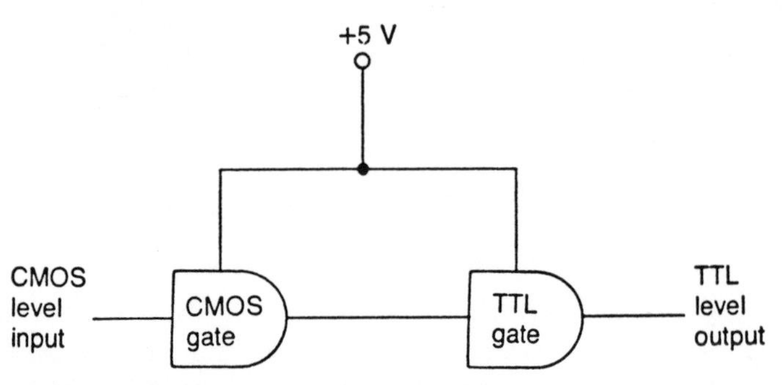

Fig. 14-11. Some CMOS circuits can drive TTL ICs directly.

although optimum performance generally requires a higher supply voltage. The value of the pull-up resistor will usually be fairly low. 1 KΩ is typically used.

Some CMOS devices can drive TTL gates directly, as shown in Fig. 14-11. The power supply must be suitable for TTL devices (5 volts).

In many cases, a CMOS driving TTL situation will involve problems stemming from the ability of the CMOS device to supply and sink the required currents for each of the logic states. Typical TTL specifications call for current no more than −1.6 mA for a logic 0, and no more than 40 μA for a logic 1. In cases where a single device doesn't have sufficient current handling capability, two or more identical gates can be paralleled, as illustrated in Fig. 14-12.

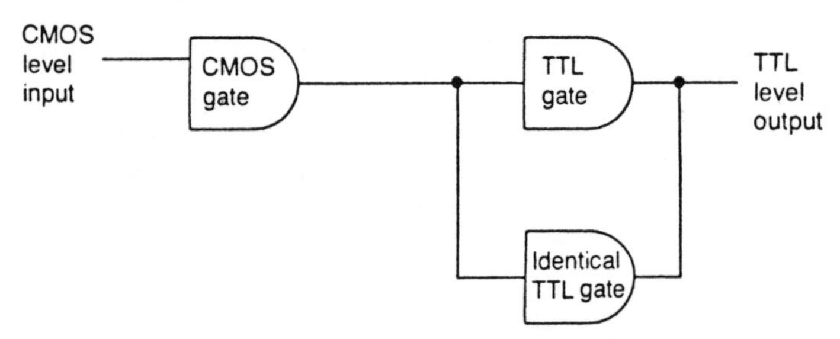

Fig. 14-12. Paralleling gates allow CMOS gates to drive TTL units.

Certain CMOS ICs are specially designed to make interfacing with TTL components easier. The CD4009A hex inverter and the CD4010A hex buffer are ideal for use as logic-level shifters. One terminal can take a high voltage (12 to 15 volts) from the CMOS portion of the circuit, while the 5 volts for TTL can be fed into the same interface chip at a second terminal. Figure 14-13 demonstrates the use of a CD4010A buffer for interfacing a CMOS gate with a TTL device.

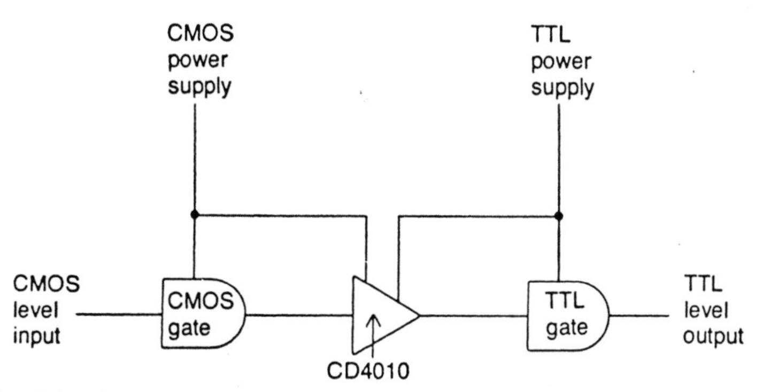

**Fig. 14-13. Special buffers, such as the CD4010, can be used to interface CMOS and TTL chips.**

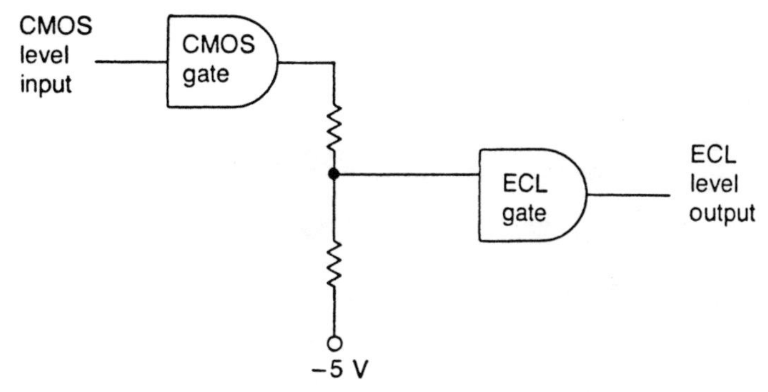

**Fig. 14-14. A simple voltage divider network permits interfacing between CMOS and ECL devices.**

Connecting CMOS and ECL ICs can be a little trickier. ECL gates tend to work best with voltages between ground potential and −5 volts. CMOS can be made to work with this voltage by using a voltage-divider resistance network, as shown in Fig. 14-14. In many cases, the CMOS and ECL chips will have to be powered separately.

# 15

# Digital Devices in Linear Applications

Digital and analog (or linear) circuits are fundamentally different from one another. Practically all ICs are clearly digital or linear and there is very little overlap. This chapter will cover a few of the exceptions.

## TIMERS

Timer ICs, such as the 555, are linear devices, but since they are used in monostable and a stable multivibrators, and produce clear rectangular or square waves, they are easily compatible with most digital gates. Most timer ICs have flexible power supply requirements so levels can be matched with most logic families. In digital circuits, timers are generally used in the astable mode as clocks.

## DIGITAL GATES AS LINEAR AMPLIFIERS

The most basic linear application is amplification. A weak signal at the input appears as a stronger, but otherwise identical signal at the output. This seems quite removed from digital operations, but a simple linear amplifier can be constructed from three inverters, two resistors, and a capacitor. The circuit is shown in Fig. 15-1.

The gain of this circuit (the amount of amplification) is determined by the ratio between the values of the two resistors. The gain can be calculated with this simple formula;

$$GAIN = R2 / R1$$

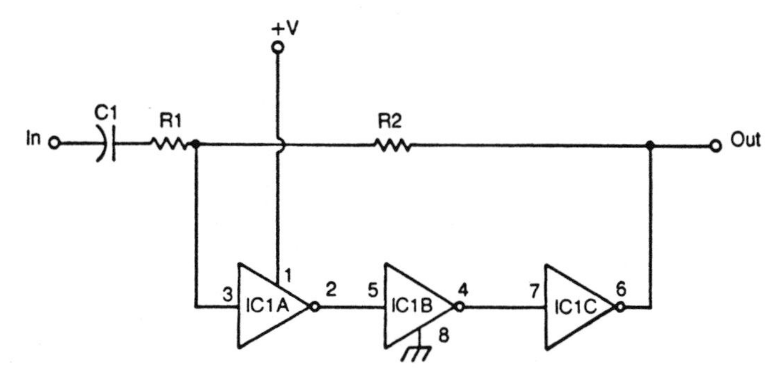

Fig. 15-1. Three digital inverters can be forced to act like a linear amplifier.

For the best results, both resistors should have relatively large values. I wouldn't recommend using anything less than about 100 KΩ.

## DIGITAL OSCILLATORS

It's not difficult to use feedback to create a digital oscillator. An example is shown in Fig. 15-2. This circuit produces square waves that can be used in either digital or analog circuits.

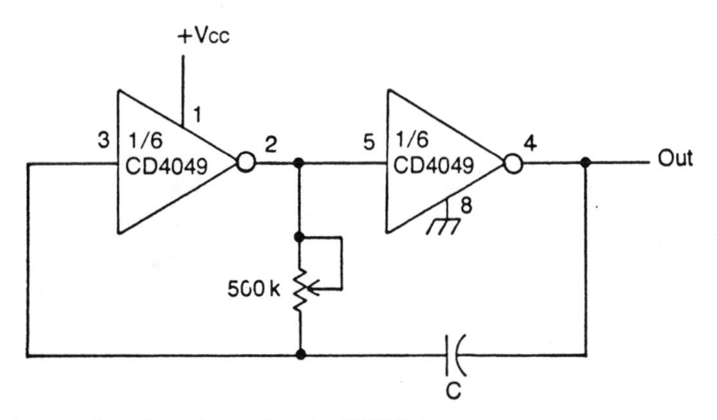

Fig. 15-2. An inverter based clock circuit using CMOS devices.

Linear oscillators can be designed to generate many other waveshapes in addition to square waves. Surprisingly, it is also possible to create other waveshapes, usually with some distortion. In many applications such distortion will be negligible.

A reasonable triangle wave can be generated by the digital circuit shown in Fig. 15-3. Resistor R1 and capacitor C1 are the frequency determining components. The formula for calculating the output frequency is;

$$F = 1/(1.4R1C1)$$

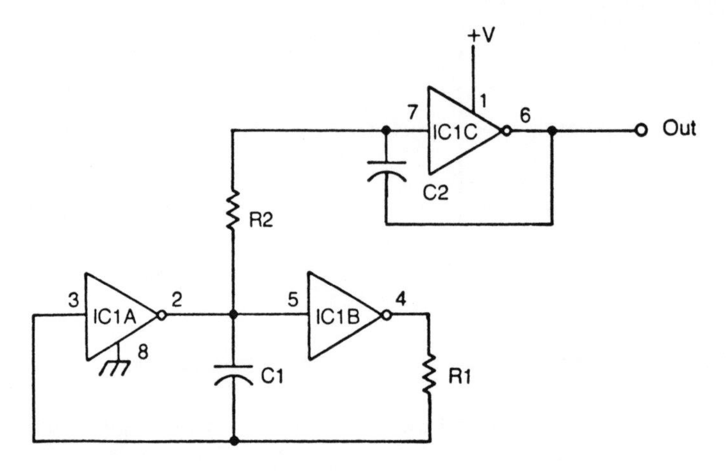


Fig. 15-3. Digital ICs can be used to generate a triangle wave signal.

Using the component values listed in Table 15-1, the circuit will generate triangle waves with a frequency of about 1000 Hz.

Sine waves are usually the most difficult waveforms to generate in pure form by digital or linear circuitry. The ideal sine wave is 100 percent pure. It consists of just the fundamental frequency, and has no harmonic content at all. Distortion adds harmonics.

Fortunately, a reasonable approximation of a sine wave, with just slight distortion, will be adequate for all but the most critical applications. We can usually settle for a signal that contains a few low-level high harmonics.

The circuit shown in Fig. 15-4 generates a fair pseudo-sine wave using the phase-shift oscillator technique. Once again, three inverters are used, along with a handful of passive components.

All three of the resistors should have identical values, as should the capacitors. The exact values used will determine the output frequency according to this formula:

$$F = 1/3.3RC$$

A sample parts list for this circuit is given in Table 15-2. Using these component values the output frequency will be about 1000 Hz.

Even better sine waves can be created from complex stopped waves. This is illustrated in Fig. 15-5. The more steps there are in the wave, the closer it will resemble

| | |
|---|---|
| IC1 | CD4049 hex inverter |
| R1 | 15 k resistor * |
| R2 | 22 k resistor |
| C1*, C2 | 0.05 $\mu$F capacitor |

* --- see text

Table 15-1. A Digital Triangle Wave Oscillator Can Be Constructed from These Parts. (See Fig. 15-3).

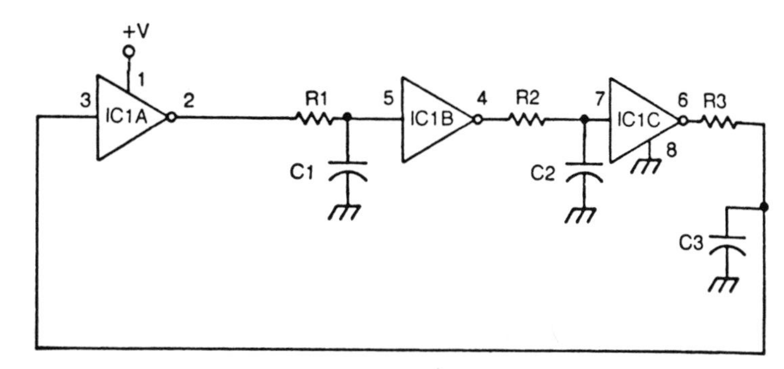


Fig. 15-4. A digital version of the linear phase-shift oscillator circuit.

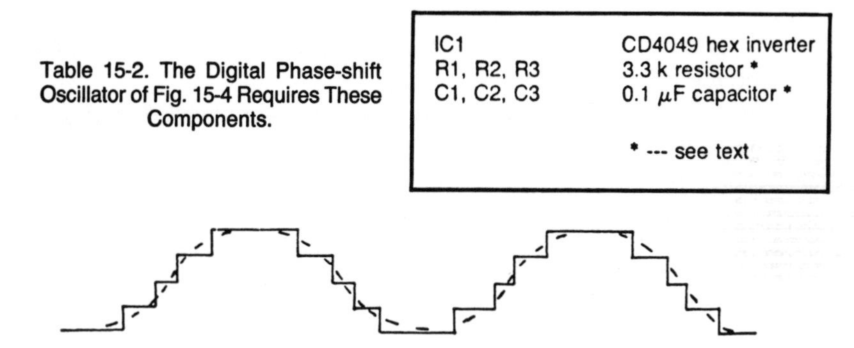

Table 15-2. The Digital Phase-shift Oscillator of Fig. 15-4 Requires These Components.

| | |
|---|---|
| IC1 | CD4049 hex inverter |
| R1, R2, R3 | 3.3 k resistor * |
| C1, C2, C3 | 0.1 $\mu$F capacitor * |

* --- see text

Fig. 15-5. A sine wave can be approximated by filtering a stepped-wave signal.

a true sine wave. The amplitude of the harmonics drops as the number of steps increases. The frequency of the harmonics also increases with the number of steps.

Most of the harmonic content of a stepped wave can be removed from the signal with a low-pass filter. This is simply a circuit that passes low frequencies, but blocks high frequencies, as shown in Fig. 15-6.

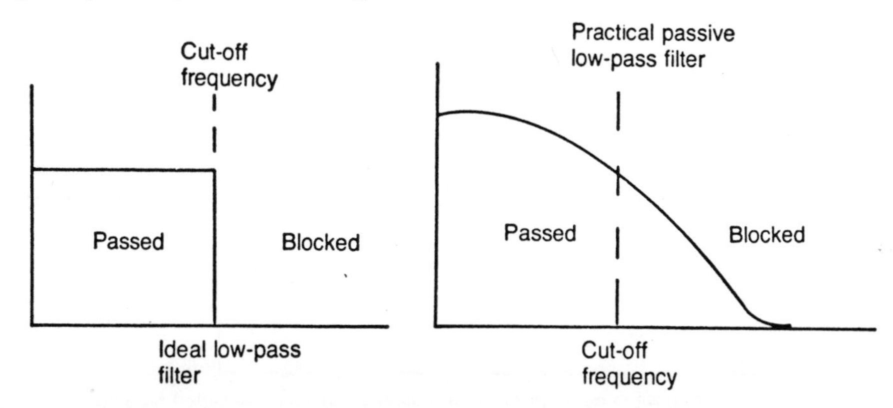


Fig. 15-6. A low-pass filter blocks high frequencies, but allows low frequencies through to the output.

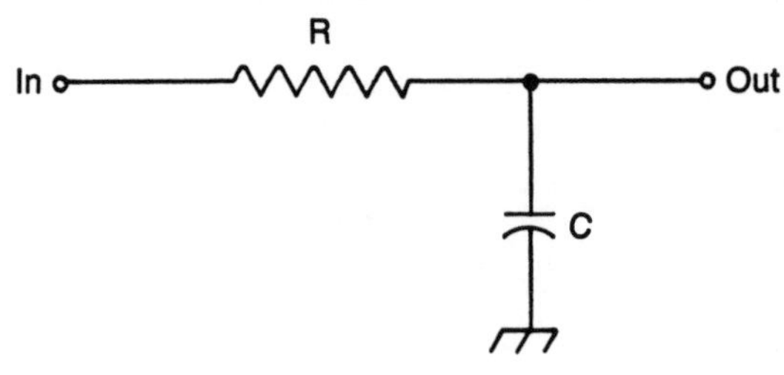


Fig. 15-7. A passive low-pass filter can be made from a resistor and a capacitor.

Depending on the component values in the filter circuit, some specific frequency is the cutoff frequency. In an ideal filter all frequencies below the cutoff frequency will be passed completely (with no attenuation), while all frequencies above this point will be completely blocked. Real circuitry cannot achieve such ideal filter response. Fortunately, in this particular application, a simple passive filter with a gradual cutoff slope will do the job reasonably well.

Only two components—a resistor and a capacitor—are needed to create a simple low-pass filter. The arrangement of the components is shown in Fig. 15-7. High frequencies are shunted through the capacitor to ground. Low frequencies can't pass through the capacitor, so they are fed through to the output. The formula for determining the cutoff frequency for this simple passive low-pass filter network is;

$$F_c = 159000 / RC$$

For a 1000 Hz cutoff you could use a 0.1 $\mu$F capacitor and a 15 KΩ resistor.

A suitable stepped waveform for digitally synthesizing a sine wave can be made from a string of D-type flip-flops. All of the outputs, except for the last one, are mixed together through a series of resistors. Determining the ideal values for these resistors can be tricky. If the ideal values are not used, the circuit will work, but the output will not be a sine wave.

The basic stepped-wave generator is illustrated in Fig. 15-8. A practical four-stage version of this circuit is shown in Fig. 15-9, with the parts list given in Table 15-3.

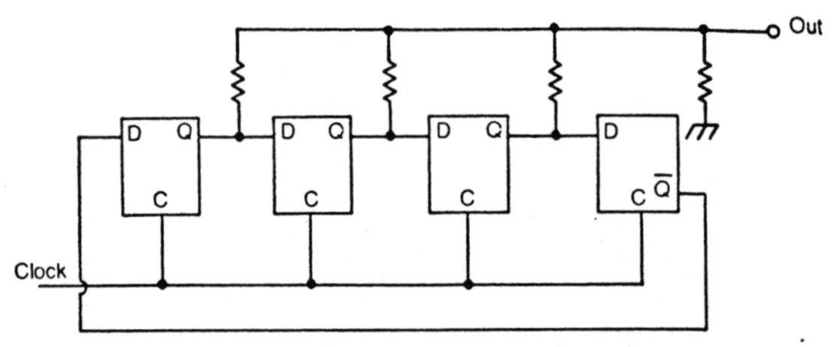


Fig. 15-8. A basic stepped-wave generator can be constructed from cascaded flip-flops.

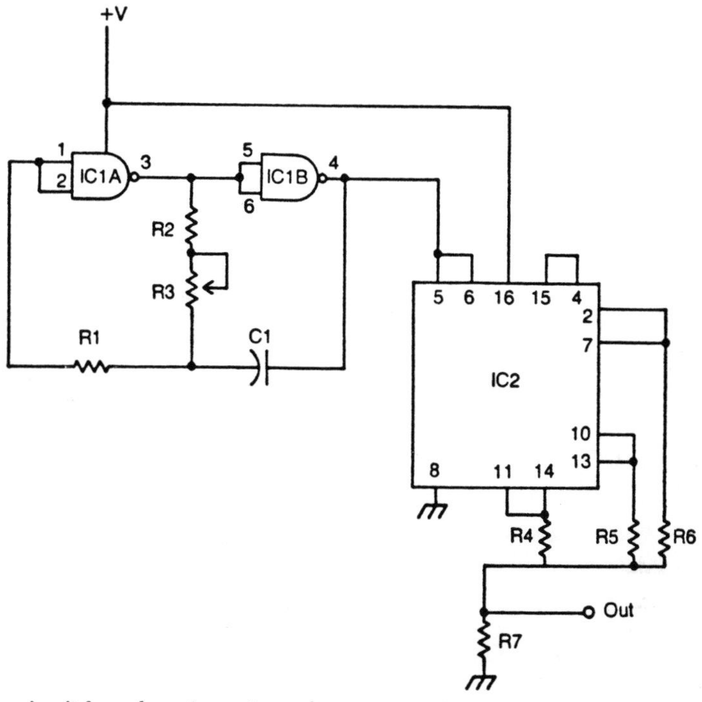

Fig. 15-9. The circuit for a four-stage stepped-wave generator.

IC1 generates a clock frequency ten times the desired output frequency. This is because ten steps are used to create the output waveform.

More steps in the digital signal increases the similarity to a sine wave, an eight-stage circuit is shown in Fig. 15-10. The parts list for this circuit is given in Table 15-4. This circuit functions in pretty much the same way as the four-stage circuit, except this time the clock frequency must be 16 times the desired output frequency.

Table 15-3. A Parts List for the Four-stage Stepped-wave Generator Circuit of Fig. 15-9.

|  |  |
|---|---|
| IC1 | CD4011 quad NAND gate |
| IC2 | CD4012 quad latch |
| R1 | 1.5 megohm resistor |
| R2, R4, R6 | 22 k resistor |
| R3 | 100 k potentiometer |
| R5 | 33 k resistor |
| R7 | 2.2 k resistor |
| C1 | 0.1 $\mu$F capacitor |

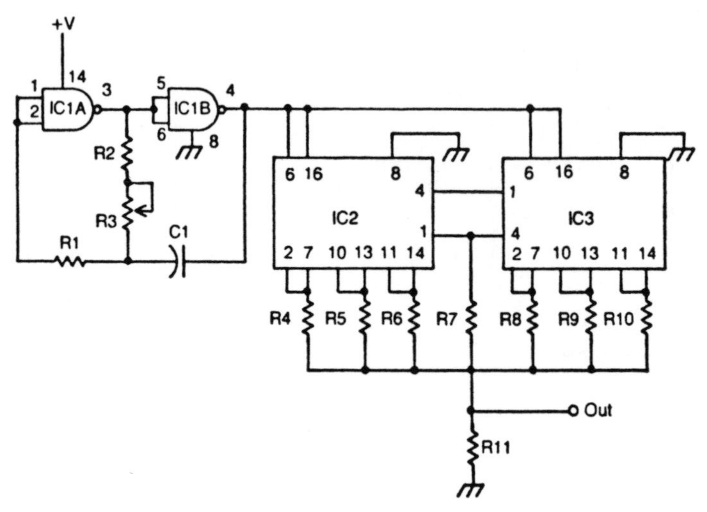


Fig. 15-10. A better sine wave approximation can be obtained by expanding the stepped-wave generator to eight stages.

Table 15-4. The Eight-stage Stepped-wave Generator Circuit of Fig. 15-10 Calls for These Parts.

| | |
|---|---|
| IC1 | CD4011 quad NAND gate |
| IC2, IC3 | CD4042 quad latch |
| R1 | 1.5 megohm resistor |
| R2 | 47 k resistor |
| R3 | 100 k potentiometer |
| R4, R10 | 22 k resistor |
| R5, R9 | 39 k resistor |
| R6, R7, R8 | 56 k resistor |
| R11 | 2.2 k resistor |

## DIGITAL FILTERS

Digital circuits can also be brought into play as filters. Digital filters are often used in modern communications systems, or in computer analysis or synthesis of sounds.

Before we turn to an actual digital filter circuit let's expand the basic passive low-pass filter circuit of Fig. 15-7 into what is known as a commuting filter. In a commuting filter, a number of identical capacitors are sequentially switched in and out of the circuit. A simplified commuting filter circuit is shown in Fig. 15-11. As shown here, the circuit has eight capacitors, so the switch grounds each in turn in this sequence;

C1 - C2 - C3 - C4 - C5 - C6 - C7 - C8 -
C1 - C2 - C3 - C4 - C5 - C6 - C7 - C8 -
C1 - C2 - C3 - and so on.

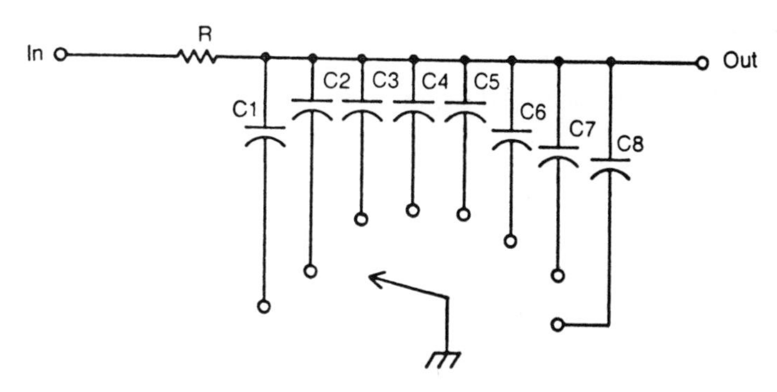

Fig. 15-11. A commuting filter is similar to a low-pass filter with multiple switched capacitors.

The capacitor switching action changes the operation of the filter from a low-pass to a band-pass type. In a band-pass filter, only a specific band, or range, of frequencies is passed. Anything outside that band will be blocked. The basic response graph for a band-pass filter is shown in Fig. 15-12. The center frequency of the pass band can be found with this formula;

$$F_c = 1 / 2nRC$$

where n is the number of capacitors, or switch positions.

Actually, there are a number of pass bands for a commuting filter. Harmonics (integer multiples) of the center frequency will also be allowed to pass through the commuting filter. For example, if the center frequency is 1000 Hz, there will be additional pass bands with center frequencies of;

2000 Hz (second harmonic)
3000 Hz (third harmonic)

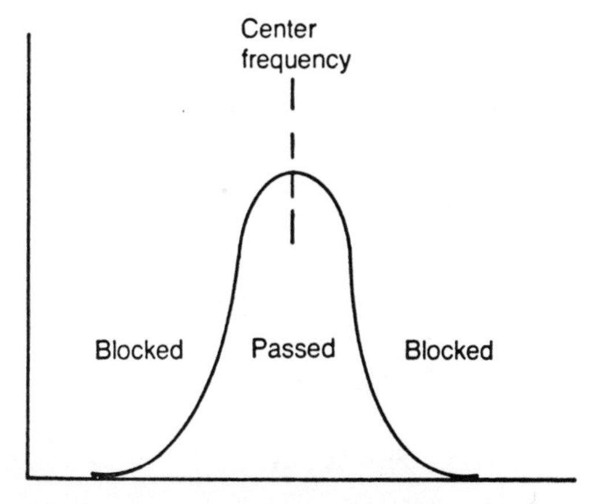

Fig. 15-12. A commuting filter has a band-pass frequency response.

4000 Hz (fourth harmonic)
5000 Hz (fifth harmonic)
6000 Hz (sixth harmonic) and so on.

Each harmonic pass band will be lower in amplitude than the next lower pass band, so the upper harmonics will eventually be filtered out. A low-pass filter added to the output of a commuting filter will eliminate the harmonic pass bands if that is required for the specific application.

Figure 15-13 shows the frequency response graph for an unmodified commuting filter. Since the multiple pass bands on this graph resemble the teeth of a comb, this type of circuit is sometimes called a comb filter.

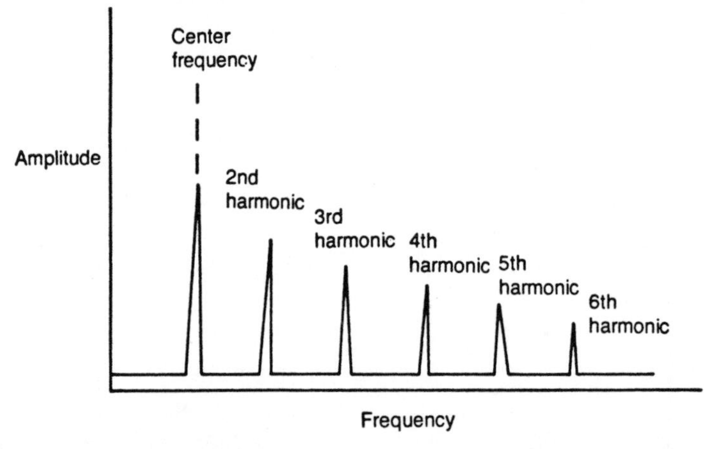


Fig. 15-13. Commuting (or "comb") filters also pass harmonics of the center frequency.

The circuit shown in Fig. 15-11 is not practical. There must be some automated means to continuously switch the capacitors at a fairly high rate. Digital circuitry can perform this function quite easily. A suitable circuit is shown in Fig. 15-14. Notice that two inputs are required—a clock signal to drive the counter, and a signal to be filtered. The clock signal should be a square or rectangular wave, as with any digital circuit. The input signal to be filtered can be any waveshape at all.

The capacitor values are fairly critical in this circuit. Components with no more than a 10 percent tolerance rating should be used here. Typical component values are listed in Table 15-5, but feel free to experiment with different resistor and capacitor values.

A more linear frequency response and narrower pass bands can be achieved by adding more decoder stages (capacitors to be switched). The number of stages and the clock frequency determine the width of the pass band, according to this formula;

$$F = X / N$$

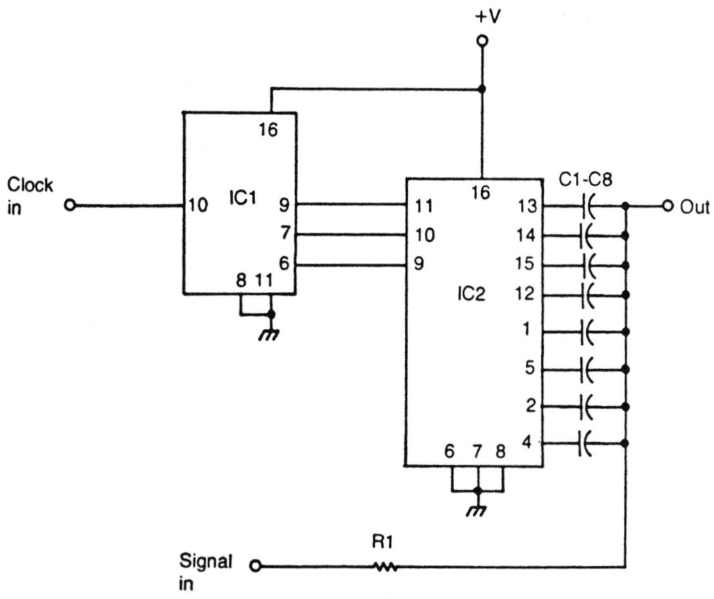

Fig. 15-14. This digital circuit performs a filtering function.

Table 15-5. Components Needed to Construct the Digital Filter Circuit of Fig. 15-14.

| | |
|---|---|
| IC1 | CD4040 BCD ripple counter |
| IC2 | CD4051 BCD to decimal decoder |
| R1 | 1 k resistor (experiment) |
| C1 - C8 | 0.01 $\mu$F capacitor (experiment) |

where F is the pass band, X is the clock frequency, and N is the number of capacitor stages.

Almost any input signal can be used for this circuit, but the input signal's peak-to-peak voltage must be less than the voltage powering the digital ICs.

## BILATERAL SWITCHES

There is at least one digital IC that was designed for use with both digital and analog signals. This is the bilateral switch. This is a switch that can be open or shut according to a logic signal. For purely digital signals, a similar function could be achieved with a complex gating circuit, but this would be awkward at best. Such a gating circuit also could not switch analog signals.

Figure 15-15 shows the pinout diagram and functional internal structure of the CD4066 quad bilateral switch. This CMOS chip consists of four independent, digitally controlled

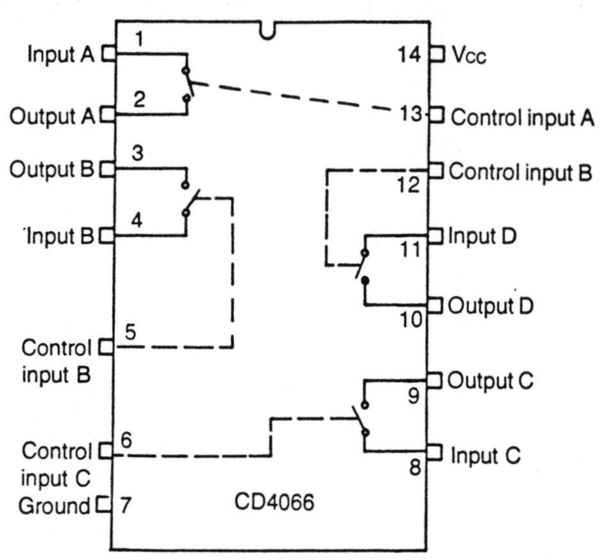


Fig. 15-15. A typical digitally controlled switch device is the CD4066 quad bilateral switch.

switches. The switches are called bilateral because they have no fixed polarity. The pins labeled as switch inputs on the diagram can be used as switch outputs, and vice versa. Which end of the switch is the input and which is the output does not matter.

Digital switch units like the CD4066 are often used in hybrid circuits that use both digital and analog devices. Analog components like resistors or capacitors can be selected or programmed via digital control signals. Digital-to-analog signal conversion is one obvious application for this type of device. Almost any switched function can be made programmable or can be automated with bilateral switches.

## LINEAR ICS

There are a great many linear ICs available. There are so many of them, it would be ridiculous to try to cover them here. A number of books have been written on various linear ICs.

Linear ICs range from simple arrays of matched components on a single chip-up, to complex systems, such as a complete radio receiver on a single chip. Popular circuit functions, such as amplifiers and voltage regulators, can be performed by many dedicated ICs.

A number of linear ICs, such as timers and op amps, are general purpose devices used in a great many different applications. The op amp, or operational amplifier, is particularly versatile. This type of circuit was originally designed to perform mathematical operations, but op amps can be used in almost any linear application, including;

- dc amplification

- ac amplification
- signal mixing
- filtering
- signal generation
- signal comparison
- voltage regulation

Almost any electronic function you can think of can be performed by some integrated circuit somewhere.

# Index

# Index

# Other Bestsellers From TAB